Dynamical System Models
in the Life Sciences
and Their Underlying Scientific Issues

Dynamical System Models in the Life Sciences

and Their Underlying Scientific Issues

Frederic Y M Wan

University of California, Irvine, USA

 **World Scientific**

NEW JERSEY · LONDON · SINGAPORE · BEIJING · SHANGHAI · HONG KONG · TAIPEI · CHENNAI · TOKYO

Published by

World Scientific Publishing Co. Pte. Ltd.

5 Toh Tuck Link, Singapore 596224

USA office: 27 Warren Street, Suite 401-402, Hackensack, NJ 07601

UK office: 57 Shelton Street, Covent Garden, London WC2H 9HE

Library of Congress Cataloging-in-Publication Data

Names: Wan, Frederic Y. M.

Title: Dynamical system models in the life sciences and their underlying scientific issues /
Frederic Y.M. Wan, University of California, Irvine.

Description: New Jersey : World Scientific, 2017. | Includes bibliographical references and index.

Identifiers: LCCN 2016051221| ISBN 9789813143333 (hardcover : alk. paper) |
ISBN 9789813143708 (softcover : alk. paper)

Subjects: LCSH: Life sciences--Mathematical models.

Classification: LCC QH323.5 .W36 2017 | DDC 570.1/5195--dc23

LC record available at https://lccn.loc.gov/2016051221

British Library Cataloguing-in-Publication Data

A catalogue record for this book is available from the British Library.

Printed in Singapore

To my wife, Julia, who saw more potential than what met the eyes.

Contents

Preface   xvii

1. Mathematical Models and the Modeling Cycle   1
   1.1 Mathematical Models   1
   1.2 The Modeling Cycle   2
        1.2.1 Formulation   3
        1.2.2 Analysis   3
        1.2.3 Validation   3
        1.2.4 Improved formulation   4
        1.2.5 The cycle   4
   1.3 Scientific Issues   4
   1.4 Scales of Biological Phenomena   6
   1.5 Let's Get Started   7

Part 1. Growth of a Population   9

2. Evolution and Equilibrium   11
   2.1 Growth without Immigration   11
        2.1.1 Population size dependent growth rates   11
        2.1.2 Solution of the model ODE   13
        2.1.3 Initial condition and the initial value problem   14
   2.2 Exponential Growth   15
        2.2.1 Linear growth rate   15
        2.2.2 The growth rate constant   16
        2.2.3 Model validation   16
        2.2.4 Prediction of future population   17
   2.3 Estimation of System Parameters   17
        2.3.1 The least squares fit   17
        2.3.2 The least squares method for exponential growth   18
   2.4 A Constant Growth Rate Model   21
   2.5 Intravenous Drug Infusion   22

Contents

	2.5.1	Questions on hypoglycemia	22
	2.5.2	Linear models with immigration	22
	2.5.3	Mathematical analysis	23
	2.5.4	Model improvements	24
2.6		Mosquito Population with Bio-Control	25
	2.6.1	The problem of mosquito eradication	25
	2.6.2	Constant rate of eradication	26
	2.6.3	Seasonal varying growth rate	26

3. Stability and Bifurcation | | | 27
3.1		The Logistic Growth Model	27
	3.1.1	Taylor polynomials for the growth function	27
	3.1.2	Normalization and exact solution	29
	3.1.3	Implication of the exact solution	29
	3.1.4	Validation and model improvement	30
3.2		Critical Points and Their Stability	30
	3.2.1	Critical points of autonomous ODE	30
	3.2.2	Stability of a critical point	32
	3.2.3	Graphical method for critical points and their stability	34
	3.2.4	Linear stability analysis	36
3.3		Over-fishing of a Commercial Fish Population	37
	3.3.1	Fish harvesting	37
	3.3.2	Critical points and their stability	39
	3.3.3	Bifurcation for fish harvesting	40
3.4		Bifurcation	41
	3.4.1	Bifurcation diagram and horizontal tangency	41
	3.4.2	Basic bifurcation types	43
3.5		Structural Stability and Hyperbolic Points	48

Part 2. Interacting Populations | | | **51**

4. Linear Interactions | | | 53
4.1		Types of Interaction	53
4.2		Red Blood Cell Production	54
	4.2.1	Model formulation	54
	4.2.2	Method of elimination	55
	4.2.3	Implication of mathematical results	56
4.3		Linear Systems with Constant Coefficients	58
	4.3.1	Vector form	58
	4.3.2	Complementary solutions	59
	4.3.3	Fundamental matrix solution	59
	4.3.4	Matrix diagonalization	63
	4.3.5	De-coupling of the linear system	63

		4.3.6	The matrix exponential	65
		4.3.7	Symmetric matrices	66
	4.4		DNA Mutation	68
		4.4.1	The double helix	68
		4.4.2	Mutation due to base substitutions	69
		4.4.3	Linear model	70
		4.4.4	Equal opportunity substitution	70
	4.5		Defective Matrices	73
		4.5.1	A simple example	73
		4.5.2	Systematic reduction to Jordan form	75
		4.5.3	Matrices with a single eigenvector	75
		4.5.4	Multiple but insufficient eigenvectors	77
		4.5.5	General defective matrix	79
5.			Nonlinear Autonomous Interactions	81
	5.1		Predator-Prey Interaction	81
		5.1.1	Rabbits and foxes	81
		5.1.2	Reduction of order	82
		5.1.3	Critical points	85
		5.1.4	Linear stability analysis	86
	5.2		Linear Autonomous Systems	87
		5.2.1	Critical point and its stability	87
		5.2.2	Phase portrait	89
		5.2.3	Stability regions	93
	5.3		Nonlinear Autonomous Systems	96
		5.3.1	Critical points	96
		5.3.2	Linear stability analysis	97
		5.3.3	Hyperbolic critical points	100
	5.4		The Phase Portrait	103
		5.4.1	Limit cycles	103
		5.4.2	Non-existence of a limit cycle	106
		5.4.3	Reversible systems	107
		5.4.4	Multiple eigenvalues	109
	5.5		Basic Bifurcation Types	110
		5.5.1	The bacteria-antibody problem	110
		5.5.2	The three basic types of bifurcation	112
		5.5.3	Hopf bifurcation	113
6.			HIV Dynamics and Drug Treatments	117
	6.1		The Human Immunodeficiency Virus (HIV)	117
		6.1.1	HIV is a retrovirus	117
		6.1.2	The immune system	117
		6.1.3	The virus life cycle	118

	6.1.4	Three phases of HIV infection	119
	6.1.5	Treatment of HIV	120
	6.1.6	The principal direction of HIV research	121
6.2		HIV Dynamics	122
	6.2.1	Model formulation	122
	6.2.2	Steady states and bifurcation for virus saturation	123
	6.2.3	Steady states and bifurcation for a small virus population	126
	6.2.4	The three-component model	127
6.3		Drug Treatments	130
	6.3.1	The activities of HIV	130
	6.3.2	Combining RT and protease inhibitor	131
	6.3.3	Effectiveness of AIDS cocktail treatments	133
	6.3.4	On the modified three-component model	134

7. Index Theory, Bistability and Feedback — **137**

7.1		Poincaré Index for Planar Systems	137
7.2		Bistability — A Dimerized Reaction	140
	7.2.1	Interaction involving a dimer	140
	7.2.2	Fixed points and bifurcation	141
	7.2.3	Linear stability analysis	141
7.3		Bistability — Two Competing Populations	143
	7.3.1	Leopards and hyenas	144
	7.3.2	Critical points	145
	7.3.3	Linear stability analysis	145
	7.3.4	Existence of bistability	146
	7.3.5	Other bistability configurations?	148
	7.3.6	Multiple bifurcation parameters	150
7.4		Cell Lineages and Feedback	151
	7.4.1	Multistage cell lineages	151
	7.4.2	Performance objectives	153
	7.4.3	Feedback control of output	153
	7.4.4	A finite steady state	154
	7.4.5	Linear stability	155
	7.4.6	Fast regeneration time	156

Part 3. Optimization — **159**

8 The Economics of Growth — **161**

8.1		Maximum Sustained Yield	161
	8.1.1	Fishing effort and maximum sustainable yield	161
	8.1.2	Depensation and other growth rates	163
8.2		Economic Overfishing vs. Biological Overfishing	164
	8.2.1	Revenue for a price taker	164

      8.2.2   Effort cost and net revenue 165

      8.2.3   Economic overfishing vs. biological overfishing 166

      8.2.4   Regulatory controls 166

  8.3   Discounting the Future 168

      8.3.1   Current loss and future payoff 168

      8.3.2   Interest and discount rate 169

9.   Optimization over a Planning Period   171

  9.1   Calculus of Variations 171

      9.1.1   Present value of total profit 171

      9.1.2   A problem in the calculus of variations 172

      9.1.3   The basic problem 173

  9.2   An Illustration of the Solution Process 174

      9.2.1   The action of a linear oscillator 174

      9.2.2   The condition of stationarity 175

      9.2.3   The Euler differential equation 176

  9.3   The General Basic Problem 177

      9.3.1   The fundamental lemma 177

      9.3.2   Euler differential equation for the basic problem 179

  9.4   Integration of Special Cases 180

      9.4.1 $F(x, y, y') = F(x)$ 180

      9.4.2 $F(x, y, y') = F(y)$ 180

      9.4.3 $F(x, y, y') = F(y')$ 181

      9.4.4 $F(x, y, y') = F(x, y')$ 182

      9.4.5 $F(x, y, y') = F(y, y')$ 183

      9.4.6 $F(x, y, y') = F(x, y)$ 183

  9.5   The Fishery Problem with Catch-Dependent Cost of Harvest . . . 184

      9.5.1   The Euler DE and BVP 184

      9.5.2   Linear growth rate 184

      9.5.3   Constant unit harvest cost $(c_1 = 0)$ 185

      9.5.4   The method of most rapid approach 187

10.   Modifications of the Basic Problem   191

  10.1 Smooth and PWS Extremals 191

  10.2 Unspecified Terminal Unknown Value 194

      10.2.1 Euler boundary conditions 194

      10.2.2 Terminal payoff 196

      10.2.3 Euler boundary conditions with terminal payoff 199

  10.3 Higher Derivatives and More Unknowns 199

      10.3.1 Higher derivatives equivalent to more unknowns 199

      10.3.2 Euler differential equations for several unknowns 200

      10.3.3 Some examples 201

  10.4 Parametric Form and Free End Problems 203

    10.4.1   Basic problem in parametric form 203
    10.4.2   A first integral and Erdmann's conditions 204
    10.4.3   Free end problems 205

11. Boundary Value Problems are More Complex   209
    11.1 Boundary Value Problems and Their Complications 209
        11.1.1   Second order equations 209
        11.1.2   No solution or too many solutions 210
    11.2 One-Dimensional Heat Conduction 211
        11.2.1   Rate of change of heat 211
        11.2.2   Fourier's law and the heat equation 213
        11.2.3   Steady state temperature distributions 214
        11.2.4   The time dependent problem 216
    11.3 Analytical Solutions for BVP 218
        11.3.1   Reduction of order for autonomous equations 218
        11.3.2   A nonautonomous nonlinear ODE 220
    11.4 The IVP Approach for Linear BVP 221
        11.4.1   Homogeneous linear ODE 221
        11.4.2   Inhomogeneous ODE 224
        11.4.3   Some possible complications 225
    11.5 The Shooting Method 225

Part 4. Constraints and Control   229

12. "Do Your Best" and the Maximum Principle   231
    12.1 The Variational Notation 231
        12.1.1   Variations of a function 231
        12.1.2   Variations of a function of functions 232
        12.1.3   Variations of the performance index 233
    12.2 Adjoint Functions 234
        12.2.1   A minimum cost problem 234
        12.2.2   Constraints on the investment rate 236
        12.2.3   The optimal control 237
    12.3 With Constraints, Do Your Best 239
        12.3.1   The augmented performance index 239
        12.3.2   The interior solution 240
        12.3.3   Do the best you can 241
    12.4 Fishery Problem without Prescribed Terminal Condition 243
        12.4.1   The "Do Your Best" approach 243
        12.4.2   Interior control is singular 244
        12.4.3   Upper corner solution near the end 245
        12.4.4   Optimal control for $t < t_s$ 246
    12.5 Optimal Allocation of Stem Cells for Medical Applications . . 248

12.6 The Maximum Principle 249
    12.6.1 The Hamiltonian 249
    12.6.2 The general problem 250
    12.6.3 Maximization of the Hamiltonian 251
    12.6.4 The stem cell problem 253
    12.6.5 The fishery problem with constant unit harvest cost 255
    12.6.6 On the Maximum Principle 259

13. Chlamydia Trachomatis   261
  13.1 A Proof of Concept Model of C. Trachomatis 261
    13.1.1 Life cycle of Chlamydia Trachomatis 261
    13.1.2 A linear growth rate model 261
    13.1.3 Hamiltonian and Maximum Principle 263
    13.1.4 Biological mechanisms for optimal strategy 265
  13.2 Chlamydia with Finite Carrying Capacity 266
    13.2.1 The model 266
    13.2.2 The singular solutions 268
    13.2.3 The upper corner control is optimal adjacent to T 269
    13.2.4 Lower corner control optimal only for $R_0 \leq 1/2$ 271
  13.3 Optimal Conversion Strategy for $R_0 \leq 1/2$ 272
    13.3.1 The optimal control is bang-bang for $u_{\max} < 1/2$ 272
    13.3.2 Sub-interval of singular solution possible for $u_{\max} \geq 1/2$ . . 273
    13.3.3 On various threshold values 277
    13.3.4 Summary for $R_0 \leq 1/2$ 279
    13.3.5 Case I — $0 < R_0 \leq R_c$ 280
    13.3.6 Case II — $R_c < R_0 < 1/2$ 280
  13.4 Optimal Conversion Strategy for $R_0 > 1/2$ 280
    13.4.1 Upper corner control for $1 - u_{\max} \geq R_0 \ (> 1/2)$ 280
    13.4.2 Optimal conversion strategies for $1 - u_{\max} < R_0$ 282
    13.4.3 The relative magnitude of t_x and t_S 285
    13.4.4 Summary of the optimal conversion strategy for $R_0 > 1/2$. 288
    13.4.5 Case III — $u_{\max} \leq 1/2$ 288
    13.4.6 Case IV — $u_{\max} > 1/2$ 288
  13.5 Mathematics and the Biology of Chlamydia 288

14. Genetic Instability and Carcinogenesis   293
  14.1 Genetic Instability is a Two-Edge Sword 293
  14.2 Activation of an Oncogene 295
    14.2.1 A one-step system 295
    14.2.2 Dependence of death rate on mutation rate 296
    14.2.3 Dimensionless formulation 297
  14.3 Shortest Time by a Constant Mutation Rate 298
    14.3.1 An alternative description of the evolving populations . . . 298

14.3.2 Numerical solution for a time-invariant control 299

14.4 The TOP 301

    14.4.1 Fastest time to cancer 301

    14.4.2 The Maximum Principle 302

    14.4.3 Some preliminary results on adjoint functions 304

    14.4.4 Vanishing Hamiltonian for our TOP 306

14.5 Strictly Concave Death Rates 306

    14.5.1 Upper corner control near the start 306

    14.5.2 Optimal mutation rate is bang-bang 308

    14.5.3 General strictly concave death rate 310

14.6 Optimal Switch Point for Bang-Bang Mutation 310

    14.6.1 A brute force scheme on $x_s = x_2(T_s)$ 310

    14.6.2 Some bounds for the optimal switch time 311

14.7 Other Types of Death Rates 314

    14.7.1 A death rate linear in mutation rate 314

    14.7.2 Strictly convex death rate 314

15. Mathematical Modeling Revisited   315

15.1 From Simple to Complex 315

    15.1.1 Simple question 315

    15.1.2 Simple model and upgrading 316

    15.1.3 Instantaneous rate of change 317

15.2 Open to Options 318

    15.2.1 Higher order rates of change? 318

    15.2.2 Mathematical effectiveness & computational efficiency . . . 319

15.3 Improvement or Alternative 321

    15.3.1 Weighted parameter estimation 321

    15.3.2 Rate limiting carrying capacity 322

15.4 Active Modeling Experience 323

Appendix A   First Order ODE   325

A.1 Separable ODE 325

    A.1.1 Reduction to a calculus problem 325

    A.1.2 Initial condition and the initial value problem 326

    A.1.3 Scale invariant first order ODE 327

A.2 First Order Linear ODE 328

    A.2.1 Integrating factor 328

    A.2.2 The Bernoulli equation 329

A.3 An Exact First Order ODE 330

    A.3.1 Test for an exact equation 330

    A.3.2 Reduction to a calculus problem 331

A.4 When an ODE Is Not Exact 333

A.5 Summary of Methods for First Order ODE 334

Appendix B   Basic Numerical Methods   335

    B.1   Simple Euler 335

    B.2   Slowly 336

    B.3   But Not So Surely 338

    B.4   No Clue 341

    B.5   Clueless? 343

    B.6   Well-Posed Problems 346

    B.7   Higher Order Numerical Schemes for IVP 348

Appendix C   Assignments   351

    C.1   Assignment I 351

    C.2   Assignment II 352

    C.3   Assignment III 354

    C.4   Assignment IV 356

    C.5   Assignment V 358

    C.6   A Typical Midterm Examination 359

    C.7   Assignment VI 360

    C.8   Assignment VII 362

    C.9   Assignment VIII 363

    C.10   A Typical Final Examination 365

Bibliography   369

Index   371

Preface

The molecular revolution in the 20th century has provided biological sciences with an abundance of factual information but with no more insight to the biochemical processes at work in biological organisms. In a 2009 report entitled *A New Biology for 21st Century* <http://www.nap.edu/catalog/php?record_id=12764>, the National Research Council (NRC) recommends a new approach to biology that "depends on greater integration within biology, and closer collaboration with physical, computational, and earth scientists, mathematicians and engineers be used to find solutions ...". Given the centrality of mathematical modeling in these areas designated for collaboration, it is implicit in the NRC recommendation that it be exploited for significant advances in biology.

Prompted by their own experience, faculty at the University of California, Irvine (UCI) independently recognized the same needs for interdisciplinary collaboration and initiated in 2007 a gateway graduate program in Mathematical and Computational Biology (MCB) <http://mcsb.bio.uci.edu/> to educate a new generation of researchers in the life sciences capable of such collaboration. The MCB program was made possible with an award from the Howard Hughes Medical Institute (HHMI)-National Institute of Biomedical Imaging and Bioengineering (NIBIB) Interface Initiative and has also been sustained by subsequent NIH training grants and a P50 Center Grant <http://ccbs.uci.edu/> for the last ten years. A new option of stand-alone interdisciplinary Ph.D. degree program in Mathematical and Computational Systems Biology was established in 2015 to provide more flexibility to our educational effort in this area. Several dozen MCB students have received their Ph.D. degrees since the Program's inception.

For the core curriculum of the MCB Program, the Mathematics Department at UCI introduced in 2007 a three-quarter course sequence on Mathematical and Computational Biology, Math 227A,B,C, to acquaint interested students with mathematical modeling of biological phenomena and the attendant mathematical and computational techniques for extracting useful information from mathematical models. This volume grew out of the course notes for the first quarter of the three-quarter course sequence, focusing on models involving *ordinary differential equations* (ODE). Models involving *partial differential equations* (PDE) are studied in

Math 227B while Math 227C is concerned with differential equation models involving uncertainty (stochastic differential equations, SDE). While differential equations (DE), ordinary, partial or stochastic, are known to be important in the quantitative understanding of life science phenomena, teaching DE modeling is much more challenging than teaching the mathematics of DE for a number of reasons including the following:

- There is no recipe or algorithm for the development of a mathematical model for any specific phenomenon.
- The undergraduate training of students in our program is diverse: Those majored in the biological sciences are often deficient in the needed training in mathematics and computing. Those majored in non-biological sciences are usually deficient in their knowledge of biology.
- Even among life science majors, the amount of background science needed for the teaching of a relatively comprehensive set of models is too extensive to have been covered in an undergraduate curriculum in biology.

In order for students to be more or less similarly prepared for the MCB curriculum, the Program requires attendance of a three-week pre-program boot camp, one week in mathematics and computational software and two weeks on basic biology. The math and computing week provides all participants a working knowledge of MatLab and Mathematica while using them to review undergraduate mathematics relevant to the MCB core curriculum. Even so, experience in teaching nearly ten groups of program students suggested that only a working knowledge of basic calculus and minimal matrix computation can be expected of the students in general. To accommodate the need of such a group of students for gradual acquisition of more mathematical knowledge and computational skills, the course starts with the simplest of growth models for a single population but forces them to face the challenge of parameter estimation in order to obtain a meaningful and realistic solution of the problem. As models evolve to provide different information, students are led to consider more and more complex models that require them to learn more sophisticated mathematics and computational skills. This volume culminates in some rather challenging discussion and applications of Pontryagin's Maximum Principle for constrained optimal control problems.

On the teaching of mathematical modeling, courses and textbooks on mathematical models typically consist of 1) case studies of specific phenomena, or 2) exposition of different mathematical methods with discussion of specific mathematical models as examples of their applications. While the MCB curriculum for mathematical modeling needs both approaches, it specifically organizes the teaching of modeling around what we want to know about the phenomenon of interest. Typically, we group the development of course material around mathematically expressed scientific issues such as

- How does a phenomenon of interest evolves with time (evolution)?

- Does it tend to a time-independent or time-varying steady state (equilibrium or periodic motion)?
- Is a steady state sensitive to small external perturbation (stability)?
- Does a small change of the system characteristics cause a catastrophic change in the steady states and/or their stability (bifurcation)?
- Does a biological organism maintain its normal functions or developments after environmental or genetic changes (robustness)?
- What is the mechanism for an organism to maintain robustness (feedback, redundancy)?
- etc.

Different mathematical models are needed for different issues about the same phenomenon. Different analytical and computational requirements relevant to the different models are then identified. Appropriate mathematical and computational tools are then introduced as needed for extracting from these models the information sought at the start.

Analysis and numerical simulations of mathematical models require knowledge of mathematical methods and computational techniques for differential equations. With the mixed background of MCB students, some having had minimal training in quantitative methods, it is necessary to introduce the mathematical and computational techniques needed for studying the models being investigated throughout the course. In terms of mathematical techniques, this first volume on ODE models is divided into the following three related parts:

(1) *Initial Value Problems* (IVP) for which **all** the auxiliary conditions supplementing the ODE are prescribed at the same value of the independent variable, usually the initial value t_0 of the interval of the independent variable t (giving rise to the term initial value problem or *IVP* for short) or the terminal value t_T (known as terminal value problem) but may be some point in between.
(2) *Boundary Value Problems* (BVP) for which some of the auxiliary conditions are prescribed at the initial point t_0 while the remaining ones prescribed at the end point t_T resulting in what is known as two-point boundary value problems.
(3) *Calculus of Variations* (CoV) *and Optimal Control* which reduce two important classes of optimization problems to solving BVP.

Phenomena in the biological sciences and the corresponding mathematical models are generally rather complex and require extensive computation. Fortunately, much of the computing involved can be relegated to user friendly mathematical software such as Mathematica, MatLab and Maple. Still, a working knowledge of basic ODE is so critical to mathematical modeling that we take every opportunity to work out the mathematical details whenever possible. For an exposure to a variety of ODE systems, we work in this volume with a wide range of mathematical models for a diverse group of biological phenomena. These range from the efficacy of cocktail treatment of AID, to the mechanism for effective regenerative biology, to the level of genetic instability for carcinogenesis. In the process, the volume introduce readers to frontier research on ODE models for biological phenomena of current interest. Nevertheless, the principal goal of this volume is not teaching mathematics of ODE or an exposition of current research in mathematical and computational biology. The principal new elements of the volume is on mathematical modeling of biological phenomena with an emphasis on the kind of scientific issues (evolution, stability, bifurcation, robustness, etc.) of interest pertaining to the particular phenomenon.

There is no text on mathematical biology that meets the intended purposes and the planned curriculum indicated above. In the absence of an appropriate text, MCB students are provided with a set of course notes for a coherent coverage of the prescribed course material. Experience from more than 40 years of teaching suggests that students are best served by continuous engagement in an active learning process. To provide an active learning environment (short of engaging students in research projects, which the MCB Program does concurrently through three quarterly lab rotations), the course notes for each quarter course are accompanied by nine sets of required weekly exercises on the course material in addition to the suggested exercises sprinkled throughout these notes. To incentivize students for punctual completion of the assigned exercises, their homework assignments are graded and their overall performance on the assignments constitutes 35% of their final course grade. The course is reputed to have a fast pace and the weekly exercises are by and large not straightforward. Optional discussion sessions are offered to help students with the exercises; however they are never provided with written solutions for the homework assignments, mid term or final examination.

Frederic Y.M. Wan
Irvine, CA
February 28, 2017

Chapter 1

Mathematical Models and the Modeling Cycle

1.1 Mathematical Models

Before architects design and construct a building or artists create a sculpture, they usually make a model of the object they have in mind to see if the original objectives are met. An architect would want to make sure that the design is structurally sound (so that the columns do not buckle under the building's own weight or the roof does not collapse under the added weight from unusual amount of snow accumulation, etc.). A sculptor on the other hand may want to be sure that the sculpture conveys to viewers the desired artistic features that generate certain esthetic reaction such as pleasure or shock. As an integral part of producing any of its lines of jet planes, an aircraft company also builds models of these planes and put them to wind tunnel and other tests to investigate their aerodynamic efficiency, structural integrity, passenger comfort and other design criteria.

Physical models often provide more than a valuable tool to gather useful qualitative information and quantitative data about an actual design. They may suggest possible general behavior or properties of a class of related phenomena. An apple is seen to drop to the ground at increasing speed when it leaves a tree. When it happens again and again to other apples as well as other objects in free fall from rest, scientists and engineers accept this observed behavior as a universal truth and apply it to predict the outcome of other falling objects with considerable success. However, there is a limitation to such inferences and extrapolations. For example, when a free falling object is dropped from great height, it can be seen to reach a maximum (terminal) velocity and does not accelerate beyond that speed. At high speed, air drag from surrounding atmosphere would provide a non-negligible resistive force that eventually balances the gravitational force. Inference from apples falling from trees (and hence not subject to significant air drag) would lead to incorrect (and possibly catastrophic) conclusions for objects falling from much greater heights. This is just one of many examples that shows any universal truth inferred from empirical data requires validation. Yet additional experiments at adequately high altitudes for substantial air drag may not be possible or sufficiently encompassing.

There are also phenomena and situations for which we simply cannot produce a physical model for testing and validation. An example is our economy. How do we build a model to test whether a stimulus package would get us out of the great recession or ruin it altogether as some critics would have it? Another is testing whether a certain drug dosage would cure a disease or kill the patient because of the excessive toxicity of the dosage. The drug dosage issue has led to the current federal policy and practice of phase testing on animal subjects before trying the drug on human. While risk to human is reduced by regulations established by the Federal Food and Drug Administration (FDA), it is not eliminated.

Even if you can build a model, it may not be advisable to do so for a number of reasons. One of these would be the cost involved in model construction and model testing. Other show-stoppers include 1) the lack of scalability of phenomena that do not behave in proportion to size, and 2) the return time of information sought is simply too long or too slow to meet the actual needs.

For all these reasons, we wish to have a different way to ensure the success of our endeavors other than building and testing physical models alone whether or not it is possible to do so. One effort in this direction is to formulate a mathematical representation of the particular phenomenon or activity in question (more conventionally known as a *mathematical model*) that can be used to generate the information we wish to know or to offer another tool for validating a general principle we may have inferred from the empirical data on the phenomenon. It is reported in the literature that mathematical modeling occurred at least as early as year 1225 when Robert Grosseteste advanced his mathematical theory for the Big Bang in cosmology. Since then, the mathematical modeling process has been refined and evolved over time and has especially flourished over the last few centuries. We are therefore in a great position to benefit from these developments.

If you have no prior experience working with mathematical modeling, it is natural to ask what does it entail and how you may proceed. Perhaps the best answer would be to see the actual development of such a model and the process of analyzing it. We do this by a simple model on the growth (and decline) of the Earth's human population in the next chapter. However, there may also be some benefits to be gained from an overview of the mathematical modeling process to provide a framework for the development of specific models. The remaining sections of this chapter provide such an overview. They summarize and expand on what have already been observed in existing text on mathematical models and their analysis such as [4, 33]. If you are a person of action and does not benefit from abstract discussion, please feel free to skip to the next chapter and see how the modeling process actually works.

1.2   The Modeling Cycle

Mathematical modeling is not one but (for our purposes) four different sets of activities that cycle through repeatedly until a reasonable goal is reached (or clearly not attainable). We sketch a broad outline of the activities of each step of this cycle in the subsections below. These activities will be fleshed out in the chapters to follow when we model specific phenomena.

1.2.1 *Formulation*

Mathematical modeling is usually prompted by questions we have about a phenomenon of interest and want to uncover the answers through mathematical relationships that characterize the phenomenon. An adequate set of mathematical relations (e.g., equations, inequalities, truth tables, etc.) based on relevant scientific principles or reasonable assumptions constitute a mathematical model for the phenomenon. It is generally advisable to begin a study of a particular phenomenon with the simplest meaningful model. By this, we mean a model with a smallest and simplest set of mathematical relations that seems adequate for the questions we have about the phenomenon. Understandably, this recommendation is rather vague and requires more clarification. However, there is no recipe for formulating the simplest model; only that it generally requires good judgement and sensible decision on the part of the modeler and any choice made is not irrevocable. Specific examples in subsequent chapters will help to clarify this recommendation.

1.2.2 *Analysis*

The next set of activities after model formulation is to analyze the mathematical relations of the model formulated and extract from them relevant information on the phenomenon of interest by known mathematical and computational methods. On occasions, new mathematics or computational techniques may have to be developed for this purpose. Since the models in this volume involve only ordinary differential equations, it is expected that the mathematical and computational methods needed are generally related to such equations. The type of information sought often helps to narrow the range of mathematical and computational techniques for this analysis phase of modeling.

1.2.3 *Validation*

The information extracted from our model in the previous step should be given a reality check. If the output from the model is consistent with known observations or reasonable expectation of the phenomenon, the model may be used to generate more information that helps to answer questions we have about the phenomenon or for other applications. If not, we would need to go to the next phase of the modeling cycle for an improved model and better model performance. Even if the model should pass whatever reality checks available, there may still be reasons to refine it for more informative or useful output.

1.2.4   *Improved formulation*

If the model being examined is found to be inadequate in the validation phase or a more informative model is called for, we would have to modify the *first model* by refining it to include more features and properties of the phenomenon being modeled. There are no recipes on what more should be included or how the model should be modified for improved performance. Fruitful modifications are often model-specific but the models examined in this volume were selected to provide some concrete approaches for *model improvement* that are applicable to many other models. Even if the first model should prove to be very representative of the modeled phenomenon, it may still be important to improve it further. A historically important example is the tiny discrepancy between the prediction of the Newtonian theory of planetary motion and the actual orbit of planet Mercury. The accumulation of a small discrepancy for each complete orbit over many centuries has become noticeable and the Newtonian model had to be modified to get a more accurate theory. This was accomplished eventually by Einstein's theory of relativity.

1.2.5   *The cycle*

With an improved model, we would repeat the earlier steps of the modeling cycle:

- Model Formulation (or modification)
- Model Analysis
- Model Validation
- Applications or/and Model Modification (for improved output)

The process continues cyclically until our objectives are met (though we may cycle the process again at a later date when a need for more information arises). We shall see in some details how these four phases of mathematical modeling at work in the chapters to follow (see also [33]).

1.3   Scientific Issues

Conventional teaching of mathematical modeling tends to focus on the "analysis and computation" phase of the modeling cycle because the discussion is generally more systematic, more universal and more suitable for classroom instruction. In contrast, there are no fixed rules or recipes for the formulation of mathematical models. The method of construction varies with scientific discipline and even from phenomenon to phenomenon within a given discipline. Furthermore, the background knowledge of science needed to investigate a particular phenomenon also varies from phenomenon to phenomenon. For this reason, the case study approach became an alternative for those who wish to emphasize the modeling aspect in teaching mathematical modeling. Our principal goal here eschews both. Instead, we wish to emphasize the fact that an appropriate mathematical model for any phenomenon to be investigated must be grounded on the scientific issue(s) needed to be addressed by the investigation: What do we want to know about the phenomenon? Are we interested in

- How does the phenomenon change and evolve with time?
- Whether it eventually evolves into a time-independent or time-varying steady state?
- Whether small perturbations drive the phenomenon away from a particular steady state?
- Whether a small change in the system's constitution causes qualitative changes in the evolution of the phenomenon?
- Does a biological organism maintain its normal functions or developments after experiencing environmental or genetic changes?
- What is the mechanism for maintaining normal development?
- How can the outcome be minimized or maximized?
- How do we steer a phenomenon toward a desired outcome?
- How does signal or information move from its spatial location?
- Does a concentration gradient induce movement and spread of signal or information toward uniformity?
- Does deterministic input always lead to the same output?
- How do we extract useful information from output in an uncertain/noisy environment?
- ...

In addition to helping the development of an appropriate mathematical model for the problem of interest, knowing the scientific issues to be addressed also help identify the relevant analytical and computational methods for extracting the desired information from the model. The different models examined herein are selected to showcase many of the following *mathematical issues* (corresponding to the questions raised above):

- Evolution
- Equilibrium and Periodicity
- Stability
- Bifurcation
- Robustness
- Feedback and Redundancy
- Optimization
- Control
- Propagation
- Diffusion
- Chaos
- Randomness (stochasticity).
- ...

Certain mathematical and computational techniques are known to be relevant to each of these scientific issues for extracting the desired information from a mathematical model. Hence, identifying the scientific issue(s) of interest often leads to a more relevant and informative model and suggests the type of analysis and computation appropriate for the model.

1.4   Scales of Biological Phenomena

Scientists and mathematicians face many challenges in mathematical modeling. Among the challenges in modeling biological phenomena is the very wide range of scales of these phenomena. Very roughly, the range in size spans

- Molecular
- Cellular
- Anatomical
- Organismic
- Population
- Species
- Ecosystem

Within each of these categories, there may also be a wide range in size. For example, at the level of organisms, their size ranges from bacteria of the order of a micron (10^{-6} meter) to the size of an elephant.

Temporal scales may also differ by orders of magnitude as we move down the list of phenomenon types below:

- Molecular
- Cellular
- Developmental
- Organismic
- Societal
- Evolutionary

Modeling phenomena that involve several scales of different orders of magnitude (such as mutation, cell proliferation and tumor growth for example) poses serious computational challenges. The high resolution required for analysis and computation for the upper part of the list is typically unnecessary and wasteful for those near the bottom. There is also the time consuming computation that may be unacceptable for the time and resources available for the problem.

One way to overcome these multi-scale phenomena is to organize our modeling investigation according to their scale and focus on components with the same length and/or time scale. Another would be to move a component up the length or time scale by focusing on its average behavior over a temporal or spatial span to result in what is known as macroscopic behavior like the others. Most models in this volume are consequences of these two ways to reduce the investigation of complicated phenomena to a manageable form. Other approaches include doing the analysis and computation for each component at its level and then integrate the output of the different components.

1.5 Let's Get Started

In the sections above, we highlighted three areas that require our attention as we begin to engage in mathematical modeling of biological phenomena. While there are more extensive and broader general discussions of the mathematical modeling process, the modeling cycle, the scientific issues and the range of spatial and temporal scales are singled out here for the readers to see how they play out in the models discussed in this volume. The discussion of these topics are also intended to sensitize present and future modelers to the need to address them in their own modeling work. However, no general discussion of the modeling process in the abstract can provide a mathematical modeler all the needed tools to do mathematical modeling. Only by seeing actual modeling at work for specific cases and hands-on experience to develop a few models can we accumulate the experience and expertise to become knowledgeable and successful mathematical modelers. Let us begin to develop in the next chapter a simple mathematical model for the familiar phenomenon of the growth of human population on Earth to see how these general notions such as the modeling cycle and scientific issues play out in practice.

Part 1

Growth of a Population

Chapter 2

Evolution and Equilibrium

2.1 Growth without Immigration

2.1.1 *Population size dependent growth rates*

As indicated earlier, the goal here is to use a simple problem to begin illustrating the mathematical modeling process. We do so by modeling the familiar growth of human population on Earth. Typically, mathematical modeling is prompted by the modeler's quest for some information about a phenomenon. The human population on Earth has grown from 4 billions in 1974 to 7 billions by 2011. Naturally, we are concerned about future growth and want to know how the population size will change with time.

To quantify the evolution of the human population, let $y(t)$ be the population size at time t. Size may be measured by the number of individuals or tonnage of total biomass. Time may be measured in seconds, minutes, hours, days or years from some reference time. To answer by mathematical modeling the question how does the population size evolves with time, we need to formulate some mathematical relation(s) about $y(t)$ on the basis of known scientific principle (such as Newton's law in physics for motion of mass particles). Unfortunately, there is no equivalence of Newton's law in biology that governs and regulates population growth. In that case, the modeler would have to adopt some reasonable assumptions (postulates) for the phenomenon on hand based on available data and general observations. For a first model, we wish to avoid plunging into highly technical area of statistical analysis of available data or the complex relations of birth and death processes. Instead, we start with a simple phenomenological model of population growth.

For this simple first model, we should not (and cannot) make assumptions on the size of the population at any future time $y(t)$, since that is what we want to deduce from the model. This forces us to focus on the next level of observable, the change of the population with time. With the human population on Earth changing in fractions of a second (as can be seen from the U.S. Department of Commerce website on the U.S. and World population clock [31]), it is not unreasonable to think of the two variables y and t as continuous variables if we measure time in years and size in tons of biomass or billions of individuals. In that case, we may work with the instantaneous rate of change of the Earth's (human) population dy/dt in developing our mathematical model. In fact, this volume is concerned mainly with phenomena that can be modeled by relations involving instantaneous rates of change. When the rate of change involved is with respect to time, such models are known as *dynamical systems*.

Extensive observations on a variety of populations suggest that the rate of change of the human population should depend on the size of the existing population: It would even be tempting to think that the more people there are, the faster the population would grow. However, this property would not hold if growth is limited by finite resources. So for the simplest model, we may reasonably limit our consideration to a growth rate that depends only on the size of the current population and adopted it as our first (and simplest) phenomenological model for the growth of a relatively small population. Mathematically, this *instantaneous rate of growth postulate* can be stated as the following mathematical relation:

$$\frac{dy}{dt} = f(y) \qquad (2.1.1)$$

where $f(y)$ is the growth rate function (dependent only on the size of the current population $y(t)$ in this first model). While we may not know the specific form of the function $f(\cdot)$ without additional information, the observable facts of the human population on Earth suggest that it has the following two general properties:

$$(i) \;\; f(0) = 0, \quad (ii) \;\; f'(0) = \left[\frac{df}{dy}\right]_{y=0} \equiv a_0. \qquad (2.1.2)$$

The first condition characterizes the fact that for the human population, we must have people to beget offsprings. The second assures a change in population size with time as long as there are people. With these two assumptions, Taylor's theorem for a twice differentiable $f(y)$ gives

$$f(y) = f(0) + f'(0)y + \frac{1}{2}\, f''(\varsigma)y^2 \simeq a_0 y$$

for a sufficiently small y.

If $a_0 = f'(0) > 0$, the relation (2.1.1) in fact reflects a population that increases initially: the larger the current population the faster the growth. However, when the population is sufficiently large, the neglected quadratic term would become important and cannot be omitted. In reality, limited resources (such as land and water) may limit further growth. We will address the issue of limited resources in a later section. Here, we focus on population sizes for which further growth is still possible.

It is also possible that $a_0 = f'(0) < 0$ for some other populations, e.g., those required encounter for mating. In that case, the population would decrease toward extinction if the initial population is sufficiently small. We will also have occasion to encounter this type of populations in later sections of this volume.

2.1.2 *Solution of the model ODE*

The relation (2.1.1) is a mathematical equation for the unknown human population $y(t)$. However, it is an equation in which the derivative of the unknown, dy/dt, also appears. As such, it is called a *differential equation* to distinguish it from the more familiar algebraic equations. With only functions of one variable and their derivatives (and no partial derivatives) in such equations, they are known generally as *ordinary differential equations* (customarily abbreviated as ODE for brevity). Some basic information on methods of solution for such equations can be found in the appendices of this volume. We also develop the most basic idea for solving this type of equations, the reduction to a calculus problem, for this first model and a few other simple models in the next few chapters.

Normally, we would have to specify the function $f(\cdot)$ in order to solve the ODE for $y(t)$. To illustrate, if $f(y) = a_0 y$ where a_0 is some known constant, we may re-arrange (2.1.1) to read

$$\frac{1}{y}\frac{dy}{dt} = a_0. \tag{2.1.3}$$

But we know from the chain rule of calculus that $d\left[\ln(y(t)\right]/dt = y^{-1}(dy/dt)$ so that the re-arranged ODE (2.1.3) becomes

$$\frac{d}{dt}\left[\ln(y(t))\right] = a_0$$

or

$$\ln(y(t)) = a_0 t + C_0$$

where C_0 is a (yet unknown) constant of integration. Exponentiate both sides of the relation above, we get

$$y(t) = Ce^{at} \tag{2.1.4}$$

where $C = e^{C_0}$ is a new constant of integration.

Evidently, the method for solving the first order ODE (2.1.1) depends on the fact that we were able to separate the unknown y from the independent variable t on the right-hand side of the ODE. This is straightforward for the simple growth rate function $f(y) = a_0 y$. But even if the constant a_0 is replaced by a function $a(t)$ and the ODE becomes

$$\frac{dy}{dt} = a(t)y = f(y;t), \tag{2.1.5}$$

the separation of quantities involving each variable can still be accomplished to get $y^{-1}(dy/dt) = a(t)$ so that

$$y(t) = Ce^{A(t)}$$

with

$$A(t) = \int^{t} a(z)dz.$$

An ODE of the form (2.1.1) where the right-hand side f does not depend on t explicitly is said to be an *autonomous* ODE. If f does depend on t explicitly as in (2.1.5), the ODE is said to be *non-autonomous*.

2.1.3   *Initial condition and the initial value problem*

In a first course in calculus, the instructor usually takes great pain to emphasize the importance of a constant of integration when we integrate. We see presently why such constants are critical in mathematical modeling. In the question we posed that led to the relation (2.1.1), we started with the knowledge that the Earth's population being 7 billions in 2011. Whatever the model we choose to study the Earth's future population, it must be consistent with that fact. Since, the calendar label for time is merely a convention, we may more conveniently let $t = 0$ be year 2011 (or more precisely, for that moment in 2011 when the human population reached 7 billions). In that case, we have

$$y(0) = 7 \text{ (billion)}.$$

(Note that such re-labeling of time would not be appropriate if the ODE system is non-autonomous unless we make a corresponding change of t in $f(y,t)$.) With the simple model (2.1.3) and its consequence (2.1.4), we can choose the constant C to meet this requirement:

$$y(0) = \left[Ce^{a_0 t}\right]_{t=0} = C = 7,$$

so that

$$y(t) = 7e^{a_0 t}. \tag{2.1.6}$$

More generally, if the initial population is Y_0 (instead of 7 billion) individuals, we would have

$$y(0) = Y_0 \tag{2.1.7}$$

and

$$y(t) = Y_0 e^{a_0 t}. \tag{2.1.8}$$

The method of reduction to a calculus problem for the special case of a linear growth rate can also be carried out for a general $f(y)$ in (2.1.1) to get

$$\frac{1}{f(y)}\frac{dy}{dt} = 1 \quad \text{or} \quad \int_0^y \frac{dz}{f(z)} = t + C_0 \tag{2.1.9}$$

where C_0 is a constant of integration. It is fixed by the initial condition (2.1.7) to give

$$\int_{Y_0}^y \frac{dz}{f(z)} = t. \tag{2.1.10}$$

Whether or not we can find an anti-derivative of $1/f(y)$, the relation (2.1.10) gives t as a function of y. Under suitable condition, we can solve (2.1.10) for y in terms of t to get $y = Y(t)$ (to be shown below for some special cases). Whether or not (2.1.10) is invertible, the relation describes how one variable evolves as the other variable changes and we can always plot t as a function of y (by integrating the right-hand side of (2.1.10) numerically if necessary). By interchanging the role of t and y, the same graph can also be viewed as y as a function of t.

The general ODE (2.1.1) and the initial condition (2.1.7) together constitute an *initial value problem* (or IVP for short). Had we forgotten to include a constant of integration in handling the calculus problem associated with (2.1.1), we would not be able to meet the requirement posed by the *initial condition, a known fact that must be a part of the model.* Since the study of the Earth's human population cannot proceed without either of these two items, the growth rate and the initial population, it is the IVP (and not just the ODE (2.1.1)) that constitutes the mathematical model for investigating the *evolution* of a population with time.

It should be noted that the choice of reference time may be changed provided that we make the corresponding change in the determination of the constant of integration. For example, if we take the time unit to be the calendar year and denote year 2011 by t_0 with $y(t_0) = Y_0$, then we have instead of (2.1.10)

$$t - t_0 = \int_{Y_0}^{y} \frac{dz}{f(z)}.$$

For the simple first model of a linear growth rate (2.1.4), the application of the initial condition led to

$$y(t_0) = Ce^{a_0 t_0} = Y_0$$

or

$$C = Y_0 e^{-a_0 t_0}$$

so that

$$y(t) = Y_0 e^{a_0(t - t_0)}.$$

2.2   Exponential Growth

2.2.1   *Linear growth rate*

In order to obtain specific consequences of the mathematical model of the last section (e.g., the size of the population by year 2050), we noted that we need to specify the function $f(y)$ in the ODE (2.1.1). To illustrate how we may obtain such consequences as well as further developments of the modeling process, we adopted the simplest and rather reasonable hypothesis that the rate of population growth is proportional to the current size of the population so that $f(y) = a_0 y$ where a_0 is the constant of proportionality. In that case, the IVP for the mathematical model adopted becomes

$$y' = a_0 y, \qquad y(0) = Y_0 \tag{2.2.1}$$

where we have used a prime, $'$, to indicate differentiation with respect to the independent variable t. The solution for this IVP has already been found to be $y(t) = Y_0 e^{a_0 t}$ in (2.1.8). With the linear ODE (2.2.1) leading to an exponentially growing population, the corresponding model is sometime referred to as the *exponential growth model*. With the right-hand side $a_0 y$ of the ODE being linear in y, the same IVP is also called the *linear growth rate model*.

2.2.2   The growth rate constant

With Y_0 specified by the initial condition (e.g., 7 billion in 2011), we still need the value of the *growth rate constant* a_0 in order for the solution (2.1.6) of the IVP to provide specific information on the Earth's population in future times. Similar to how we chose the constant of integration Y_0, we may fix the value of a_0 by requiring the model be consistent with another piece of information from census data.

Suppose we continue to take $Y_0 = 7$ (billion) and measure time in years with $t = 0$ as year 2011 (without being too precise about the moment of the population reaching Y_0). We know from available data that the Earth's human population was 4 billion in the year 1974. For (2.1.6) to be consistent with this second piece of data, we would want $y(-37) = 4$ (since $t = 0$ being year 2011) so that

$$4 = 7 \times e^{-37a_0} \qquad \text{or} \qquad a_0 = -\frac{1}{37}\ln\left(\frac{4}{7}\right) = 0.01512475\cdots. \qquad (2.2.2)$$

The corresponding solution of the IVP is then

$$y(t) = 7e^{(0.01512475\cdots)t}. \qquad (2.2.3)$$

It should be noted that $a_0 = 0.015\cdots$ is positive so that, consistent with our expectation, the rate of growth of the population size is an increasing function of the population size at that instant in time. It is important to keep in mind that the positivity of a_0 is a consequence of the data available for the human population and not an assumption. In particular, the sign may not be the same for a different population.

2.2.3   Model validation

While we can now use (2.2.3) to answer more concretely the questions we asked at the start of the modeling discussion and others, it would be prudent to see how consistent is the simple exponential growth model with available census data beyond those used up for determining the two parameters Y_0 and a_0 in the linear IVP (2.2.1). In Table 2.1, we show a representative set of census data from the past on the last line and the prediction by (2.2.3) on the third line.

Table 2.1

$Year$	1804	1927	1959	1974	1987	1999	2011	2023	2056
t_n	−207	−84	−52	−37	−24	−12	0	23.58	45.83
$y(t_n)$	0.306	1.97	3.19	4	4.87	5.84	7	10	14
$\bar{y}_n = data$	1	2	3	4	5	6	7	?	?

Evidently, the predictions by the solution (2.2.3) of the linear (*aka* exponential growth) model seem reasonable for data shown back to 1927 (over a span of 84 years) with a maximum percentage error less than 7%. The adequacy of exponential growth deteriorates dramatically for the remaining data point displayed in the table. Still, it is reasonable to think that the predictive value of the model should be adequate for the next 50 years (after 2011) given its accuracy for an even longer period of the recent past.

2.2.4 Prediction of future population

If we are satisfied with the level of accuracy shown by the exponential growth model when compared with available census data, we can use (2.2.3) to answer questions about the human population of interest:

- **When will the Earth's population reach 10 billion?** Suppose it is in year t_1. Then the solution (2.2.3) requires

$$10 = 7e^{(0.01512475\cdots)t_1}$$

or

$$t_1 = \frac{1}{0.01512475\cdots} \ln\left(\frac{10}{7}\right) = 23.5822 \ldots \quad \text{(years after 2011)}$$

The answer to our question is by (mid) 2034.

- **What is the Earth's population in year 2050?** With year 2050 corresponding to $t_2 = 39$ (years after 2011), the solution (2.2.3) gives

$$y(39) = 7e^{(0.01512475\cdots)(39)} = 12.626\cdots \; billion$$

- **When will the Earth's population double (to 14 billion)?** Let that year be t_3. Then the solution (2.2.3) requires

$$14 = 7e^{(0.01512475\cdots)t_3}$$

so that

$$t_3 = \frac{1}{0.01512475\cdots} \ln(2) = 45.8286671\cdots \text{ (years after 2011)}$$

The Earth's human population is predicted to double by late 2056 according to our exponential growth model.

With the model predicting the Earth's population growing exponentially without bound, it is expected intuitively that it ceases to be appropriate beyond some year in the distant (or not so distant) future as growth should eventually be limited by finite resources (such as food, water, land, etc.). We should therefore consider possible improvement of our model to remedy this deficiency for application over a broader range of population size. This we will do in the next chapter.

2.3 Estimation of System Parameters

2.3.1 The least squares fit

Before embarking on modifying the model (2.2.1) for possible improvements, we note that there are ways to improve on the predictive power of this linear growth rate model without changing the actual model. One of these is a better use of the data available from the U.S. Census Bureau record for determining the two parameters Y_0 and a_0 in (2.1.8). Up to now, we have made use of only two data points to fix these two parameters. In this first effort, the census data for the population of 2011 was used as the initial condition to determine Y_0 and the data point for 1974 was used for determining the growth rate constant a_0. It would seem that we could get better accuracy if we use more of the available data (even if not all of them). However, the model (2.2.1) which gave rise to (2.1.8) contains only two parameters that can be used to fit the available data. As such, only two data points can be fitted exactly. To make use of more data points, we would have to either modify the model to contain more parameters (possibly as many as the number of data points we wish to fit exactly) or consider alternative ways to use the data. We will do the first of these in the next and later chapters. In this section, we illustrate how the second approach may be implemented by the *method of least squares*.

Suppose the evolution of the human population is given by

$$y(t) = Y(t; a_1, a_2, \cdots, a_m)$$

where $\{a_1, a_2, \cdots, a_m\}$ are m parameters at our disposal for fitting data and $Y(t; a_1, a_2, \cdots, a_m)$ is a known function of time t and these m parameters. An example is the exponential growth function (2.1.8) with $y(t) = Y(t; Y_0, a_0) = Y_0 e^{at}$ that has two parameters $a_1 = Y_0$ and $a_2 = a_0$. For any given choice of values for these m parameters, the value $y(t_k) = Y(t_k; a_1, a_2, \cdots, a_m)$ would generally not be in agreement with the census data $\bar{y}_k$ for $t = t_k$ (such as those given in the last line of Table 2.1) leaving us with a residual

$$E_k = Y(t_k; a_1, a_2, \cdots, a_m) - \bar{y}_k.$$

If we want to make use of N census data points $\{\bar{y}_1, \bar{y}_2, \bar{y}_3, \cdots, \bar{y}_N\}$, there would be N such residuals $\{E_1, E_2, E_3, \cdots, E_N\}$. If $N \leq m$, we can choose $\{a_1, a_2, \cdots, a_m\}$ to make all $E_k = 0, k = 1, \ldots, N$. (In fact, we can do it with just N of the m parameters and set the remaining $m - N$ parameters equal to zero.)

On the other hand, if $m < N$, then only m of the $\{E_k\}$ can be reduced to zero with the m parameters at our disposal, leaving the other $N - m$ residuals generally with a positive value. To the extent that we do not wish to favor some of the N residuals and ignore the others for the case $m < N$, one option is to choose the m parameters to render some combinations of the N residuals to zero. We can do this for as many as m different (independent) combinations, with all m parameters specified by m different (appropriate) conditions. Instead of implementing either of these two approaches for the $m < N$ case, we describe in the next subsection one of the most popular methods for choosing the m unknown parameters $\{a_1, a_2, \cdots, a_m\}$ known as the *least squares method* (a special case of *regression analysis* in statistics).

2.3.2   *The least squares method for exponential growth*

To illustrate the *method of least square fit*, suppose we want to make use of the four pieces of census data from the U.S. Census Bureau for $1974, 1987, 1999$ and 2011 in

Table 2.1 to determine the two parameters Y_0 and a_0: Clearly, we cannot choose Y_0 and a_0 to match all four data points exactly. Whatever choice we make, there would generally be a difference $y_n - \bar{y}_n$ between the value $y_n = y(t_n) = Y_0 e^{a_0 t_n}$ and the Census Bureau data $\bar{y}_n$ for l_n for two or more of the data points. Since we cannot make all the differences zero, we will do "the best we can." To do that, we need to specify what do we mean by "best"? One possible criterion is to minimize the sum of the squares of the discrepancies:

$$S_N^2 = (y_1 - \bar{y}_1)^2 + \ldots + (y_N - \bar{y}_N)^2 = \sum_{k=0}^{N} (y_k - \bar{y}_k)^2. \qquad (2.3.1)$$

Note that we should not minimize just the sum of all discrepancies as that would allow cancellations of large discrepancies with opposite signs.

With the four data points for years 1974 ($t_1 = -37$), 1987 ($t_2 = -24$), 1999 ($t_3 = -12$) and 2011 ($t_4 = 0$), the quantity y_n in (2.3.1) is given by (2.1.8) to be $Y_0 e^{a_0 t_n}$. As such, S_N^2 is a function of Y_0 and a_0. Our (least squares) minimization criterion is to choose Y_0 and a_0 so that S_N^2 is (at least) a (local) minimum. From elementary calculus, we know that this minimum is attained at a stationary point of $S_N^2(Y_0, a_0)$ determined by

$$\frac{\partial S_N^2}{\partial Y_0} = 2 \sum_{k=1}^{4} \left(Y_0 A^{t_k} - \bar{y}_k \right) A^{t_k} = 0 \qquad (2.3.2)$$

$$\frac{\partial S_N^2}{\partial a_0} = 2 \sum_{k=1}^{4} \left(Y_0 A^{t_k} - \bar{y}_k \right) t_k Y_0 A^{t_k} = 0 \qquad (2.3.3)$$

where $A = e^{a_0}$. The first equation can be solved for Y_0 to get

$$Y_0 = \frac{\sum_{k=1}^{4} \bar{y}_k A^{t_k}}{\sum_{k=1}^{4} A^{2t_k}} = \frac{\bar{y}_1 A^{t_1} + \bar{y}_2 A^{t_2} + \bar{y}_3 A^{t_3} + \bar{y}_4 A^{t_4}}{A^{2t_1} + A^{2t_2} + A^{2t_3} + A^{2t_4}}$$

$$= \frac{4A^{-37} + 5A^{-24} + 6A^{-12} + 7}{A^{-74} + A^{-48} + A^{-24} + 1}. \qquad (2.3.4)$$

The expression (2.3.4) can be used to eliminate Y_0 from the second stationary condition (2.3.3) to get

$$\frac{\sum_{m=1}^{4} \bar{y}_m A^{t_m}}{\sum_{m=1}^{4} A^{2t_m}} \sum_{k=1}^{4} t_k A^{2t_k} = \sum_{i=1}^{4} t_k A^{t_k} \bar{y}_k \qquad (2.3.5)$$

or, with $t_4 = 0$ corresponding to 2011,

$$\frac{7 + 6A^{-12} + 5A^{-24} + 4A^{-37}}{1 + A^{-24} + A^{-48} + A^{-74}} = \frac{(6 \cdot 12) A^{-12} + (5 \cdot 24) A^{-24} + (4 \cdot 37) A^{-37}}{12A^{-24} + 24A^{-48} + 37A^{-74}}. \qquad (2.3.6)$$

After cancellation of the A^{-111} terms, we are left with

$$x^{12} P_{86}(x) = 0$$

where $x = 1/A = e^{-a_0}$ and $P_{86}(x)$ is a polynomial of ($x = 1/A$ of) degree 86. It can be solved for x using *NSolve* of Mathematica (or similar software in Maple or MatLab). Remarkably, there is only one real root corresponding to a real a_0 value:

$$e^{-a_0} = x = 0.9852906761558474 \cdots \quad \text{or} \quad a_0 = 0.0148186 \cdots . \tag{2.3.7}$$

We then get from (2.3.4)

$$Y_0 = 7.06244 \cdots \tag{2.3.8}$$

so that (2.1.8) becomes

$$y(t) = 7.06244 \cdots \; e^{(0.0148186 \cdots)t}. \tag{2.3.9}$$

As found in (2.3.7) and (2.3.8), the values for Y_0 and a_0 determined by the least squares method are nearly the same as those with $Y_0 = \bar{y}_4 = 7$ and $a_0 = 0.01512475 \cdots$ determined by only the two data points for 1974 and 2011.

We can now use (2.3.9) to compute the corresponding y_k for different past and future years:

Table 2.2   Solution by Least Squares.

$Year$	1974	1987	1999	2011
$y_n = y(t_n)$	4.08...	4.95...	5.91...	7.08...
$\bar{y}_n = data$	4	5	6	7

The agreement between the actual data and (2.3.9) based on the *least squares method* using four data points is evidently "better" than the exact fit with two data points of the previous section for the period of 1974–2011. The error in the predictions for 1974 is well below 3% while the errors for the other years after 1974 are smaller than what we got previously using only the data for 1974 and 2011. However, the improvements are not so significant and, similar to the previous (two data points) solution, the accuracy of the prediction deteriorates as we move further away from the years involved in the least squares solution (as shown in Table 2.3). While the prediction for the population in 1804 is slight better than that by using only two data points, the least squares fit solution predicts only about 1/3 of the 1 billion population reported by U.N. census. As approximations of the actual data, the prediction of (2.3.9) is just as unacceptable as (2.2.3) sufficiently far away from the 1974–2011 range.

Table 2.3

$Year$	1804	1927	1959	1974	1987	1999	2011
n	−207	−84	−52	−37	−24	−12	0
$y_n = y(t_n)$	0.33	2.03	3.27	4.08	4.95	5.91	7.06
$\bar{y}_n = data$	1	2	3	4	5	6	7

2.4   A Constant Growth Rate Model

Can we improve on the accuracy of a linear growth rate model by using more known data points from the U.S. Census Bureau data in a least squares fit? Or must we change the mathematical model to get a better fit. Such questions take us again to the model modification and improvement phase of the modeling cycle. An answer to the first will be left to the reader who have now been provided with a framework to seek it (or to learn more about statistical methods before attempting). As for the second option, there are definitely other models different from (2.2.1); some are quite competitive and credible. To illustrate, we consider a model related to a linear fit in statistics. If we replace the linear growth rate assumption by a constant growth rate hypothesis, i.e., $y'(t) = b$ (a constant), the population size at time t can immediately be obtained by simple integration:

$$y(t) = c + bt, \qquad\qquad (2.4.1)$$

where c and b are two parameters to be chosen to fit a set of available data by some criterion (including the least squares fit). Fitting data with a straight line is one of the simplest and often employed statistical tools (linear regression) especially when no characterization appropriate for the phenomenon such as (2.1.1) suggests itself. (In the context of evolution of the human population, adoption of the linear regression amounts to an assumption of a constant growth rate.)

Exercise 1 *a) Determine the values of b and c in (2.4.1) by a least squares fit using the same 4 data points (for 1974, 1987, 1999 and 2011). b) Use the result from a) to generate the predicted population $y(t)$ for the six years spanning 1927–2011 in Table 2.1. c) Compare with the corresponding results in Tables 2.1 and 2.2.*

The following exercise relates the exponential growth parameters to those for the linear growth rate model.

Exercise 2 *a) Take the natural logarithm on both sides of (2.1.8) to get $z(t) = \alpha + \beta t$ and relate $z(t)$, α and β to $y(t)$, a_0 and Y_0.*

b) Determine Y_0 and a_0 of (2.1.8) for the four data points of Table 2.2 but now using $z(t) = \alpha + \beta t$ instead of (2.1.8).

With the results from the exercises above, we can compare the prediction from the fit by (2.4.1) with the linear growth rate model. Whatever the comparison may show, it is evident that the new model (2.4.1) still predicts unbounded growth with time. A different model is needed to capture the growth limiting effects of the Earth's finite resources. Such a model will be developed in the next chapter. In the rest of this chapter, we continue to examine populations with a linear growth rate but now allow for the possibility of immigration.

2.5   Intravenous Drug Infusion

2.5.1   *Questions on hypoglycemia*

Intravenous infusion of glucose into the bloodstream is an important medical procedure for individuals suffering from hypoglycemia (abnormally low level of glucose
in blood stream). In order to raise the concentration of glucose in the patient's
blood stream to an adequate level and maintain it at that level, a common practice is to infuse glucose solution continuously into the patient's body intravenously.
Suppose that glucose is infused at the constant rate c grams per minute while it is
concurrently converted into energy as well as removed from the bloodstream at a
rate proportional to the amount of glucose in the body. Of interest is the amount
of glucose in the blood stream at any future time $t > 0$. More specifically, if the
patient needs to maintain glucose in the blood stream at the level of g grams, what
does the infusion rate c have to be in order to attain and maintain that glucose
level?

2.5.2   *Linear models with immigration*

Let $G(t)$ be the amount of glucose (in units of grams) in the blood stream at time t
and the initial level of glucose be G_0. For a simple model that includes the essential
features of the patient's changing glucose concentration, it seems reasonable to
assume that glucose is lost from the blood stream (through conversion and removal)
in proportion to the amount present with a rate constant a_0 that can be determined
from measurement record. If glucose is infused into the blood stream at a rate $c(t)$,
the instantaneous time rate of change of $G(t)$ is taken to be the difference between
the rate of loss (through conversion and removal) and the infusion rate so that we
have the following linear ODE model for glucose concentration:

$$\frac{dG}{dt} = -a_0 G + c. \tag{2.5.1}$$

At the inception of glucose infusion at $t = 0$, we have the initial glucose concentration:

$$G(0) = G_0. \tag{2.5.2}$$

This completes the model formulation part of the mathematical modeling which
consists of an accounting of the different rates of change contributing to total rate of
change of the glucose concentration. The next step is to extract useful information
from the IVP (2.5.1)–(2.5.2) to check for the model's adequacy in answering the
questions we asked at the start of this section.

    Before embarking upon finding a solution of the IVP, we note that the math-
ematical problem here is similar but not identical to the growth of the human
population with the glucose concentration taking the place of population size. On
the one hand, the rate of change of $G(t)$ with time depends on the concentration
$G(t)$ again in the form $G' = f(G)$ if c is a constant. For the model (2.5.1) with

$c = 0$, we have $f(0) = 0$ but now $df/dG < 0$ instead of $df/dy > 0$ in the case of the human population. This reflects the fact that the rate of human population *growth* increases with size while the rate of *reduction* of glucose increases with concentration level.

Even with reduction taken as negative growth, there is still a substantive difference between the present model and that of human population growth previously considered. For the latter, we have $f(y = 0) = 0$ reflecting the fact that the Earth is a closed system; there is no immigration or emigration of human population. In contrast we have for the glucose problem $f(G = 0) = c > 0$ with the positive rate of glucose infusion c corresponding to a rate of immigration when view as population growth.

2.5.3 *Mathematical analysis*

The numerical value of the natural loss rate constant a_0 is now for the glucose concentration. It is to be determined from available data on glucose concentration recorded in the patient's blood stream. We will leave it as a project for the readers to determine the value for this constant from whatever data there may be (available from medical record). The value of a_0 will be assumed to have been determined so that we can proceed to the next phase of the modeling process.

The ODE (2.5.1) is not separable (unless $c(t)$ does not vary with t). However, it is linear and known (from an elementary ODE course) to be solvable with the help of an *integrating factor*. For this method, we rearrange the first order linear ODE for G to read $G' + a_0 G = c$ or after multiplication by a factor $e^{a_0 t} > 0$,

$$e^{a_0 t}\left(\frac{dG}{dt} + a_0 G\right) = \frac{d}{dt}\left(e^{a_0 t} G\right) = c e^{a_0 t}. \tag{2.5.3}$$

In the equivalent form (2.5.3), we have reduced the original ODE to one with a known right-hand side so that we have just a calculus problem. Upon integrating both sides of (2.5.3), we obtain

$$e^{a_0 t} G = C + \int_0^t c(\tau) e^{a_0 \tau} d\tau.$$

The very important constant of integration C allows us to satisfy the initial condition (2.5.2):

$$G(0) = C = G_0$$

so that

$$G(t) = e^{-a_0 t}\left[G_0 + \int_0^t c(\tau) e^{a_0 \tau} d\tau\right]. \tag{2.5.4}$$

Exercise 3 *There is an obvious integrating factor for*

$$y' = -\frac{1}{t} y + t.$$

Find it and use it to determine the solution for this ODE.

2.5.3.1 *Consequences and implications of the model*

In order to obtain specific consequences of our mathematical model such as the amount of glucose in the blood stream at some future time, we consider first the special case of a constant glucose infusion rate $c(t) = c_0$. In that case, we can carry out the integration in the expression (2.5.4) for $G(t)$ to get

$$G(t) = e^{-a_0 t} \left[G_0 - \frac{c_0}{a_0} \left(1 - e^{a_0 \tau} \right) \right] \tag{2.5.5}$$

$$= \left(G_0 - \frac{c_0}{a_0} \right) e^{-a_0 t} + \frac{c_0}{a_0}.$$

Evidently, the glucose concentration evolves from the initial concentration G_0 toward c_0/a_0 as $t \to \infty$ whatever the initial concentration G_0 may be.

Every individual naturally acquires glucose at some rate $c_0 = c_n$ (through normal food ingestion for example) and the glucose in his/her blood stream is removed (through conversion into energy or secretion) from the body at a rate $a_0 G(t)$. In that case, the body reaches a steady state of c_n/a_0 amount of glucose in the bloodstream. If c_n/a_0 is not less than g grams, the level of glucose normally needed by the body, the individual is in normal health as far as supply of glucose is concerned and nothing needs to be done. However, if c_n/a_0 should be less than g, then the individual needs to raise his/her glucose level. In some cases, this is done by intravenous glucose infusion. If the infusion is at a constant rate c_i gms/sec., then our model gives a steady state glucose level of $(c_n + c_i)/a_0 \equiv c_0/a_0$. For the individual suffering from hypoglycemia to maintain a normal level of glucose in the blood stream, c_0/a_0 should be at least g gms so that the steady infusion rate should be no less than

$$c_i = a_0 g - c_n.$$

2.5.4 *Model improvements*

Glucose deficiency in the blood stream means the glucose concentration is far from saturation. Hence, a reduction rate in proportion to the existing concentration should be adequate for the purpose of modeling and analysis. The issue of model validation is more appropriately focused on how intravenous infusion is administered. While a steady and continuous infusion at a constant rate is possible and practiced sometimes, another form of infusion is at constant rate in periodic intervals. Mathematically, this would take the form of

$$c_i(t) = \bar{c}_i \sum_{n=0} \left[H(t - n\Delta) - H(t - \tau - n\Delta) \right]. \tag{2.5.6}$$

The necessary analysis similar to that of the previous subsection should be repeated for this particular form of intravenous infusion but will be left as an exercise. Instead, we consider here a simpler problem of a different kind of periodic infusion rate.

For some insight to the problem of periodic infusion, we consider here the infusion rate

$$c_i(t) = \bar{c}_i \{1 + \sin(\omega t)\} \tag{2.5.7}$$

for which the ODE (2.5.1) becomes

$$\frac{dG}{dt} = -a_0 G + [c_n + \bar{c}_i \{1 + \sin(\omega t)\}] . \tag{2.5.8}$$

The same integrating factor $e^{-a_0 t}$ again yields

$$G(t) = e^{-a_0 t} \left[G_0 + \int_0^t \{(c_n + \bar{c}_i) + \bar{c}_i \sin(\omega t)\} e^{a_0 \tau} d\tau \right]$$

$$= \frac{1}{a_0} (c_n + \bar{c}_i) + e^{-a_0 t} \left[G_0 - \frac{1}{a_0} (c_n + \bar{c}_i) + \bar{c}_i \int_0^t \sin(\omega \tau) e^{a_0 \tau} d\tau \right]$$

$$= \frac{1}{a_0} (c_n + \bar{c}_i) + e^{-a_0 t} \left[G_0 - \frac{1}{a_0} (c_n + \bar{c}_i) - \frac{\bar{c}_i \omega}{a_0^2 + \omega^2} \right]$$

$$+ \frac{\bar{c}_i}{a_0^2 + \omega^2} [a_0 \sin(\omega t) - \omega \cos(\omega t)] .$$

From this exact solution, it appears that even in steady state the glucose concentration in the patient blood stream resulting from this periodic infusion rate continues to fluctuate with time. Though the level of fluctuation varies with the magnitude of the various system parameter values, the result of this kind of medical intervention is qualitatively inferior to that of the steady and continuous infusion at constant rate.

Though the infusion rate (2.5.7) is a highly idealized version of periodic infusion rates such as (2.5.6), its consequences offer valuable insight to more realistic infusion rates such as (2.5.6) since these consequences are expected to be quite representative of those of more realistic practices. Even more significantly, highly idealized models are valuable as they often provide insight to the real phenomenon they model.

Another benefit in a different direction is how the simple form of the forcing term $\sin(\omega t)$ has led to a simpler method of solution for the ODE (2.5.8) known as the method of undetermined coefficients for a particular solution of an inhomogeneous linear ODE with constant coefficients (see [1] for example). It is simpler to execute as the solution process involves mainly algebraic calculations that are considerable simpler than what had to be done to obtain $G(t)$ above. Readers are urged to become more familiar with this method for subsequent applications.

2.6 Mosquito Population with Bio-Control

2.6.1 *The problem of mosquito eradication*

In the discussion of the human population on Earth, we can assume that the growth rate function $f(y)$ has the property $f(0) = 0$ since the Earth is a closed system; there is no immigration/emigration of human beings to/from Earth. This is not the case for many other kind of populations, notably those of any sovereign country such as the United States. It is of some interest to investigate a population that allows immigration. Instead of studying of a human population with immigration (with emigration treated as negative immigration), we consider in this section the problem of eradicating mosquitoes in some neighborhood around a large pond infested with mosquito larvae. The larvae become mosquitoes which in turn breed more larvae, etc. The residents around the pond would like to get rid of the larvae and mosquitoes that have been causing considerable aggravation and potentially health problems. They can obtain from Vector Control, the county or state agency dedicated to protecting public health by controlling rats, flies, mosquitoes, and other vector related problems, some bio-control agents (such as mosquito fish, dragonflies and/or other natural predators of mosquitoes) to control and preferably eradicate the mosquito population.

2.6.2   *Constant rate of eradication*

Exercise 4 *a) We can begin with the simpler problem of bio-agent eradicating the mosquito larvae population $M(t)$ at a constant rate b_0 and formulate a linear growth rate model on that basis.*

b) Obtain the exact solution of the IVP with an initial larvae population M_0.

c) Analyze the consequences and implication of the model for different ranges of M_0.

2.6.3   *Seasonal varying growth rate*

In reality, the growth rate constant of the mosquito population is not uniform in time but seasonal as larvae are spawn and hatched in the spring and summer months. Females emerged in late summer seek sheltered areas where they "hibernate" until spring. Warm weather brings them out in search of water on which to lay eggs. As an approximate characterization of this time-varying cyclical growth, we take the "growth rate constant" to be $a(t) = a_0 \cos(2\pi t)$ where a_0 is a constant. In that case, the evolution of the mosquito larvae is modeled by the *non-autonomous ODE*

$$\frac{dM}{dt} - a_0 \cos(2\pi t)\, M = -b(t) \tag{2.6.1}$$

which can be seen to be not separable unless $b(t) = 0$. While it is linear, an integrating factor does not readily suggest itself. Given their frequent occurrence in applications, a systematic method for finding an integrating factor for linear first order ODE with variable coefficient(s) such as (2.6.1) as well as methods for other kinds of first order ODE (taught in any elementary course on ODE and available in any book on elementary ODE) are summarized in Appendix A of this volume. It is left as an exercise for the reader to apply the method of integrating factor to obtain the solution for (2.6.1) and deduce the implications of the solution.

Chapter 3

Stability and Bifurcation

3.1 The Logistic Growth Model

3.1.1 Taylor polynomials for the growth function

The linear rate of growth hypothesis leading to the model (2.2.1) and (2.1.5) of Chapter 2 was found to result in a monotone increasing population size that fits the census data well over a period of nearly a hundred years. Yet, exponential growth or any unbounded growth is not consistent with an environment of limited resources. Recall that our exponential growth based on four data points of the last chapter predicts the world population to reach 10 billion by 2023 and 14 billion by 2056. On the other hand, United Nation's best estimate is that population would only reach 9.7 billion by 2050 and exceed 10.1 billion by the end of the century (with the qualification that these estimates could be higher, if birth rates do not continue to drop as they have in the last half-century). It appears that the awareness or real effects of limited resources is being felt by an Earth bound human population and this rate limiting factor should be included in our mathematical model. This means we should modify our linear growth rate model (2.2.1) if the model is to be applicable for a longer period of time into the future. The question is how do we make the needed modification?

Seasoned mathematical modelers may be able to use their knowledge in biology and experience with other related modeling work to formulate an appropriate model by ad hoc reasoning. However, a novice modeler is generally encouraged to take a more systematic approach by returning to where we began to specialize the general growth model (2.1.1) to a linear growth rate in Chapter 2. More specifically, we deduced from the following Taylor series of the growth rate function:

$$f(y) = f(0) + f'(0)y + \frac{1}{2}f''(0)y^2 + \cdots$$
$$= f(0) + f'(0)y + \frac{1}{2}f''(0)y^2 + R_3 \qquad (3.1.1)$$

where the remainder term R_3 is proportional to y^k, $k \geq 3$. From this, we obtain

- the constant growth rate model if we terminate the series (3.1.1) after the constant term to get $y' = f(y) \simeq f(0) \equiv b$; and

- the linear growth rate model $y' = f(y) \simeq f'(0)y \equiv a_0 y$ by terminating a term later and assuming no immigration.

It is natural then to retain at least one more term of the Taylor series to get a different model that may meet our interest in a rate limiting growth:

$$f(y) \simeq f(0) + f'(0)y + \frac{1}{2}f''(0)y^2. \tag{3.1.2}$$

With $f(0) = 0$ (by the *no immigration hypothesis*) and by setting

$$f'(0) = a_0, \quad \frac{1}{2}f''(0) = -\frac{a_0}{Y_c} \tag{3.1.3}$$

we can take the modified growth rate differential equation in the form

$$\frac{dy}{dt} = a_0 y \left(1 - \frac{y}{Y_c}\right). \tag{3.1.4}$$

The least squares estimation based on the census data for the period of 1974–2011 has already shown $a_0 > 0$. While we can always write the quadratic approximation (3.1.2) as (3.1.4) since Y_c may be of either sign depending on the sign of $f''(0)$, we take here $Y_c > 0$ for our new model since the rate limiting finite resources would force a diminishing return as the population increases. In fact, Y_c is known as the *carrying capacity* of the environment, a population size not to be exceeded in steady state. As long as the quadratic growth rate model (3.1.4) is applicable, the rate of growth actually becomes negative for a population size greater than the carrying capacity so that population self-corrects to adjust to what the environment would support. The quadratically nonlinear differential equation (3.1.4) is known as the *logistic growth* equation in the field of population dynamics.

The ODE (3.1.4) and an initial condition

$$y(t_0) = Y_0$$

constitute an IVP used extensively in different fields of population studies. For an autonomous ODE, the initial time t_0 may be relocated to the origin by a change of the independent variable $t = \tau + t_0$. In terms of τ, the ODE becomes

$$\frac{dy}{d\tau} = f(y) = a_0 y \left(1 - \frac{y}{Y_c}\right)$$

which is the same as the original equation. As such we can work with t as the independent variable with the initial condition

$$y(0) = Y_0 \tag{3.1.5}$$

as we did in (2.1.5) of Chapter 2.

3.1.2   Normalization and exact solution

The ODE (3.1.4) for logistic growth is again *separable* and an exact solution is assured (and is actually informative in this case). Before we embark on the solution process, it is often advantageous to reduce the number of parameters, such as a_0 and Y_c, in the model whenever possible. In the present case, we have three parameters, a and Y_c in (3.1.4) and Y_0 in the initial condition (3.1.5). We can eliminate two of them by introducing two new variables

$$\tau = a_0 t, \qquad z(\tau) = \frac{y(t)}{Y_c}. \tag{3.1.6}$$

In terms of the two new variables, (3.1.4) takes the form

$$\frac{dz}{d\tau} = z(1 - z). \tag{3.1.7}$$

Correspondingly, the initial condition $y(0) = Y_0$ becomes

$$z(0) = z_0 \equiv \frac{Y_0}{Y_c}. \tag{3.1.8}$$

Note that the new and equivalent ODE (3.1.7) is parameter-free and hence universally applicable to populations with a different natural growth rate constant and a different carrying capacity. The process for its exact solution is to reduce it to a calculus problem by re-arranging (3.1.7) to read

$$\frac{1}{z(1 - z)} \frac{dz}{d\tau} = 1. \tag{3.1.9}$$

Upon carrying out the integration (by partial fractions), applying the initial condition (3.1.8) and solving the resulting relation for $z(t)$, we obtain the following exact solution of the IVP:

$$z(\tau) = \frac{z_0}{z_0 + (1 - z_0)e^{-\tau}}.$$

In terms of the original variables y and t, we have

$$y(t) = \frac{Y_c Y_0}{Y_0 + (Y_c - Y_0)e^{-a_0 t}}. \tag{3.1.10}$$

Exercise 5 *Obtain the exact solution (3.1.10) for (3.1.7) and (3.1.8). (Hint: apply partial fractions to (3.1.9).)*

3.1.3   Implication of the exact solution

When an exact solution such as (3.1.10) for the IVP (3.1.7)–(3.1.8) is available, nearly all information of interest can be obtained from the explicit solution. For example, we see from (3.1.10) that the population would increase with time tending to the carrying capacity Y_c from below if $Y_0 < Y_c$. On the other hand, if $0 < Y_c < Y_0$, the population would decrease with time and tend to the carrying capacity from above. If $Y_0 = Y_c$, then $y(t) = Y_0 = Y_c$ for all $t \geq 0$ so that the population remains at the initial size indefinitely. The situation is similar if $Y_0 = 0$ with $y(t) = 0$ for all $t \geq 0$. (Note that the remaining case $Y_0 < 0$ is not relevant since population size cannot be negative.) These conclusions are certainly more consistent with our understanding of growth and decline of the human population than the linear (or constant) growth rate model (leading to exponential growth). Instead of unlimited growth, the new model predicts that the human population should tend to a limiting population size eventually.

3.1.4 *Validation and model improvement*

As such, the logistic model is seen to be more appropriate and reliable for a longer time interval for the human population on Earth. More generally, the logistic growth model should be adequate for population growths that are autonomous with the following three properties for its growth rate function:

$$i) \ \ f(0) = 0, \quad ii) \ \ f(y) > 0, \quad iii) \ \ f''(y) < 0. \tag{3.1.11}$$

For more quantitative information beyond the qualitative observations in the preceding paragraph, we would need numerical values for the three parameters Y_0, a and Y_c. These can be obtained by the method of least squares or other methods for parameter estimation using census data. We will not pursue such a discussion here. Instead, we note that, unlike Newton's laws of motion, the properties (3.1.11) for the growth rate function do not hold for all evolving populations. A growth rate that is not monotone increasing even for small y is the "depensation growth rate" for some fish population. These are known to have a threshold population size for extinction Y_e; their growth is more adequately modeled by the separable equation

$$y' = ay(y - Y_e)(Y_c - y), \tag{3.1.12}$$

where a, Y_e and Y_c are three constant parameters characterizing the instantaneous rate of change of the population in addition to the initial population Y_0. From the perspective of Taylor's theorem, the right-hand side of the new model ODE (3.1.12) is a four-term Taylor polynomial approximation of $f(y)$ (that is cubic in y).

While we can obtain an exact solution for the corresponding IVP (again using partial fractions), the result is not particularly informative, unlike the solution for the logistic growth case. It turns out that there are methods for deducing qualitative results such as those extracted from (3.1.10) without knowing the explicit solution or even the explicit form of the growth rate function. We will begin a discussion of this geometrical method of solution for autonomous ODE in the next section.

3.2 Critical Points and Their Stability

3.2.1 *Critical points of autonomous ODE*

Up to now, our approach to extracting information from a mathematical model is analytical and quantitative in nature. In contrast, the new approach to be discussed below is geometrical and qualitative. It is principally useful for autonomous ODE such as

$$y' = f(y) \qquad (3.2.1)$$

where the growth function f does not depend explicitly on the independent variable t (corresponding to time until further notice). The exponential growth (2.2.1), the logistic growth (3.1.4) and the depensation growth (3.1.12) are examples of such growth rates. On the other hand, the ODE for intravenous glucose infusion (2.5.1) of the previous chapter is non-autonomous when the infusion rate c is a function of time.

For autonomous ODE, it is possible to have (one or more) solutions that do not change with time, i.e., $y(t) = y^c$, where y^c is a constant. For such time independent solutions, we have

$$0 = \frac{dy^c}{dt} = f(y^c)$$

so that y^c must be a *zero* of the function $f(y)$ on the right-hand side of first order autonomous ODE (3.2.1). Such a time independent solution is called a *critical point* of the ODE.

Definition 1 *A critical point y^c of the autonomous ODE $y' = f(y)$ is a zero of the growth rate function $f(y)$, i.e., $f(y^c) = 0$.*

A critical point corresponds to a time independent steady state of the population (or whatever the quantity y measures). We note in passing that the definition of a critical point applies to the more general case of a system of first order ODE written in vector form with y being a vector function of t.

Depending on the form of the growth rate, $f(y)$ may have any number of critical points: zero, one or many. The exponential growth model has only one critical point at the origin and the logistic growth model (3.1.4) has two:

$$y_c^{(1)} = 0, \qquad y_c^{(2)} = Y_c$$

while the growth rate function $f(y) = \sin(y)$ has an infinite number of critical points at $\pm n\pi$, $n = 0, 1, 2, 3, \ldots$. Not all critical points can be found explicitly. However, there are several fast root finding routines on any of the popular computational software such as MatLab, Mathematica and Maple for accurate numerical solutions of the single (scalar or vector) equation $f(y) = 0$.

On the other hand, to locate the critical points quickly without getting on the computer or hand held calculator, graphical means are often very helpful. For example,

$$f_e(y) = 1 - y - e^{-2y}$$

can be seen graphically to have only two critical points at

$$y_c^{(1)} = 0, \qquad 0 < y_c^{(2)} \lesssim 1$$

by plotting $f_e(y)$ vs. y and looking for locations where the graph crosses the abscissa (the horizontal y-axis). Such a plot shows the second intersection (other than $y_c^{(1)} = 0$) is smaller than (but very close to) 1.

For very complicated $f(y)$, it may still be possible to locate the zeros approximately without resorting to computing software. Successes are often achieved by breaking up the function into two pieces $f(y) = g(y) - h(y)$, plotting the two simpler functions $g(y)$ and $h(y)$ and locating their intersections as shown in Figure 3.1:

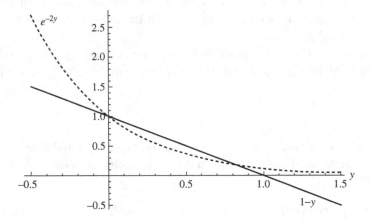

Fig. 3.1 Splitting $1 - y - e^{-2y}$ and look at intersections.

It may not be so easy to visualize the graph of $f_e(y)$ or estimate some of its zeros without actually plotting it accurately using Mathematica or MatLab. But we have no problem visualizing $g(y) = 1 - y$ and $h(y) = e^{-2y}$ and estimating the location y^c of their intersection(s). Note that there are usually alternative ways to split $f(y)$ as we could have taken $g(y) = 1 - e^{-2y}$ and $h(y) = y$ for $f_e(y)$ instead. While there is not much difference in the two ways in splitting the particular function $f_e(y)$, we generally should be flexible in exploring more than one options to find an effective splitting for a more complicated $f(y)$.

By changing $f_e(y)$ slightly to

$$f_2(y) = 2 - y - e^{-y},$$

neither critical point can be determined explicitly. But graphing $g(y) = 2 - y$ and $h(y) = e^{-y}$, it is easy to see that the positive critical point must be less than (but very close to) 2 since $g(2) = 0 < e^{-2} = h(2)$ and the negative critical point must be less than -1 (since $g(-1) = 3 > e = h(-1)$) as shown in Figure 3.2.

3.2.2 *Stability of a critical point*

In population growth models, critical points correspond to possible steady state populations which do not change with time. They correspond to equilibrium states

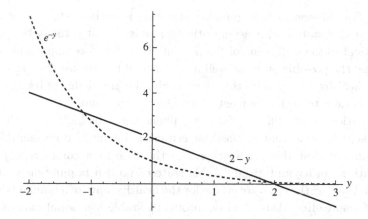

Fig. 3.2   Splitting $2 - y - e^{-y}$ and look at intersections.

analogous to a rigid ball sitting at the (convex) bottom of a trough or perching (precariously) on a (concave) hill top. These two particular equilibrium configurations show that critical points can be significantly different in nature. At the slightest disturbance, the ball on top of the hill would roll down the hill, running away from its equilibrium position. In contrast, the ball at the bottom of a trough would, upon a relatively small displacement from its critical point, merely oscillate about its original position. We characterize the trough bottom equilibrium configuration as a *stable* equilibrium or a *stable critical point* and the hill top configuration as an *unstable* equilibrium or an *unstable critical point*. Depending on the surface of the trough, whether it is smooth or rough (causing frictional resistance), the oscillating ball at the trough bottom may continue to oscillate with the same maximum amplitude indefinitely or the oscillation may diminish and eventually return to the resting configuration. In the latter case with the ball returning to equilibrium, the critical point is said to be *asymptotically stable*. In the former case when there is no friction and the ball continues to oscillate indefinitely, the critical point is *stable* but not an *attractor* that is *asymptotically stable*.

For many phenomena in science and engineering, we very much want to know whether a critical point is stable or unstable and, if stable, whether it is asymptotically stable. An example from medical science is an individual not yet infected by a spreading virus. The individual is in a steady state of good health until infected by the virus. Depending on the constitution of the individual, a small population of virus particles (virions) may be completely eliminated by the individual's immune system, or may multiply unchecked and eventually overwhelm the individual and cause his/her demise. In the first instance, the original healthy state is an asymptotically stable critical point; in the latter case, the critical point is unstable. There are other possible scenarios. An example is a small population that grows to a certain size and, without further external stimulations, remains at that size thereafter. For this case, the original healthy state is again unstable; but it evolves into a new steady state that is asymptotically stable. What actually happens is obviously critical to the well being of the patient. It is therefore important to be able to determine the possible outcome well in advance of the eventuality. In the case of an initially healthy steady state that is unstable, the knowledge of its stability may enable the patient to seek treatment to avoid eventual demise.

The situation is not different for other phenomena, biological or otherwise. It should be kept in mind that an unstable critical point may be undesirable for one phenomenon but desirable for another. In the case of a commercially valuable fish population, an asymptotically stable state of no fish population would be an economical disaster; the opposite is true for the healthy state of a human being in the presence of a spreading virus. An asymptotically stable fish population of *adequate* size would probably be good for sustaining fishery harvest but an asymptotically stable high virus population would be disastrous for a sick patient. For this reason, human intervention to modify the stability of a critical point is usually to meet an identified beneficial objective even if it is only the avoidance of potential disaster. For some applications, this may be to have an asymptotically stable critical point, for others, it may mean to ensure an unstable critical point.

3.2.3   *Graphical method for critical points and their stability*

For phenomena that can be modeled by the single first order ODE $y' = f(y)$, we may take advantage of the graphical approach already used in the previous subsection to find the critical points of the ODE and their stability as a first step toward the ultimate objective of possible strategic intervention. This is illustrated by the following example with $(\ )' = d(\ )/dt$ throughout this chapter:

Example 1 $y' = y(1 - y), \qquad y(0) = Y_0.$

For this normalized logistic population growth model with $f(y) = y(1 - y)$, the graph of $f(y)$ is the upside down parabola crossing the abscissas at the critical points $y^{(1)} = 0$ and $y^{(2)} = 1$ as shown in Figure 3.3. Evidently, the two critical points of this ODE are *isolated*.

Definition 2 *A critical point y^c of the ODE $y' = f(y)$ is an isolated critical point if $f(y^c) = 0$ and there is no other critical point for the ODE inside a neighborhood $0 < |y - y^c| < \delta$ for some $\delta > 0$.*

To see the possibility of non-isolated critical points, plot the growth rate function:

$$f(y) = \begin{cases} y(0.1 - y)(0.9 - y)(1 - y) & (y < 0.1) \\ 0 & (0.1 < y < 0.9) \\ y(0.1 - y)(0.9 - y)(1 - y) & (y > 0.9) \end{cases}$$

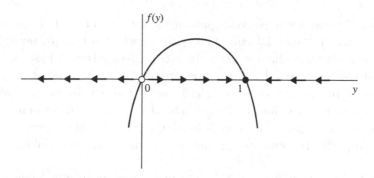

Fig. 3.3   Locating critical points of logistic growth.

and conclude that all initial values Y_0 in the range $[0.1, 0.9]$ are critical points and they are all *stable but not asymptotically stable*. (Any sufficiently small deviation ε from a critical point y^c in the interval $0.1 < y < 0.9$ would be at another critical point so that $y(t)$ would change from $y(t) = y^c$ for all t thereafter to $y(t) = y^c + \varepsilon$ for all later time and not returning to y^c.) For this example, there are also two other critical points at 0 and 1; each lies within a small interval where there is no other critical point. The two critical points 0 and 1 are *isolated* critical points. In contrast, those in the interval $[0.1, 0.9]$ are *non-isolated*. In this volume, we will be mainly concerned with isolated critical points.

Returning to the logistic growth Example 1, suppose the initial condition for y is $y(0) = Y_0 = y^{(1)} + \varepsilon > y^{(1)} (= 0)$, will $y(t)$ stay close to the critical point? approach it with increasing time? or will it run away from $y^{(1)}$? To answer these questions by the graphical method for determining the stability, we simply examines the graph of $f(y)$ near the origin to see if $f(y)$ is above or below the y-axis.

For our example, $f(y)$ is above the y-axis to the right of the origin; hence we have $y' = f(y) > 0$ for $y > 0$ so that $y(t)$ is increasing (see Figure 3.3). Graphically, that $y(t)$ (is positive and) increases with t is indicated by arrows on the positive y-axis pointing to the right and away from $y^{(1)}$. For any Y_0 to the right of the origin, $Y_0 > 0$, the direction of the arrows shows that $y(t)$ moves further to the right away from the origin as t increases. On the other hand, $f(y)$ is below the negative y-axis so that $y(t)$ is decreasing for any Y_0 to the left of the origin, $Y_0 < 0$ (which is not relevant for population growth problems). The left-pointing arrows along the negative y-axis show that $y(t)$ becomes more negative and moves further to the left of the origin with increasing time. The critical point $y^{(1)} = 0$ is seen graphically to be *unstable* as arrows on both sides (indicating the direction of increasing time of the state y at time t) are pointing away from the critical point. Whenever possible, we use a hollow circle "o" to indicate an unstable critical point.

Similar observations on $f(y)$ in the neighborhood of the other isolated critical point $y^{(2)} = 1$ lead to arrows along both sides of the y-axis pointing toward the critical point $y^{(2)}$. Hence, the second critical point is seen graphically to be asymptotically stable. Asymptotically stable critical points are indicated by a solid round dot ".". Together, we know everything about the solution of the IVP except for the numerical details. If the initial value Y_0 is positive (whether it is greater or less than $y^{(2)} = 1$), then the ODE characterizing the rate of change (be it physical, biological or social) drives $y(t)$ toward $y^{(2)} = 1$ monotonically. If Y_0 is negative, then $y(t)$ would decrease further and monotonically head toward $-\infty$ in the limit as $t \to \infty$. There are no other possibilities. This graphical method for determining stability of an isolated critical point of the logistic ODE applies to other autonomous ODE $y' = f(y)$. In particular, we have the following result from such applications:

Theorem 1 *Starting with an initial state, the evolution of any temporal process governed by a single first order autonomous ODE does not oscillate.*

Proof For the first order ODE $y' = f(y)$, the graph of $f(y)$ vs. y must be of one sign between any two consecutive zeros of $f(y)$, i.e., any two consecutive isolated critical points. For any initial value Y_0 in between, $y(t)$ must be increasing toward the larger zero if $f(y) > 0$ and decreasing toward the smaller zero if $f(y) < 0$. In other words, starting from an initial value Y_0 the solution must be either monotone increasing or monotone decreasing until it reaches an asymptotically stable critical point, ∞ or $-\infty$.   □

It follows from this theorem that the solution $y(t)$ of $y' = f(y)$ must be a monotone function away from its critical points. Damped or undamped oscillatory motions (such as those of the mass–dashpot-spring systems) do not occur; there are no periodic solutions for such equations.

It is worth repeating that $Y_0 < 0$ is not relevant as an initial condition for the ODE $y' = f(y)$ if it is intended to be a mathematical model for population growth. However, it may be necessary and appropriate to discuss the behavior of $y(t)$ for $Y_0 < 0$ either to complete the mathematical analysis or for other applications of the mathematical model which allow for negative values of $y(t)$.

3.2.4 *Linear stability analysis*

The geometrical approach introduced in the previous subsection determines completely the qualitative behavior of an evolving population modeled by $y' = f(y)$. Its effectiveness however is mostly for one or two first order autonomous ODE. It is less effective for three such ODE and not particularly useful for more equations except in special cases since visualization becomes more difficult or impossible in higher dimensions. To complement the graphical method for phenomena modeled by several autonomous ODE, we note here that the stability of the two critical points for logistic growth can also be determined analytically without the exact solution for the relevant IVP. Suppose the initial condition is not exactly at a critical point y^c but at some value near it, say $Y_0 = y^c + \epsilon$. For sufficiently small $t > 0$, the response $y(t)$ can be written as $y(t) = y^c + \epsilon x(t)$ to a good approximation with $x(0) = 1$.

Given $f(y^c) = 0$ for a critical point y^c and by Taylor's theorem (assuming that $f(y)$ is at least twice continuously differentiable with respect to its argument)

$$f(y) = f(y^c + \epsilon x(t)) = f(y^c) + f'(y^c)\epsilon x(t) + \frac{1}{2}f''(\xi)[\epsilon x(t)]^2,$$

where $f'(y) = df/dy$, the ODE $y' = f(y)$ may be accurately approximated by

$$x' = f'(y^c)x(t) \tag{3.2.2}$$

for sufficiently small ϵ and t so that $\frac{1}{2}|\epsilon f''(\xi)|[x(t)]^2 \ll |f'(y^c)x(t)|$. It follows that the critical point y^c is asymptotically stable if $f'(y^c) < 0$ and unstable if $f'(y^c) > 0$. (The case $f'(y^c) = 0$ will be discussed below.) Thus, for the stability of the critical point, we need only to know the sign of $f'(y^c)$.

From the viewpoint of the graphical approach, this analytical result tells us that we only need to sketch f near y^c to see the slope of $f(y)$ at that critical point; we do not even need to compute the derivative of $f(y)$ if it should be too complicated to do so. In addition, it saves us from performing the chore of plotting the entire graph of $f(y)$. On the other hand, the result we get is local in nature as the "linearized" equation (3.2.2) is adequate only for the range of ϵ and t when the neglected terms remain small. Global results for solutions of IVP are generally difficult to get for higher order systems; we often have to settle for limited local results by this method of linearization, known as *linear stability analysis*. For a single first order autonomous ODE however, changes are limited to along the y-axis, the local results are sufficient for a complete qualitative characterization of the phenomenon being modeled.

The two critical points of the logistic growth ODE (3.1.4) exhibit two distinctly different kinds of stability, $y^{(1)} = 0$ is unstable while $y^{(2)} = 1$ is asymptotically stable. They correspond to $f'(0) > 0$ and $f'(1) < 0$, respectively. For a general $f(y)$, we noted earlier a third possibility $f'(y^c) = 0$. In this case, stability is determined by the sign of $f''(y^c)$ or the next nonvanishing derivative at y^c. Near the critical point, the graph of the growth function is qualitatively as indicated in Figure 3.4(a) if $f''(y^c) > 0$, and as in Figure 3.4(b) if $f''(y^c) < 0$. In the first case, $f(y)$ is increasing on both sides of the critical point and hence stable for initial values slightly to the left of y^c, i.e., $Y_0 < y^c$, and unstable for those slightly to the right, i.e., $Y_0 > y^c$. In the second case, $f(y)$ is decreasing on both sides of the critical point and hence unstable for initial values $Y_0 < y^c$ and asymptotically stable for $Y_0 > y^c$. In either case, stability is different qualitatively from the two types exhibited by the critical points of logistic growth with $f'(y^c) \neq 0$. A critical point with $f'(y^c) = 0$ and $f''(y^c)$ not vanishing is said to be *semi-stable*.

3.3 Over-fishing of a Commercial Fish Population

3.3.1 *Fish harvesting*

When left alone, a fish population grows within the limit of the *carrying capacity* of the body of water and the available nutrients. Typically, the growth process may

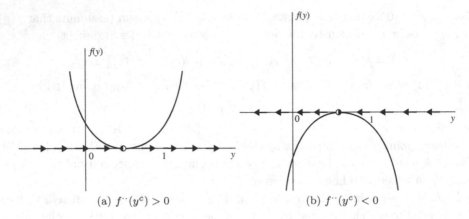

(a) $f''(y^c) > 0$ (b) $f''(y^c) < 0$

Fig. 3.4 $f'(y^c) = 0$.

be modeled by the first-order ODE

$$\frac{dy}{dt} = f(t, y)$$

where $y(t)$ is the total tonnage of fish biomass at time t. For $y(t)$ small compared to the region's carrying capacity Y_c, $f(t, y)$ may be taken to be linear in y, i.e., $f(t, y) = f_1(t)y + f_0(t)$; the undisturbed fish population growth is exponential if $f_0(t) = 0$ and $f_1(t)$ is a positive constant a. A growth (rate) function adequate for a much wider range of $y(t)$ is the *logistic growth* (rate)

$$f(t, y) = ay\left(1 - \frac{y}{Y_c}\right).$$

Other more complex growth functions (such as the depensation model (3.1.12) mentioned earlier) have been used in fishery management research. Here, we work with the logistic growth model with a time-invariant natural growth rate to illustrate the modeling and analysis of fishery. Given the initial population $y(0) = Y_0$, the corresponding IVP for an unharvested fish population is just the logistic growth analyzed previously.

A commercially valuable fish population (whether it is in a fish farm, a river or an ocean) is not likely to be left alone to grow naturally. More often, it is harvested continually at a rate $h(t) \geq 0$. For a seasonal fishery, $h(t)$ may be a periodic function which vanishes over regular intervals. In that case, the actual growth rate of the fish population would be determined by the growth rate model

$$y' = f(t, y) - h, \quad y(0) = Y_0 \tag{3.3.1}$$

where $(\ )' = d(\ )/dt$. A fishery with access to the fishing ground would naturally want to choose a harvesting strategy $h_{op}(t)$ that is most advantageous for its purpose.

3.3.2 Critical points and their stability

To illustrate the issues and process related to the harvesting of a fish population, we limit our discussion here to the special case of a logistic model for the natural growth of the fish population and work with a constant harvest rate h. The evolution of the fish population is then determined by

$$y' = y(1 - y) - h, \qquad y(0) = Y_0. \tag{3.3.2}$$

As the ODE involved is autonomous, considerable information about the evolution of the fish population can be extracted by a critical point analysis. The locations of the two critical points (corresponding to the two steady state fish population sizes) would now depend on h. We can determine them by solving the quadratic equation $F(y; h) \equiv f(y) - h = y(1 - y) - h = 0$ (with $F(y; h)$ being the growth rate for the harvested fish population) to get

$$\begin{pmatrix} y^{(1)} \\ y^{(2)} \end{pmatrix} = \frac{1}{2} \begin{pmatrix} 1 - \sqrt{1 - 4h} \\ 1 + \sqrt{1 - 4h} \end{pmatrix}$$

with $0 < y^{(1)} < y^{(2)} < 1$ for $0 < h < 1/4$. The two critical points coalesce into one at $h = 1/4$ and the ODE has no critical point for $h > 1/4$.

The same conclusions can be seen readily from the graph of $y(1-y)$ and the position of the horizontal line h as shown in Figure 3.5. The figure tells us that there are two critical points $y^{(1)}(h)$ and $y^{(2)}(h)$ for our fish harvesting model corresponding to the two intersections between the horizontal line and the upside down parabola $y(1 - y)$. The graph of $y(1 - y)$ is seen to be above the horizontal line h for y in the interval between the two critical points at the location "o" and "$\cdot$", respectively. We conclude immediately that $y^{(1)}(h)$ is unstable while $y^{(2)}(h)$ is asymptotically stable. As h increases from 0 to 1/4, the horizontal line moves up toward the peak of the natural fish growth rate graph and the two critical points move toward each other. The critical points coalesce when the line is tangent to the peak of the natural growth rate curve at $h = 1/4$ and $y^{(1)} = y^{(2)} = 1/2$. For larger values of h, the line does not intersect the natural growth rate curve and the ODE does not have any critical point in this range of harvest rate.

For $h > 1/4$, the growth rate $F(y; h) = y(1 - y) - h < 0$ so that $y(t)$ decreases to zero for any initial condition Y_0. Any harvest rate $h > 1/4$ would overfish the fish stock and the fish population would dwindle toward extinction.

Evidently, the value of the parameter h not only affects the location of the critical point of the model but also their count. While the effect of changing h is qualitatively insignificant as long as h remains less than $\frac{1}{4}$, the change in the number of critical points as h passes through the threshold value $h_c = \frac{1}{4}$ is abrupt and qualitatively significant, changing from 2 to 0 as h increases from $\frac{1}{4} - \varepsilon$ to $\frac{1}{4} + \varepsilon$. In this volume, we adopt the following (less formal than the usual) definition of bifurcation:

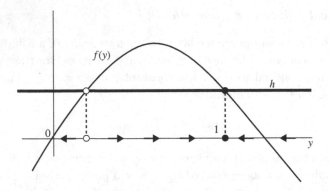

Fig. 3.5   Critical points for fish harvesting.

Definition 3 *Let the evolution of a dynamical system be adequately modeled by an autonomous ODE*

$$y' = f(y; \mu) \qquad (3.3.3)$$

where μ is a parameter (with $\mu = h$ in the fishery model (3.3.2)). The dynamical system is said to undergo a **bifurcation** *at a specific value μ_c of the parameter μ if there is an abrupt change in the count of critical points and/or the type of stability of any one of its critical points as μ passes through the threshold value μ_c. The threshold value μ_c ($= 1/4$ in the fishery case) itself is said to be a* **bifurcation point** *of the model ODE.*

3.3.3   Bifurcation for fish harvesting

It is evident from the fishery model that the location of critical point of an autonomous ODE may depend on the values of the various system parameters. In the case of the temporally uniform harvest rate model (3.3.2), the only system parameter is the constant harvest rate h. If we had kept the original form of the logistic growth model and do not re-scale the ODE, we would have

$$y' = ay\left(1 - \frac{y}{Y_c}\right) - H$$

(with $h = H/\left(aY_c\right)$ after normalization). In that case, there would be three system parameters a, Y_c and H. The un-scaled ODE still has two critical points, but they are at the two roots

$$\begin{pmatrix} y^{(1)} \\ y^{(2)} \end{pmatrix} = \frac{Y_c}{2}\begin{pmatrix} 1 - \sqrt{1 - 4h} \\ 1 + \sqrt{1 - 4h} \end{pmatrix}, \qquad h = \frac{H}{aY_c} \qquad (3.3.4)$$

of

$$y^2 - Y_c y + \frac{Y_c H}{a} = 0. \qquad (3.3.5)$$

Though there are now three parameters in the ODE, only one combination of the three, $h = H/(aY_c)$ constitutes the bifurcation parameter. In other words, possible occurrence of bifurcation depends on the value of only one combination of the three parameters.

The particular abrupt qualitative change in the fishery problem caused by a bifurcation at the bifurcation point $h_c = 1/4$ involves a change in critical point count from two to none as μ passes through the bifurcation point is known as a *saddle-node bifurcation*. This change in critical point count along with changes in the stability of the critical points can be succinctly summarized by the *bifurcation diagram* in Figure 3.6.

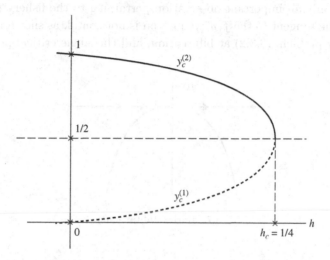

Fig. 3.6   Bifurcation diagram for fishery problem (saddle-node).

3.4   Bifurcation

3.4.1   *Bifurcation diagram and horizontal tangency*

The fishery example suggests that we need to better understand possible bifurcation in the more general growth model (3.3.3),

$$y' = f(y; \mu), \qquad y(0) = Y_0, \tag{3.4.1}$$

where μ is a bifurcation parameter corresponding to h in the fishery problem. For most problems, the function f is at least twice continuously differentiable in y with f, $f_{,y}$, $f_{,\mu}$ and $f_{,yy}$ continuous in μ. We assume these properties of f in the discussion below.

Similar to the fishery problem, the qualitative features of $y(t)$ as a function of time may be ascertained by determining the critical points $\{y^{(k)}(\mu)\}$ and analyzing their stability. The model should also be examined for possible bifurcation to anticipate possible catastrophic changes. As illustrated by the fishery problem, this is often clearly delineated by way of a bifurcation diagram which displays the graphs of all the critical points of $f(y; \mu)$ as functions of the bifurcation parameter μ. For the fish harvesting problem, there are two critical points (and hence two different graphs corresponding to an upper and lower branches of the side-way parabola in Figure 3.6); the bifurcation parameter is the harvest rate $(\mu =) h$. In general, there are as many graphs in the bifurcation diagram as the number of critical points for the model ODE.

Bifurcation analysis for (3.4.1) will be discussed in more detail below. We note at this point only an important observation pertaining to the fishery problem: At bifurcation, the tangent to the $f(y^c; \mu_c)$ graph is horizontal, as shown in Figure 3.7 for the fishery problem (3.3.2) at bifurcation, and the single critical point involved is semi-stable.

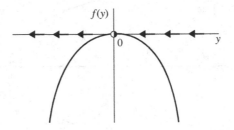

Fig. 3.7   Horizontal tangency.

This phenomenon of horizontal tangency at bifurcation is seen to be true generally from the Taylor expansion of $f(y, \mu)$:

$$f(y; \mu) = f(y^c; \mu) + f_{,y}(y^c; \mu)(y - y^c) + \frac{1}{2!} f_{,yy}(y^c; \mu)(y - y^c)^2 + \cdots ,$$

where any of the relevant critical point $y^c = y^c(\mu)$ generally varies with μ. As long as $f_{,y}(y^c; \mu) \neq 0$, continuity of $f(y^c, \mu)$ and $f_{,y}(y^c, \mu)$ in both y and μ requires that $f(y + \Delta y; \mu + \Delta \mu)$ continue to be of the same sign as $f(y; \mu)$ for sufficiently small $|\Delta y|$ and $|\Delta \mu|$. Hence, there is no bifurcation in that case. This proves the following useful theorem:

Theorem 2 *If bifurcation occurs at the bifurcation point μ_c, the slope of $f(y; \mu)$ as a function of y is necessarily horizontal at a relevant critical point y^c so that*

$$f_{,y}^{(c)} \equiv f_{,y}(y^c; \mu_c) = 0, \qquad f_{,y} = \frac{\partial f}{\partial y}. \tag{3.4.2}$$

Remark 1 *Theorem 2 shows that the phenomenon of horizontal tangent at the critical point at bifurcation is not unique to the fishery problem but also persists in the other models as well. The theorem only indicates that horizontal tangency is necessary at bifurcation. That it is also sufficient will be shown in a later section.*

Theorem 2 is useful when graphical method is not practical. We illustrate this use of the theorem with the fishery problem in normalized form (3.3.2). The critical point is determined by the condition

$$f(y^c; h_c) = y^c(1 - y^c) - h_c = 0,$$

while the condition of horizontal tangency for the example takes the form

$$f_{,y}(y^c; \mu_c) = 1 - 2y^c = 0.$$

As h_c does not appear explicitly in the second condition, we can solve it to get

$$y^c = \frac{1}{2}.$$

Upon substituting this result into the first condition, we obtain

$$h_c = \frac{1}{4}$$

identical to the bifurcation point found previously by graphical method.

3.4.2 *Basic bifurcation types*

Saddle-node bifurcation is not the only type of bifurcation possible. In the following subsections, we describe the three most frequently encountered bifurcation types: the *transcritical* bifurcation and the *pitchfork* bifurcation along with the already familiar *saddle-node* bifurcation. For a single first order ODE, other seemingly different bifurcations may be just a combination of these three basic types or one of the them in disguise possibly with some cosmetic modifications or extensions. The exceptions occur so rarely that readers are referred to [27] and references therein for further reading.

3.4.2.1 *Saddle node bifurcation*

We saw an example of this type of bifurcation in (3.3.2). A canonical form of a first order autonomous ODE with this type of bifurcation is

$$z' = \mu - z^2 \qquad\qquad (3.4.3)$$

where μ is the bifurcation parameter. Equation (3.3.2) may be reduced to this form by setting $z = y - \frac{1}{2}$ and $\mu = \frac{1}{4} - h$.

For the ODE (3.4.3), there are two (real) critical points $z = \pm\sqrt{\mu}$ if $\mu > 0$ ($h < \frac{1}{4}$), one critical point $z = 0$ if $\mu = 0$ ($h = \frac{1}{4}$) and no (real) critical points if $\mu < 0$ ($h > \frac{1}{4}$). These features are evident from the zeros of $\mu - z^2$ by plotting μ and z^2 separately and locate their intersections. This type of changes in the count of critical point occurs frequently in applications. It is known as a saddle-node bifurcation and is captured by a bifurcation diagram similar to Figure 3.6.

3.4.2.2 Transcritical bifurcation

The canonical ODE

The simplest ODE exhibiting a transcritical bifurcation is

$$y' = f(y; \mu) = y(\mu - y). \tag{3.4.4}$$

For $\mu \neq 0$, the graph of $f(y; \mu)$ is an upside down parabola intersecting the abscissa (y-) axis at 0 and μ. As such, the ODE has two critical points at $y^{(1)} = 0$ and $y^{(2)} = \mu$. For a positive value μ, $y^{(2)} = \mu$ is located on the positive y axis (to the right of the origin) as shown in Figure 3.8(a). From the graph of $f(y; \mu)$ for this case, we see that $y^{(1)} = 0$ is unstable and $y^{(2)} = \mu$ is asymptotically stable. As μ decreases but remaining positive, the location of $y^{(2)} = \mu$ moves toward the origin while the stability of both critical points remain unchanged. All the while, the maximum $f(y; \mu)$ decreases until $\mu = \mu_c = 0$. For the particular value $\mu_c = 0$, the entire graph of $f(y, 0) = -y^2$ lies below the y-axis but tangent to the abscissa (the y-axis) at the only critical point at the origin as shown in Figure 3.7. Further reduction of μ (to negative values) moves the graph further to the left and raises a portion of it above the y-axis so that it again crosses the abscissa at two locations. As we can see from Figure 3.8(b), one critical point is again at the origin $y^{(1)} = 0$ but now the other one, $y^{(2)} = \mu$ (< 0), is on the negative (horizontal) y-axis, illustrating graphically how the location of critical points may again depend on μ, the only system parameter in this case.

In addition to the location of (at least one of) the critical points, changing μ also affects the stability of the critical points for the present problem. For $\mu > 0$, $y_c^{(1)} = 0$ is unstable and $y_c^{(2)} = \mu$ is asymptotically stable. But for $\mu < 0$, $y_c^{(1)} = 0$ is asymptotically stable and $y_c^{(2)} = \mu$ is unstable. The stability switch of the critical point at the origin takes place at $\mu_c = 0$ where the two critical points coalesce into one semi-stable critical point. The pivotal value μ_c of a system parameter μ associated with such abrupt changes is the only bifurcation point of the ODE (3.4.4). Note that at bifurcation, the growth rate function also has a horizontal tangent at the critical point so that (3.4.2) again applies.

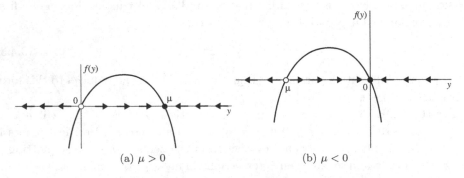

(a) $\mu > 0$ (b) $\mu < 0$

Fig. 3.8 $f(y; \mu) = y(\mu - y)$.

For (3.4.4), the number of critical points changes from two to one to two as μ passes through $\mu_c = 0$; the two critical points also exchange their stability type. This type of bifurcation is quite different from the saddle-node bifurcation for the fishery problem with a constant harvest rate; it is called a *transcritical* bifurcation. The behavior of the ODE system with this type of bifurcation is succinctly summarized by its bifurcation diagram that graphs the two critical points $\{y_c^{(1)}(\mu), y_c^{(2)}(\mu)\}$ as functions of μ as shown in Figure 3.9. Again, it is customary to use thick solid curves to indicate (asymptotically) stable critical points and dashed or dotted curves for unstable critical points.

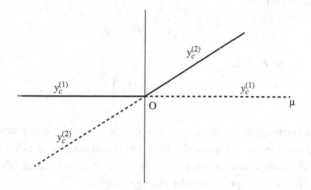

Fig. 3.9   Bifurcation diagram for transcritical bifurcation.

Fixed effort harvesting

On the open sea, the prescription of a uniform harvesting rate is not realistic. More typical is a fixed fishing effort E encapsulating the total fishing boats, equipment and fishermen available for the fishing period, and the catch rate h not only depends on E, but also on the fish population $y(t)$. Evidently, the larger the fish population at the fishing ground covered by the fishing boat(s), the larger the catch would be for the same effort E. At the other extreme, if there is no fish, there would be no fish caught however large the effort may be. As a first approximation, we consider the case that the harvest rate is proportional to both the effort E and the fish population size y with

$$h(t) = qEy(t)$$

where q is a constant of proportionality whose unit is "per unit time per unit effort." Assuming a natural growth rate of the logistic type (when it is not being harvested), the growth rate of the fish population being harvested by a fixed fishing effort is given by

$$y' = y(1 - y) - qEy = y\left[(1 - qE) - y\right] \equiv f_E(y). \tag{3.4.5}$$

It is easy to see from $f_E(y; E) = 0$ that there are two critical points (steady state fish populations) for this growth model:

$$y^{(1)} = 0, \qquad y^{(2)} = 1 - qE. \tag{3.4.6}$$

Starting from $E = 0$ for which the fish has its two natural (normalized) equilibrium states (0 and 1), the non-zero critical point $y^{(2)} = 1 - qE$ decreases with increasing fishing effort but remains positive as long as $qE < 1$, while the trivial state $y^{(1)} = 0$ does not change with E. The stability of the critical points can be deduced from a graph of the growth rate function $f_E(y)$ but can also be obtained from a linear stability analysis. With

$$\frac{\partial f_E}{\partial y} = (1 - qE) - 2y$$

we have for the range $E < 1/q$

$$\left[\frac{\partial f_E}{\partial y} \right]_{y=y^{(1)}} = (1 - qE) > 0, \qquad \left[\frac{\partial f_E}{\partial y} \right]_{y=y^{(2)}} = -(1 - qE) < 0$$

so that $y^{(1)} = 0$ is unstable while $y^{(2)} = 1 - qE$ is asymptotically stable.

For $E > 1/q$ on the other hand, the nontrivial critical point is now negative and the other critical point remains zero. But with excessive fishing effort ($E > 1/q$), the stability of the critical point would change since

$$\left[\frac{\partial f}{\partial y} \right]_{y=y^{(1)}} = (1 - qE) < 0, \qquad \left[\frac{\partial f}{\partial y} \right]_{y=y^{(2)}} = -(1 - qE) > 0.$$

While $y^{(2)} = 1 - qE < 0$ is not biologically meaningful, the other critical point remaining at the origin has changed its stability and becomes asymptotically stable. (This is bad news for those interested in re-invigorating the depleted fish population.) The two critical points coalesce at $E = 1/q \equiv E_c$ with $y^{(1)} = 0 = y^{(2)}$ being semi-stable at bifurcation.

The bifurcation diagram is similar to that for the canonical problem (3.4.4) showing that the bifurcation at $E_c = 1/q$ is transcritical.

3.4.2.3 *Pitchfork bifurcation*

Saddle-node bifurcation involves a change of the number of critical points from two to (one to) none. (Since all critical points disappear after bifurcation, nothing can be said about stability type changes after bifurcation.) In contrast, the number of critical points remains the same but the stability of the zero and nonzero critical point changes type after a transcritical bifurcation. The example

$$y' = y(\mu - y^2) \tag{3.4.7}$$

shows a third kind of bifurcation.

For $\mu > 0$, the graph of $f(y; \mu) = y(\mu - y^2)$ is as shown in Figure 3.10(a). We see from the graph that there are three isolated critical points at $y^{(1)} = -\sqrt{\mu}, y^{(2)} = 0,$

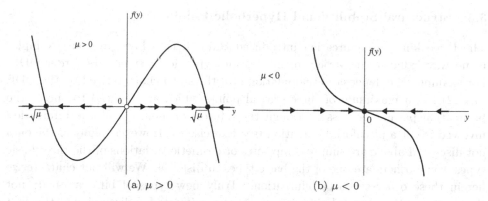

Fig. 3.10 Graph of (3.4.7) for $f(y; \mu)$.

and $y^{(3)} = \sqrt{\mu}$ with the one at the origin being unstable while the other two are asymptotically stable.

For $\mu < 0$, we have $f(y; \mu) = -y(|\mu| + y^2)$ so that its graph crosses the y-axis only at the origin as shown in Figure 3.10(b). Hence, there is only one critical point $y_c = 0$ and it is asymptotically stable. The transition from three to one critical point occurs at the bifurcation point $\mu_c = 0$ for which the graph of $f(y; \mu_c) = -y^3$ is similar to the one for $\mu < 0$ except that its tangent to the y-axis at the critical point is horizontal with $f_{,y}(y_c; \mu_c) = 0$ (and $f_{,yy}(y_c; \mu_c) = 0$ as well).

The various changes associated with bifurcation for this example are different from those of the two previous examples and are succinctly summarized by the bifurcation diagram in Figure 3.11. One common denominator permeates through all three types of bifurcation is that, at bifurcation, the growth rate function also has a horizontal tangent at the critical point as required by Theorem 2.

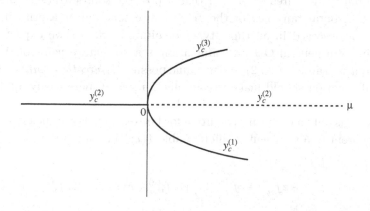

Fig. 3.11 Bifurcation diagram for pitch fork bifurcation.

3.5   Structural Stability and Hyperbolic Points

The three kinds of bifurcation introduced through the three previous examples in no way exhaust the variety of bifurcation types for a single first order ODE. For example, the depensation population growth model characterized by the ODE (3.1.12) has a maximum of three critical points which are reduced to one as the bifurcation parameter passes through the bifurcation point. But the bifurcation involved is not a pitchfork bifurcation (see Exercise 6). However, many of the ones not discussed above are simply composites or cosmetic variations of the three basic types; some others are one of the basic types in disguise. We will not characterize herein these other types of bifurcation. Truly new types of bifurcation do not occur sufficiently frequently in applications to merit detailed discussion in this first exposure to bifurcation. Those interested are referred to [17,27] for further reading.

Exercise 6 *Sketch the bifurcation diagram for the ODE below with μ as the bifurcation parameter*

$$z' = z(z - \mu)(2 - z).$$

Even if we restrict our attention to the three basic types of bifurcation discussed above, there is still the issue of how to uncover such *structural instability* at bifurcation (changing critical point counts and stability) if the growth rate function should be too complex for the graphical method that works well for the simple ODE. It may be too difficult or too tedious to investigate how the critical points change as system parameters range over a high dimensional parameter space. Imagine what needs to be done even for the fishery problem with the three system parameters Y_c, a and H if we did not know that a change of only one combination of the three is involved in triggering bifurcation. We need an alternative characterization of bifurcation that does not rely on knowing the solution behavior or having an explicit solution of the critical points in terms of the parameter(s).

In Theorem 2, we already learned that a possible solution lies in the distinct feature of horizontal tangency of the growth rate function at a critical point at bifurcation as observed in all three types of bifurcation. As we expect that the critical point of a general ODE at bifurcation is a *stationary point* of the growth rate function, we can use (3.4.2) to determine possible *bifurcation points*. If nothing else, the information should make the graphical method more easily applicable to general growth rate functions by localizing the search for possible bifurcation points.

Theorem 2 is only a necessary condition for bifurcation. For sufficiency, we apply Taylor's theorem to a sufficiently differentiable $f(y; \mu)$ about (y^c, μ_c),

$$f(y; \mu) = f(y^c + z; \mu_c + r) \tag{3.5.1}$$

$$= f^{(c)} + z f_{,y}^{(c)} + r f_{,\mu}^{(c)} + \frac{1}{2}\left[z^2 f_{,yy}^{(c)} + r^2 f_{,\mu\mu}^{(c)} + 2zr f_{,y\mu}^{(c)} \right] + \cdots$$

where

$$z = y - y^c, \qquad r = \mu - \mu_c, \qquad g^{(c)} = g(y^c; \mu_c).$$

If $\alpha \equiv f_{,y}^{(c)} = f_{,y}(y^c; \mu_c) \neq 0$, then in a small neighborhood of (y^c, μ_c) with $|z| \ll 1$ and $|r| \ll 1$, the ODE for $y(t)$ is accurately approximated by

$$z' = \alpha z + \beta r, \qquad \beta = f_{,\mu}^{(o)}.$$

Evidently, $z(t)$ grows exponentially if $\alpha \equiv f_{,y}^{(c)} > 0$ and decays exponentially if $\alpha < 0$. It follows that the (isolated) critical point y^c is unstable or asymptotically stable depending only on α being positive or negative. In particular, no small change in $r = \mu - \mu_c$, could affect the nature of the stability of the critical point y^c of the ODE. In other words, the ODE is structurally stable as μ passes through μ_c as long as the critical point y_c is not a stationary point of $f(y; \mu_c)$.

Definition 4 *A critical point $y^c(\mu)$ of the ODE*

$$y' = f(y; \mu) \tag{3.5.2}$$

*is called a **hyperbolic (critical) point** of the ODE if it is not a stationary point of the growth rate function $f(y; \mu)$ and a **non-hyperbolic point** if it is.*

From the development in the preceding paragraphs, we conclude:

Proposition 1 *A first order ODE (3.5.2) is structurally stable for a particular value of the system parameter μ if all critical points (for that μ value) are hyperbolic.*

The proposition is merely a more formal re-statement of Theorem 2 in terms of hyperbolic critical points. For the chosen model of a phenomenon to be structurally unstable at a particular value of the bifurcation parameter, the relevant critical point must be non-hyperbolic.

On the issue of sufficiency, suppose the critical point y^c is non-hyperbolic at $\mu = \mu_c$ so that $\alpha = f_{,y}(y^c; \mu_c) = 0$. In that case, the Taylor expansion (3.5.1) simplifies to

$$f(y; \mu) = \frac{1}{2}\left[z^2 f_{,yy}^{(c)}\right] + r f_{,\mu}^{(c)} + O(z^n)$$

for $n \geq 3$ so that $f(y; \mu)$ is of one sign for sufficiently small $|z|$ to render the critical point semi-stable. We have then the following sufficiency theorem for bifurcation:

Theorem 3 *The autonomous ODE model $y' = f(y; \mu)$ is structurally unstable at $\mu = \mu_c$ if some critical point $y^c(\mu_c)$ is non-hyperbolic.*

We leave the topic of bifurcation with the following more intriguing example as an exercise:

Exercise 7 $\qquad\qquad y' = 1 - \sin(y) - e^{-\mu y}\cos(y).$

Part 2

Interacting Populations

Chapter 4

Linear Interactions

4.1 Types of Interaction

A population of a biological species rarely exists in isolation. While any species may be investigated without consideration of others when the issue of interest permits such exclusion, the evolution of populations becomes even more interesting when they interact with each other. Among the features that make such interactions interesting is the different types of interaction. Even if we restrict our discussion to interaction of two populations with linear growth/declination rates, typified by

$$x' = a_{11}x + a_{12}y, \qquad y' = a_{21}x + a_{22}y, \qquad (4.1.1)$$

we can see different relations between them by the signs of the coefficients $\{a_{ij}\}$. Below are some possible types of interaction:

- Predator-Prey: $a_{11} > 0$, $a_{12} < 0$ while $a_{21} > 0$, $a_{22} < 0$
- Competitive: $a_{11} > 0$, $a_{12} < 0$ while $a_{22} > 0$, $a_{21} < 0$
- Cooperative: $a_{11} < 0$, $a_{12} > 0$ while $a_{21} > 0$, $a_{22} < 0$
- Parasitic: $a_{11} > 0$, $a_{12} = 0$ (neglecting nonlinear interaction terms) while $a_{21} > 0$, $a_{22} < 0$
- Nihilistic: $a_{ij} < 0$ for all i, j.

In the case of the parasitic interaction, a linear ODE model is generally not adequate except for a very small population of parasites. Typically, more complex and interesting interactions often require nonlinear models. However, there is certainly enough substance among the linear models for a separate discussion of such models. Mathematical techniques for analyzing linear models will also be needed in a significant way for nonlinear models in the next chapter.

4.2   Red Blood Cell Production

4.2.1   *Model formulation*

4.2.1.1   *The question*

Humans and other vertebrate organisms live on oxygen and red blood cells (RBC) are their principal means of delivering oxygen to the body tissues via blood flow. Proper supply and functioning of RBC are therefore critical to their survival. Abnormalities of RBC occur in many forms, including various types of anemias (e.g., the Sickle-cell disease), hemolysis (a term for various form of RBC breakdown), and polycythemias (a general expression for excessive supply of RBC). Most of them are life threatening and require drug treatment or other forms of clinical intervention to deal with their possible consequences. When the question is how to restore and maintain an adequate supply of functioning RBC, we need to know what is the normal supply of RBC.

4.2.1.2   *Known facts*

RBC are produced by bone marrow at the rate of about 200 billion (2×10^{11}) per day or about 2.4 million/sec. Each cell has a half-life of about 55–60 days. At any one time, a human adult has about 30 trillions (3×10^{13}) RBC. An average adult has about 8–10 lbs of bone marrow (about 5% of body weight). Bone marrow of mature adult degrades very slowly (rapidly replenished through some feedback mechanism) until old age when the degradation rate becomes high. However, significant loss of RBC (due to disease or abnormal environmental changes) induces production of additional bone marrow at a faster rate to replenish the excessive loss of RBC. Of interest is the adequacy of the replacement and how it may vary with system parameters that characterize the individual's constitution.

4.2.1.3   *A linear model*

Let $R(t)$ and $M(t)$ be the biomass of RBC and bone marrow, respectively, at time t. We take here the unit of time to be a day, given the available data on replacement and production of RBC mentioned above. The large size of the RBC population allows us to think of R and M as continuously differentiable real-valued functions. As a first effort to understand the supply of RBC in a human being, the information on degradation and replenishment activities of the RBC population is modeled by the following pair of linear ODE on the rate of change of RBC and bone marrow biomass, respectively:

$$\frac{dR}{dt} = -\alpha R + \beta M, \qquad \frac{dM}{dt} = \gamma(R_c - R) - \varepsilon M, \qquad (4.2.1)$$

where $\alpha > 0$, $\varepsilon \geq 0$, $\beta > 0$, $\gamma > 0$ and where $R_c > 0$ is a maintenance level of RBC.

Exercise 8 *Describe the role of each term on the right-hand side of the two ODE in (4.2.1).*

From the known facts listed earlier, we may assign the following values to the various parameters in the ODE:

- $\alpha = \ln(2)/60$ from the definition of half-life and a RBC half-life of 60 days.
- $\beta = 1.73 \times 10^8$ cells/lbs/day in order for 10 lbs of bone marrow to replenish the daily loss of about 1.73 billion cells (but should be considerably less since bone marrow is also responsible for producing many other cells as well).
- $\gamma = 5 \times 10^{-14}$ lbs//cell/day on the basis of a production of 0.1 lbs of additional bone marrow needed to replenish the loss of 2×10^{11} RBC (about one day's normal production) in about 10 days.
- $\varepsilon = 0$ for mature adults and $\varepsilon > 0$ for senior citizens.

The linear system of two differential equations is supplemented by two initial conditions,

$$R(0) = R_0, \qquad M(0) = M_0, \tag{4.2.2}$$

to form an IVP as a mathematical model for the evolution of RBC and bone marrow biomass in a typical individual.

4.2.2 Method of elimination

Since the model involves only two linear ODE with constant coefficients, they can be solved by any number of methods found in a typical text on elementary differential equations [1]. One of these is the method of elimination consisting of differentiating one of the equation and use the result to eliminate one unknown from the other equation leaving a second order linear ODE with constant coefficients. The resulting second order linear equation is of the same form as the usual equation of motion for a spring mass system and its solution is well known to most students in science and engineering.

To apply this method to the RBC problem, we differentiate the first equation in (4.2.1) to get

$$\frac{d^2 R}{dt^2} = -\alpha \frac{dR}{dt} + \beta \frac{dM}{dt} = -\alpha \frac{dR}{dt} + \beta \left[\gamma (R_c - R) - \varepsilon M \right],$$

where we have used the second ODE in (4.2.1) to eliminate dM/dt. The first equation is then used to eliminate M from the second order ODE above, leaving us with a single second order ODE for R alone:

$$R'' + (\alpha + \varepsilon) R' + (\beta \gamma + \varepsilon \alpha) R = \beta \gamma R_c. \tag{4.2.3}$$

The ODE is augmented by two initial conditions:

$$R(0) = R_0, \qquad R'(0) = -\alpha R_0 + \beta M_0. \tag{4.2.4}$$

The first is already given in (4.2.2) while the second results from the use of both conditions of (4.2.2) in the first ODE of (4.2.1).

With the right-hand side of (4.2.3) taken to be independent of time, we seek a (particular) solution $R^{(p)}$ independent of t. Since $dR^{(p)}/dt = d^2R^{(p)}/dt = 0$, the ODE (4.2.3) and the first ODE in (4.2.1) give immediately

$$R^{(p)} = \frac{\beta\gamma R_c}{\beta\gamma + \varepsilon\alpha}, \qquad M^{(p)} = \frac{\alpha\gamma R_c}{\beta\gamma + \varepsilon\alpha},$$

provided $\beta\gamma + \varepsilon\alpha \neq 0$. This last requirement is met since all four parameters are non-negative and at least β and γ are positive though the latter may be small for senior citizens. To satisfy the initial conditions, we need two linearly independent complementary solutions of the form $e^{\lambda t}$. Upon substitution of the assumed solution into the homogeneous ODE associated with (4.2.3), we obtain the characteristic equation

$$\lambda^1 + (\alpha + \varepsilon)\lambda + (\beta\gamma + \varepsilon\alpha) = 0 \qquad (4.2.5)$$

for λ. With the two roots of (4.2.5) given by

$$\binom{\lambda_1}{\lambda_2} = \frac{1}{2}\left[-(\alpha + \varepsilon) \pm \sqrt{(\alpha + \varepsilon)^2 - 4(\beta\gamma + \varepsilon\alpha)}\right], \qquad (4.2.6)$$

we may now complete the solution process by setting

$$R(t) = c_1 e^{\lambda_1 t} + c_2 e^{\lambda_2 t} + R^{(p)} \qquad (4.2.7)$$

for any two constants c_1 and c_2 and verifying directly that (4.2.7) is a solution of the ODE (4.2.3). (The superposition principles that led to (4.2.7) will be formulated and proved in Exercise 11 below.) The two constants can then be determined by the two initial conditions in (4.2.4).

Instead of carrying out the details for this explicit solution, we note that both characteristics roots λ_1 and λ_2 are either both negative or have a negative real part since all the parameters involved are nonnegative and at least α, β and γ are positive. It follows that $R(t) \to R^{(p)}$ as $t \to \infty$. As such, the RBC in the system should tend to $R^{(p)}$ with increasing time and the bone marrow concentration should tend to $M^{(p)}$. If we are only interested in the RBC count after a long time (and not the transient behavior), we do not need an explicit solution for the two constants c_1 and c_2.

Exercise 9 *Solve the IVP (4.2.3)–(4.2.4) to obtain $R(t)$ and $M(t)$.*

4.2.3   Implication of mathematical results

4.2.3.1   The mature adult case

Even without an explicit solution for the IVP, a number of observations can be made from the steady state solution on the evolving RBC population and the bone marrow mass. For mature adult, bone marrow remains more or less unchanged with time so that we can take $\varepsilon = 0$. This simplifies the eigenvalues to

$$\begin{pmatrix} \lambda_1 \\ \lambda_2 \end{pmatrix} = \frac{1}{2}\left[-\alpha \pm \sqrt{\alpha^2 - 4\gamma\beta}\right], \qquad \begin{pmatrix} R^{(p)} \\ M^{(p)} \end{pmatrix} = R_c \begin{pmatrix} 1 \\ \alpha/\beta \end{pmatrix}.$$

Again with $\alpha > 0$, the two eigenvalues are either both negative or have a negative real parts so that the transient (complementary) solutions decay to zero as $t \to \infty$ and become negligibly small for sufficiently large t. Actually, we can be more specific for values of the parameters given earlier. From the definition of half-life, we take $\alpha = \ln(2)/60$ (and hence $\alpha^2 = 1.335 \times 10^{-4}$) and $4\gamma\beta = 3.46 \times 10^{-5}$ so that $\alpha^2 > 4\gamma\beta$. In that case, we have $0 < \lambda_2 < \lambda_1 < 1$ and

$$\mathbf{y}(t) = \begin{pmatrix} R(t) \\ M(t) \end{pmatrix} \to \frac{R_c}{\beta\gamma + \alpha\varepsilon} \begin{pmatrix} \beta\gamma \\ \alpha\gamma \end{pmatrix} = R_c \begin{pmatrix} 1 \\ \alpha/\beta \end{pmatrix}$$

as $t \to \infty$. Thus, the RBC count $R(t)$ tends to its maintenance count R_c as we would like to see. However, the steady state bone marrow mass is much too high; it probably means that we have underestimated β. After all, not all available bone marrow is dedicated to replenishing daily RBC losses. For example, the same bone marrow tissues are also responsible for producing white blood cells and platelets that help with blood clotting. Having gone through in some detail the model formulation, mathematical analysis and model validation phase for the RBC problem, the fourth stage of the modeling cycle, the refinement of our first model to address these deficiencies, will be left to the readers as an active learning opportunity.

4.2.3.2 The aging adult case

Aging adults are known to suffer a high rate of bone marrow degradation corresponding to a non-negligible loss fraction ε. As a consequence, less RBC are produced to replenish the degraded RBC. In most case, the usual mechanism for replenishing the lost bone marrow through the $(R_c - R)$ differential is less effective for the older folks which means a smaller value of γ than normal.

We already know that both eigenvalues have a negative real part for the general problem. For γ sufficiently small (whether it is due to aging or some illness such as leukemia, lymphoma, etc.) so that $(\alpha - \varepsilon)^2 > 4\gamma\beta$, the two eigenvalues

$$\begin{pmatrix} \lambda_1 \\ \lambda_2 \end{pmatrix} = \frac{1}{2}\left[-(\alpha + \varepsilon) \pm \sqrt{(\alpha - \varepsilon)^2 - 4\gamma\beta}\right] \qquad (4.2.8)$$

continues to be negative with $\lambda_2 < \lambda_1 < 0$. In that case, the critical point of the model system remains asymptotically stable with

$$\mathbf{y}(t) = \begin{pmatrix} R(t) \\ M(t) \end{pmatrix} \to \frac{R_c}{\beta\gamma + \alpha\varepsilon} \begin{pmatrix} \beta\gamma \\ \alpha\gamma \end{pmatrix}.$$

For γ sufficiently small, the common factor $R_c/(\beta\gamma + \alpha\varepsilon)$ may be approximated by $R_c/\alpha\varepsilon$ to give

$$\lim_{t \to \infty} [\mathbf{y}(t)] \simeq \frac{\gamma R_c}{\varepsilon} \begin{pmatrix} \beta/\alpha \\ 1 \end{pmatrix}.$$

In that case, both RBC (R) and bone marrow (M) tend to zero with γ and some clinical intervention such as drug treatment, blood infusion and/or bone marrow transplant is needed to arrest further deterioration of the individual's health.

Exercise 10 *Formulate a linear model that allows for daily blood infusion to replenish lost RBC.*

4.3   Linear Systems with Constant Coefficients

4.3.1   *Vector form*

For more complex phenomena with more evolving unknown states, the method of elimination becomes tedious and often impractical. Even for linear systems of small or moderate size such as the RBC problem, the inhomogeneous (forcing) terms may be time dependent and for which a useful particular solution is not easy to find by elementary methods (such as the method of undetermined coefficients in elementary texts for differential equations). For the latter, more general method such as reduction of order and variation of parameters are available and may be applicable, we choose to offer in this section a single unified method of solution for this class of problems.

For large linear systems, it is generally more efficient and practical to work with the vector form of linear systems. For the two equations RBC problem, we may set

$$\mathbf{y}(t) = \begin{pmatrix} R(t) \\ M(t) \end{pmatrix}, \quad \mathbf{h}(t) = \begin{pmatrix} 0 \\ \gamma R_c \end{pmatrix}, \tag{4.3.1}$$

and write the system (4.2.1) as

$$\frac{d\mathbf{y}}{dt} = A\mathbf{y} + \mathbf{h}(t) \tag{4.3.2}$$

where

$$A = \begin{bmatrix} -\alpha & \beta \\ -\gamma & -\varepsilon \end{bmatrix} \tag{4.3.3}$$

is the *coefficient matrix* of the linear ODE system. Correspondingly, the initial conditions can be written as

$$\mathbf{y}(0) = \mathbf{y}^o \equiv \begin{pmatrix} R_0 \\ M_0 \end{pmatrix}. \tag{4.3.4}$$

A working knowledge of basic operations with matrices, including *row echelon reduction, matrix inversion* and *matrix eigenvalue problem*, is assumed (and can be acquired from an elementary text on linear algebra or matrix theory). A working knowledge of using mathematical software such as MatLab, Mathematica or Maple for matrices, though not required for this volume, would generally take the drudgery from working with large linear systems.

4.3.2 Complementary solutions

Whenever the number of unknowns in a linear system is not too large and the forcing term $\mathbf{h}(t)$ is in terms of elementary functions, a practical method of solution takes advantage of the superposition properties of linear systems, whether or not we use the method of elimination or work directly with the vector formulation. We describe in this section the kind of superposition principles most useful in this context.

Definition 5 A complementary solution $\mathbf{y}^{(k)}(t)$ of a general linear ODE system of the form (4.3.2) is a solution of the corresponding homogeneous linear system so that

$$\frac{d\mathbf{y}^{(k)}}{dt} = A\mathbf{y}^{(k)}. \qquad (4.3.5)$$

Exercise 11 Verify the following two superposition principles

a) Superposition Principle I: If $\mathbf{y}^{(i)}(t)$ and $\mathbf{y}^{(j)}(t)$ are two (vector) complementary solutions of a general linear system of the form (4.3.2), so is $c_i\mathbf{y}^{(i)}(t) + c_j\mathbf{y}^{(j)}(t)$ for any constants c_i and c_j. (Recall that a complementary solution is a solution for the ODE (4.3.2) with $\mathbf{h}(t) = \mathbf{0}$.)

b) Superposition Principle II: If $\mathbf{y}^{(i)}(t)$ is a complementary solution of a general linear system of the form (4.3.2) and $\mathbf{y}^{(p)}(t)$ is any particular solution of the inhomogeneous equation (4.3.2) for a prescribed $\mathbf{h}(t)$ so that

$$\frac{d\mathbf{y}^{(p)}}{dt} = A\mathbf{y}^{(p)} + \mathbf{h}(t), \qquad (4.3.6)$$

then $c_i\mathbf{y}^{(i)}(t) + \mathbf{y}^{(p)}(t)$ is a solution of (4.3.2) for any constant c_i.

Exercise 12 a) For the RBC problem, the vector forcing term $\mathbf{h}(t)$ in (4.3.2) is given in (4.3.1) and does not depend on t. Show that

$$\mathbf{y}^{(p)}(t) = \mathbf{z}^{(p)} = -A^{-1}\begin{pmatrix} 0 \\ \gamma R_c \end{pmatrix} = \frac{\gamma R_c}{\beta\gamma + \alpha\varepsilon}\begin{pmatrix} \beta \\ \alpha \end{pmatrix} \qquad (4.3.7)$$

is a (time-invariant) particular solution for that problem (with A of the model ODE (4.3.2) given in (4.3.3)). (Hint: Compute A^{-1}.)

b) Use the superposition principles to obtain the solution of the IVP defined by (4.3.1)–(4.3.4). (Recall that a complementary solution consists of an eigenvector $\mathbf{v}^{(j)}$ associated with the eigenvalue λ_j of A with magnitude $e^{\lambda_j t}$. Readers not familiar with complementary solutions for an ODE with constant coefficients should read the next subsection in preparation for this exercise.)

4.3.3 Fundamental matrix solution

Since complementary solutions play an essential role in determining the solution of linear ODE systems, we review briefly the method for finding them with the help of the eigen-pairs of the coefficient matrix A. Suppose λ_k is an eigenvalue of A and $\mathbf{v}^{(k)}$ is the corresponding eigenvector. Then it is straightforward to verify that

$$\mathbf{y}^{(k)}(t) = \mathbf{v}^{(k)}e^{\lambda_k t} \tag{4.3.8}$$

is a complementary solution given by

$$\frac{d\mathbf{y}^{(k)}(t)}{dt} = \frac{d}{dt}\left[\mathbf{v}^{(k)}e^{\lambda_k t}\right] = \mathbf{v}^{(k)}\frac{de^{\lambda_k t}}{dt} = \left[\lambda_k\mathbf{v}^{(k)}\right]e^{\lambda_k t} = \left[A\mathbf{v}^{(k)}\right]e^{\lambda_k t} = A\mathbf{y}^{(k)}(t).$$

If an $n \times n$ matrix A has a full set of (n linearly independent) eigenvectors, we can form the *fundamental matrix solution*

$$Z(t) = \left[\mathbf{v}^{(1)}e^{\lambda_1 t}, \mathbf{v}^{(2)}e^{\lambda_2 t}, \cdots, \mathbf{v}^{(n)}e^{\lambda_n t}\right] = VE_\lambda(t) \tag{4.3.9}$$

where

$$V = \left[\mathbf{v}^{(1)}, \mathbf{v}^{(2)}, \cdots, \mathbf{v}^{(n)}\right], \qquad E_\lambda(t) = \begin{bmatrix} e^{\lambda_1 t} & 0 & \cdot & \cdot & \cdot & 0 \\ 0 & e^{\lambda_2 t} & 0 & \cdot & \cdot & 0 \\ \cdot & & 0 & \cdot & \cdot & \cdot \\ \cdot & & & \cdot & & \cdot \\ 0 & \cdot & \cdot & \cdot & \cdot & 0 \\ 0 & 0 & \cdot & \cdot & 0 & e^{\lambda_n t} \end{bmatrix} \tag{4.3.10}$$

with $VE_\lambda(0) = V$. Evidently, $Z(t)$ is a complementary solution since

$$Z'(t) = VE_\lambda'(t) = VE_\lambda(t)\Lambda = Z(t)\Lambda$$

with

$$\Lambda = \begin{bmatrix} \lambda_1 & 0 & & 0 \\ 0 & \lambda_2 & 0 & \\ & 0 & \cdot & \\ & & \cdot & 0 \\ 0 & & 0 & \lambda_n \end{bmatrix},$$

or

$$Z'(t) = \left[\lambda_1\mathbf{v}^{(1)}e^{\lambda_1 t}, \ldots, \lambda_n\mathbf{v}^{(n)}e^{\lambda_n t}\right]$$

$$= A\left[\mathbf{v}^{(1)}e^{\lambda_1 t}, \ldots, \mathbf{v}^{(n)}e^{\lambda_n t}\right] = AZ.$$

By the superposition principle I, the general solution for a general (homogeneous) linear system (4.3.2) with $\mathbf{h}(t) = \mathbf{0}$ is then

$$\mathbf{y}(t) = c_1\mathbf{v}^{(1)}e^{\lambda_1 t} + \cdots + c_n\mathbf{v}^{(n)}e^{\lambda_n t}$$

$$= \left[\mathbf{v}^{(1)}e^{\lambda_1 t}, \ldots, \mathbf{v}^{(n)}e^{\lambda_n t}\right]\mathbf{c} = Z(t)\mathbf{c}.$$

Proposition 2 *If A has a full set of n eigenvectors $\{\mathbf{v}^{(k)}\}$ associated with the n (not necessarily distinct) eigenvalues $\{\lambda_i\}$, then $\mathbf{z}(t) = Z(t)\mathbf{c}$, where $\mathbf{c}$ is a constant vector with components $\{c_1, \ldots, c_n\}$, solves the IVP*

$$\mathbf{z}' = A\mathbf{z}, \qquad \mathbf{z}(0) = \mathbf{z}^0.$$

Proof With $z(0) = Z(0)c = \left[v^{(1)}, \ldots, v^{(n)}\right] c = Vc = z^0$, the linear system has a unique solution since the n eigenvectors $\left\{v^{(k)}\right\}$ are linearly independent so that V is invertible. □

The same fundamental matrix solution also solves the inhomogeneous IVP for (4.3.2):

Corollary 1 *Under the same hypothesis as the proposition above and suppose* $y^{(p)}(t)$ *is a particular solution for the inhomogeneous ODE (4.3.2), then* $y(t) = Z(t)c + y^{(p)}(t)$, *where* c *is a constant vector with components* $\{c_1, \ldots, c_n\}$, *solves the IVP*

$$y' = Ay + h, \qquad y(0) = y^0.$$

Proof With $y(0) = Z(0)c + y^{(p)}(0) = z^0$ or

$$Vc = z^0 - y^{(p)}(0),$$

the linear system for $\{c_1, \ldots, c_n\}$ has a unique solution since the n eigenvectors $\left\{v^{(k)}\right\}$ are linearly independent so that V is invertible. □

Some simple examples below show different combinations of complementary solutions. These solutions and the corresponding eigen-pairs will also be useful later in our study of stability of nonlinear systems in the next chapter.

Example 2 $A = \begin{bmatrix} -2 & 1 \\ 1 & -2 \end{bmatrix}$.

For this matrix, we have $\lambda_1 = -1$ and $\lambda_2 = -3$. The associated eigenvectors are $v^{(1)} = (1,1)^T$ and $v^{(2)} = (1,-1)^T$. The fundamental matrix solution of the ODE (4.3.2) or (4.3.5) is given by (4.3.9)–(4.3.10) in terms of these eigen-pairs in the form

$$Z(t) = \begin{bmatrix} 1 & 1 \\ 1 & -1 \end{bmatrix} \begin{bmatrix} e^{-t} & 0 \\ 0 & e^{-3t} \end{bmatrix}.$$

Example 3 $A = \begin{bmatrix} 5 & -1 \\ 3 & 1 \end{bmatrix}$.

For this matrix, we have $\lambda_1 = 2$ and $\lambda_2 = 4$. The associated eigenvectors are $v^{(1)} = (1,3)^T$ and $v^{(2)} = (1,1)^T$, respectively. The fundamental matrix solution for the corresponding linear ODE is

$$Z(t) = \begin{bmatrix} 1 & 1 \\ 3 & 1 \end{bmatrix} \begin{bmatrix} e^{2t} & 0 \\ 0 & e^{4t} \end{bmatrix}.$$

Example 4 $A = \begin{bmatrix} 1 & 1 \\ 4 & -2 \end{bmatrix}$.

For this matrix, we have $\lambda_1 = 2$ and $\lambda_2 = -3$ with $\mathbf{v}^{(1)} = (1,1)^T$ and $\mathbf{v}^{(2)} = (1,-4)^T$. The fundamental matrix solution is

$$Z(t) = \begin{bmatrix} 1 & 1 \\ 1 & -4 \end{bmatrix} \begin{bmatrix} e^{2t} & 0 \\ 0 & e^{-3t} \end{bmatrix}.$$

Example 5 $A = \begin{bmatrix} \mu & 1 \\ -1 & \mu \end{bmatrix}.$

For this matrix, we have a pair of complex conjugate eigenvalues $\lambda_1 = \mu + i$ and $\lambda_2 = \mu - i = \lambda_1^*$ with an asterisk denoting *complex conjugate* $(a + ib)^* = (a - ib)$. The corresponding eigenvectors, $\mathbf{v}^{(1)} = \mathbf{p} = (1,i)^T$ and $\mathbf{v}^{(2)} = \mathbf{p}^* = (1,-i)^T$, are also complex. The complementary solutions may be written in vector form as

$$\mathbf{y}(t) = e^{\mu t}\left(c_1 e^{it}\mathbf{p} + c_2 e^{-it}\mathbf{p}^*\right) = e^{\mu t}\left(c_1 e^{it}\mathbf{p} + c_1^* e^{-it}\mathbf{p}^*\right)$$
$$= e^{\mu t}\,[\mathbf{p},\ \mathbf{p}^*]\begin{bmatrix} e^{it} & 0 \\ 0 & e^{-it} \end{bmatrix}\begin{pmatrix} c_1 \\ c_1^* \end{pmatrix}.$$

The fundamental matrix solution is then

$$Z(t) = [\mathbf{p},\ \mathbf{p}^*]\begin{bmatrix} e^{(\mu+i)t} & 0 \\ 0 & e^{(\mu-i)t} \end{bmatrix}$$
$$= e^{\mu t}\begin{bmatrix} \cos t + i\sin t & \cos t - i\sin t \\ i\cos t - \sin t & -i\cos t - \sin t \end{bmatrix}.$$

For most applications, the solution should be real. In that case, it is customary to take $c_1 = c_r + ic_i$ and c_2 to be $c_1^* = c_r - ic_i$, the complex conjugate of the complex constant c_1 in which c_r and c_i are both real unknown constants. In terms of real valued functions and constants, we have

$$\mathbf{y}(t) = e^{\mu t}\begin{bmatrix} \cos(t) & \sin(t) \\ -\sin(t) & \cos(t) \end{bmatrix}\begin{pmatrix} 2c_r \\ -2c_i \end{pmatrix} = e^{\mu t}\begin{bmatrix} \cos(t) & \sin(t) \\ -\sin(t) & \cos(t) \end{bmatrix}\begin{pmatrix} y_1^o \\ y_2^o \end{pmatrix} \qquad (4.3.11)$$

so that

$$Z(t) = e^{\mu t}\begin{bmatrix} \cos(t) & \sin(t) \\ -\sin(t) & \cos(t) \end{bmatrix}. \qquad (4.3.12)$$

Upon differentiating $Z(t)$, we obtain

$$Z'(t) = e^{\mu t}\begin{bmatrix} \mu\cos(t) - \sin(t) & \mu\sin(t) + \cos(t) \\ -\mu\sin(t) - \cos(t) & \mu\cos(t) - \sin(t) \end{bmatrix}$$
$$= .e^{\mu t}\begin{bmatrix} \mu & 1 \\ -1 & \mu \end{bmatrix}\begin{bmatrix} \cos(t) & \sin(t) \\ -\sin(t) & \cos(t) \end{bmatrix} = AZ(t).$$

4.3.4 *Matrix diagonalization*

For a very large linear system (4.3.2), neither the method of elimination nor the method of superposition is practical. Matrix diagonalization offers a unified, effective and practical method of solution for general linear systems of ODE with constant coefficients.

Let λ_j be an *eigenvalue* of an $n \times n$ matrix A and $\mathbf{v}^{(j)}$ the corresponding *eigenvector* so that

$$A\mathbf{v}^{(j)} = \lambda_j \mathbf{v}^{(j)} \tag{4.3.13}$$

with $\{\lambda_j, \mathbf{v}^{(j)}\}$ known as an *eigen-pair* of the matrix.

Definition 6 *An $n \times n$ matrix A is said to be **nondefective** if it has a full set of n linearly independent eigenvectors. Otherwise, it is **defective**.*

For a nondefective matrix A, let V be the matrix with its columns consisting of n linearly independent eigenvectors

$$V = \left[\mathbf{v}^{(1)}, \mathbf{v}^{(2)}, \mathbf{v}^{(3)}, \cdots, \mathbf{v}^{(n)} \right]. \tag{4.3.14}$$

As noted previously, V is invertible since the columns are linearly independent. With (4.3.13), it is simple to verify

$$AV = V\Lambda \quad \text{or} \quad V^{-1}AV = \Lambda$$

where Λ is the diagonal matrix of eigenvalues:

$$\Lambda = \begin{bmatrix} \lambda_1 & 0 & \cdot & \cdot & 0 \\ 0 & \lambda_2 & \cdot & \cdot & \cdot \\ \cdot & 0 & \cdot & 0 & \cdot \\ \cdot & \cdot & \cdot & \cdot & 0 \\ 0 & \cdot & \cdot & 0 & \lambda_n \end{bmatrix}.$$

With the equality $\Lambda = V^{-1}AV$, we have effectively transformed A into the diagonal matrix Λ by a *similarity transformation* $V^{-1}AV$ using the *modal matrix V, the matrix of eigenvectors of A.*

4.3.5 *De-coupling of the linear system*

Upon pre-multiplying both sides of the ODE (4.3.2) by V^{-1}, we obtain (in view of V being a constant matrix)

$$\frac{d\mathbf{x}}{dt} = \Lambda \mathbf{x} + \mathbf{g}(t) \tag{4.3.15}$$

where

$$\mathbf{x} = V^{-1}\mathbf{y}, \quad \mathbf{g} = V^{-1}\mathbf{h}. \tag{4.3.16}$$

Correspondingly, the initial conditions (4.3.4) become

$$\mathbf{x}(0) = V^{-1}\mathbf{y}(\mathbf{0}) = V^{-1}\mathbf{y}^0 \equiv \mathbf{x}^0. \tag{4.3.17}$$

Since Λ is diagonal, the linear system of ODE for the new independent variable $\mathbf{x} = (x_1, \cdots, x_n)^T$ is uncoupled to give

$$x'_k = \lambda_k x_k + g_k(t), \qquad (k = 1, 2, 3, \cdots, n).$$

Each of these scalar first order ODEs is linear and can be solved by an integrating factor $e^{\lambda_k t}$ to give

$$x_k(t) = e^{\lambda_k t} \left\{ c_k + \int_0^t e^{-\lambda_k \tau} g_k(\tau) d\tau \right\} \tag{4.3.18}$$

where c_k is a constant of integration to be determined by the initial conditions. Recall that the n solutions $\{x_k(t)\}$ are the n components of the vector $\mathbf{x}(t) = (x_1(t), x_2(t), \cdots, x_n(t))^T = V^{-1}\mathbf{y}(t)$ (see (4.3.16)). Application of the initial condition $\mathbf{x}(0) = \mathbf{x}^0$ gives

$$\mathbf{x}(0) = \mathbf{c} = (c_1, c_2, \cdots, c_n)^T = \mathbf{x}^0 = (x_1^0, x_2^0, \cdots, x_n^0)^T = V^{-1}\mathbf{y}^0.$$

The solution (4.3.18) becomes

$$\mathbf{x}_k(t) = e^{\lambda_k t} \left\{ \left(V^{-1}\mathbf{y}^0 \right)_k + \int_0^t e^{-\lambda_k \tau} g_k(\tau) d\tau \right\}.$$

It may then be written in vector form as

$$\mathbf{x}(t) = e^{\Lambda t} \left[\mathbf{c} + \mathbf{x}^{(p)}(t) \right] = e^{\Lambda t} \left[V^{-1}\mathbf{y}^0 + \int_0^t e^{-\Lambda \tau} \mathbf{g}(\tau) d\tau \right]$$

where the matrix $e^{\Lambda t}$ is the diagonal matrix

$$e^{\Lambda t} = \left[e^{\lambda_i t} \delta_{ij} \right] = \begin{bmatrix} e^{\lambda_1 t} & 0 & \cdot & \cdot & 0 & 0 \\ 0 & e^{\lambda_2 t} & 0 & \cdot & \cdot & 0 \\ 0 & 0 & \cdot & \cdot & \cdot & \cdot \\ \cdot & \cdot & \cdot & \cdot & \cdot & \cdot \\ \cdot & \cdot & \cdot & \cdot & 0 & \cdot \\ 0 & \cdot & \cdot & 0 & e^{\lambda_{n-1} t} & 0 \\ 0 & 0 & \cdot & \cdot & 0 & e^{\lambda_n t} \end{bmatrix}$$

and

$$\mathbf{x}^{(p)}(t) = \left(x_1^{(p)}(t), x_2^{(p)}(t), \cdots, x_n^{(p)}(t) \right)^T, \qquad x_k^{(p)}(t) = \int_0^t e^{-\lambda_k \tau} g_k(\tau) d\tau$$

or more compactly

$$\mathbf{x}^{(p)}(t) = \int_0^t e^{-\Lambda \tau} \mathbf{g}(\tau) d\tau.$$

To recover the solution for the original problem, we recall from (4.3.16)

$$\mathbf{y}(t) = V\mathbf{x}(t) = V e^{\Lambda t} \left[V^{-1}\mathbf{y}^0 + \int_0^t e^{-\Lambda \tau} \left(V^{-1}\mathbf{h}(\tau) \right) d\tau \right]. \tag{4.3.19}$$

In the form of the more familiar superposition of complementary and particular solutions, we have

$$\mathbf{y}(t) = V\mathbf{x}(t) = X(t)\mathbf{y}^0 + \mathbf{y}^{(p)}(t)$$

where $X(t) = V e^{\Lambda t} V^{-1}$ is the fundamental matrix solution for A and $\mathbf{y}^{(p)}(t)$ is a particular solution for the general vector forcing function $\mathbf{h}(t)$:

$$\mathbf{y}^{(p)}(t) = V e^{\Lambda t} \mathbf{x}^{(p)}(t) = \int_0^t \left(V e^{\Lambda(t-\tau)} V^{-1} \right) \mathbf{h}(\tau) d\tau. \tag{4.3.20}$$

4.3.6   The matrix exponential

The solution (4.3.19) may be written more compactly if we denote $\mathbf{V}e^{\Lambda t}V^{-1}$ by e^{At} so that

$$\mathbf{y}(t) = e^{At}\mathbf{y}^0 + \int_0^t e^{A(t-\tau)}\mathbf{h}(\tau)d\tau, \quad \text{with} \quad e^{Ax} = Ve^{\Lambda x}V^{-1}. \qquad (4.3.21)$$

Note that e^{Ax} is well-defined and can be computed for any (non-defective) square matrix from its n eigen-pairs. As such, we have a complete solution of any IVP of the form (4.3.2)–(4.3.4) in the same familiar form as we know for a single linear ODE with constant coefficient (at least for the case when the $n \times n$ coefficient matrix A is non-defective).

While we can designate by fiat e^{At} to be any matrix including the one in (4.3.21), it is desirable for the designation/definition to reflect the known features of its scalar counterpart e^{at}. One obvious desirable feature is the Taylor series representation. In the matrix exponential adopted in (4.3.21), we have

$$e^{\Lambda} = \begin{bmatrix} \sum\limits_{k=0}^{\infty} \frac{\lambda_1^k}{k!} & 0 & \cdot & \cdot & \cdot & \cdot & \cdot & 0 \\ 0 & \sum\limits_{k=0}^{\infty} \frac{\lambda_2^k}{k!} & \cdot & \cdot & \cdot & \cdot & \cdot & 0 \\ \cdot & \cdot & \cdot & \cdot & \cdot & \cdot & \cdot & \cdot \\ \cdot & \cdot & \cdot & \cdot & \cdot & \cdot & \cdot & 0 \\ 0 & 0 & \cdot & \cdot & \cdot & \cdot & 0 & \sum\limits_{k=0}^{\infty} \frac{\lambda_n^k}{k!} \end{bmatrix} \qquad (4.3.22)$$

$$= \sum_{k=0}^{\infty} \frac{1}{k!} \begin{bmatrix} (\lambda_1)^k & 0 & \cdot & \cdot & \cdot & \cdot & \cdot & \cdot & 0 \\ 0 & (\lambda_2)^k & \cdot & \cdot & \cdot & \cdot & \cdot & \cdot & 0 \\ \cdot & \cdot & \cdot & \cdot & \cdot & \cdot & \cdot & \cdot & \cdot \\ \cdot & \cdot & \cdot & \cdot & \cdot & \cdot & \cdot & \cdot & \cdot \\ 0 & 0 & \cdot & \cdot & \cdot & \cdot & \cdot & \cdot & (\lambda_n)^k \end{bmatrix} = \sum_{k=0}^{\infty} \frac{1}{k!} \Lambda^k.$$

Application of (4.3.22) in (4.3.21) gives

$$e^A = Ve^{\Lambda}V^{-1} = \sum_{k=0}^{\infty} \frac{1}{k!} V\Lambda^k V^{-1}$$

$$= \sum_{k=0}^{\infty} \frac{1}{k!} (V\Lambda V^{-1})^k = \sum_{k=0}^{\infty} \frac{1}{k!} A^k. \qquad (4.3.23)$$

The relation (4.3.23) is in fact analogous to the infinite series definition of e^A when A is a scalar. Similar to the scalar case, we need to have a way of ensuring the infinite series of matrix terms converges and can be calculated. Convergence can in fact be shown with the help of the introduction of a *matrix norm* as a measure of magnitude of matrices (similar to the Euclidean norm measuring the magnitude of vectors) but will not be pursued here. With the infinite matrix series well defined, the series representation (4.3.23) can be used as an alternative (and more natural) definition of matrix exponential (instead of $Ve^{\Lambda x}V^{-1}$):

Definition 7 *The matrix exponential e^A of a square matrix A is defined to be the matrix series*

$$\mathbf{e}^A = \sum_{k=0}^{\infty} \frac{1}{k!} A^k. \qquad (4.3.24)$$

The expression (4.3.24) is an all encompassing definition since there is no requirement on A to have simple eigenvalues or a full set of eigenvectors (see [17]). However, it is not attractive for the purpose of calculating e^A as it requires, among others, a large number of matrix multiplications if a large number of terms in the series are needed for an accurate approximation. When A is diagonalizable by a similarity transformation, the representation in (4.3.21), taken in the form $\mathbf{e}^A = V\mathbf{e}^\Lambda V^{-1}$, provides a simple and elegant method to compute $\mathbf{e}^A$. (When A is not diagonalizable, it is desirable to have a way to evaluate (4.3.24) in a well-defined number of steps, e.g., something proportional to n^m for some m where n is the dimension of the square matrix. Such algorithms can be found in most intermediate texts on matrices or ODE.)

4.3.7 *Symmetric matrices*

When some of the eigenvalues of A are not simple, the following example show that we may still have a full set of eigenvectors and that A is still diagonalizable.

Example 6 $A = \begin{bmatrix} 3 & 2 & 4 \\ 2 & 0 & 2 \\ 4 & 2 & 3 \end{bmatrix}.$

From $|A - \lambda I| = 0$, we get $\lambda_1 = 8$, $\lambda_2 = \lambda_3 = -1$ so that -1 is an eigenvalue of multiplicity 2. It is straightforward to find $\mathbf{v}^{(1)} = (2, 1, 2)^T$. For $\lambda_2 = \lambda_3 = -1$, the set of equations in vector form for the components of the corresponding eigenvector is

$$[A - \lambda I]\,\mathbf{v} = \begin{bmatrix} 4 & 2 & 4 \\ 2 & 1 & 2 \\ 4 & 2 & 4 \end{bmatrix} \mathbf{v} = \mathbf{0}. \qquad (4.3.25)$$

All three equations of (4.3.25) are the same:

$$4v_1 + 2v_2 + 4v_3 = 0 \quad \text{or} \quad v_2 = -2v_1 - 2v_3$$

from which we get two linearly independent solution vectors for the homogeneous system (4.3.25). These can be taken to be:

$$\mathbf{v}^{(2)} = (1, -2, 0)^T, \qquad \mathbf{v}^{(3)} = (0, -2, 1)^T.$$

The first is obtained by taking $v_1 = 1$ and $v_3 = 0$ and the second by taking $v_1 = 0$ and $v_3 = 1$. By forming $\alpha \mathbf{v}^{(2)} + \beta \mathbf{v}^{(3)} = 0$, we see immediately that we must have

$\alpha = \beta = 0$ and the two vectors are linearly independent. We have then a full set of three eigenvectors for the given matrix and we can take its modal matrix to be

$$V = \begin{bmatrix} 2 & 1 & 0 \\ 1 & -2 & -2 \\ 2 & 0 & 1 \end{bmatrix}. \tag{4.3.26}$$

The modal matrix V of a square matrix is not unique given that all eigenvectors are determined only up to a multiplicative factor. In the case of a multiple eigenvalue with more than one associated eigenvectors, there is additional flexibility in choosing different combinations of the available eigenvectors for that eigenvalue. To see how we may take advantage of this flexibility, we note that for the example above $\mathbf{v}^{(1)}\cdot\mathbf{v}^{(2)} = \mathbf{v}^{(1)}\cdot\mathbf{v}^{(3)} = 0$ with a dot ("$\cdot$") denoting scalar product of two vectors. Geometrically, $\mathbf{v}^{(1)}$ is orthogonal to both $\mathbf{v}^{(2)}$ and $\mathbf{v}^{(3)}$. However, $\mathbf{v}^{(2)} \cdot \mathbf{v}^{(3)} = 4$ so that $\mathbf{v}^{(2)}$ is not orthogonal to $\mathbf{v}^{(3)}$. For many applications, there is considerable advantage to work with a mutually orthogonal set of unit eigenvectors. For one thing, the inverse of the corresponding modal matrix is just the transpose of the matrix, $V^{-1} = V^T$, in that case. (Readers already familiar with the orthogonalization process for vectors can skip the material between the two parallel lines below.)

To get an orthogonal set, we continue to take $\mathbf{v}^{(2)} = (1, -2, 0)^T$; but instead of our previous choice of $v_1 = 0$ and $v_3 = 1$ for $\mathbf{v}^{(3)}$, we take $v_3 = 1$ to ensure linear independence (and of course $v_2 = -2v_1 - 2v_3 = -2v_1 - 2$) but choose $v_1 \neq 0$ for a new $\mathbf{v}^{(3)}$ that is orthogonal to $\mathbf{v}^{(2)}$:

$$\mathbf{v}^{(2)}\cdot(v_1, -2v_1 - 2, 1)^T = 0$$

giving

$$v_1^{(3)} = -\frac{4}{5}, \qquad v_2^{(3)} = -\frac{2}{5}, \qquad v^{(3)} = \frac{1}{5}\begin{pmatrix} -4 \\ -2 \\ 5 \end{pmatrix}.$$

The corresponding modal matrix is now

$$V = \begin{bmatrix} 2 & 1 & -4/5 \\ 1 & -2 & -2/5 \\ 2 & 0 & 1 \end{bmatrix} \tag{4.3.27}$$

the columns of which are mutually orthogonal. However, the eigenvectors are not unit vectors; consequently, $V^T V$ is only diagonal but not the identity matrix I (so that V^{-1} is not quite V^T). Normalization of the eigenvectors to unit length so that $V^T V = I$ for the new modal matrix will be left as an exercise.

For an IVP $\mathbf{y}' = A\mathbf{y}$, $\mathbf{y}(t_0) = \mathbf{y}^\circ$, with the matrix A in this example, the exact solution can now be given in terms of the eigenvalues $\{8, -1, -1\}$ and the modal matrix V, either as in (4.3.26) or (4.3.27):

$$\mathbf{y}(t) = V e^{\Lambda(t-t_0)} V^{-1}\mathbf{y}^\circ = e^{A(t-t_0)}\mathbf{y}^\circ,$$

where

$$\Lambda = \begin{bmatrix} 8 & 0 & 0 \\ 0 & -1 & 0 \\ 0 & 0 & -1 \end{bmatrix}, \qquad e^{\Lambda t} = \begin{bmatrix} e^{8t} & 0 & 0 \\ 0 & e^{-t} & 0 \\ 0 & 0 & e^{-t} \end{bmatrix}.$$

A special feature of the matrix A in this Example 6 is its symmetry about the main diagonal so that the transpose of the matrix is the matrix itself: $A^T = A$. A square matrix with this property is called a *symmetric matrix*. The following is a well-known fact from linear algebra:

Theorem 4 *Symmetric matrices (and matrices with simple eigenvalues) are diagonalizable.*

Proof See [26]. $\square$

As such, all symmetric matrices have a full set of eigenvectors and are always nondefective. This is true even if some of their eigenvalues are not simple as illustrated by Example 6. In the next section, we discuss an application of this fundamental result in linear algebra in the modeling of the important biological phenomenon of DNA mutation.

4.4   DNA Mutation

4.4.1   *The double helix*

The central dogma of biology is "DNA to RNA to proteins," with protein constituting the fundamental units for all parts and functions of living organisms DNA (Deoxyribonucleic acid) are therefore the basic building blocks for these organisms. DNA molecules encode genetic information and these molecules (with their genetic information) are copied and passed on from parents to an offspring. Though highly accurate, the copying process is not immune from errors that lead to genetic mutation and consequently the evolution of living organisms. To study biological evolution, we need to have an understanding how errors are incurred in the DNA copying process, especially copying errors that result in genetic mutation.

The 1962 Nobel Prize in Physiology or Medicine was awarded to James Watson, Francis Crick and Maurice Wilkins for their discovery of the double helix structure of DNA molecules. In 1953, Watson and Crick saw an x-ray pattern of a crystal of the DNA molecule made by Rosalind Franklin and Maurice Wilkins; it gave Watson and Crick enough information to make an accurate model of the DNA molecule. Their model showed a twisted double helix with little rungs connecting the two helical strands. At each end of a rung of the double helix ladder is one of four possible molecular subunits (called *nucleobases* or simply *bases*): *adenine* (A), *guanine* (G), *cytosine* (C), and *thymine* (T).

The shape of *adenine* is complementary to *thymine*; they are bound together consistently at opposite ends of a rung of the DNA ladder through a hydrogen bond to form a *nucleotide*. The nucleobase *guanine* is structurally similar to *adenine* and complementary to *cytosine*. The *guanine-cytosine* pair are also bound together consistently at the opposite ends of a DNA ladder rung. In other words, we always find either A paired with T or G paired with C (but neither A nor T with C or G). Thus, once we know the base at one side of a rung, we can deduce the base at the other end of the same rung. For example, if along one strand of the DNA ladder, we have a base sequence

$$..... ATTAGAGCGCGT .....$$

then the corresponding sequence along the other strand opposite to the same stretch must be

$$..... TAATCTCGCGCA .....$$

(Because of their structural similarity, cytosine and thymine are called *pyrimidines* while adenine and guanine are called *purines*.) As such, a DNA molecule (or a segment of it) is specified by a sequence of the four letters A, T, C and G.

Once the model was established, its structure hinted that DNA was indeed the carrier of the genetic code and thus the key molecule of heredity, development and evolution. Heredity requires genetic information be passed on to offspring. This is accomplished during cell division when the twisted double-helix DNA molecule ladder unzips into two separate strands. One new molecule is formed from each half-ladder and, due to the required pairings, this gives rise to two identical daughter copies from each parent molecule. Though elaborate safeguards are in place to ensure fidelity in the replication, errors still occur, though infrequently.

4.4.2 *Mutation due to base substitutions*

The most common type of errors during the replication process is a replacement of one base by another at a certain site of the strand sequence. For instance, if the sequence in the parent DNA molecule *ATTAGAGC* should become *ATTACAGC* in the offspring DNA, then there is a *base substitution* $G \to C$ at the fifth site of the sequence. The base substitution replaces a pyrimidine by another pyrimidine (or one replaces a purine by another purine) is known as *transition*. A base substitution of a purine by a pyrimidine (or a pyrimidine by a purine) is called a *transversion*. Error types other than base substitution are also possible. These include deletion, insertion and inversion of a section of the sequence. Their occurrence are much less frequent than base substitutions and will be ignored in the discussion herein to focus only on mutations due to base substitutions.

For the restricted problem of mutations by base substitutions only (and for more general version of mutations), one issue of interest is the amount of mutation that has occurred after a number of generations of offspring from a DNA sequence, which may be quite long especially when we are talking about an evolutionary time scale. If data for all generations involved are not available or not used, then the difference between the original sequence of generation n and the sequence of generation $n + m$ does not necessarily give an accurate estimate of the amount of mutation that has occurred since one or more back mutation might have taken place. If the sequence in the parent DNA molecule $ATTAGAGC$ should become $ATTACAGC$ four generations later, then one base substitution of $G \to C$ at the fifth site of the sequence is only one possible kind of mutation. It may also be a sequence of base substitutions such as $G \to C \to G \to G \to C$ or $G \to C \to T \to G \to C$, with the former having one back mutation while the latter involves no back mutation at all. A more sophisticated approach is needed to determine the amount of mutation involved between an ancestral DNA site and the same site of its generation n descendent DNA molecule.

4.4.3   *Linear model*

Given a particular distribution of nucleobases for (one strand of) an initial (ancestral) DNA sequence, knowing the probability that the fraction of a particular nucleobase would mutate to another nucleobase after replication would allow us to estimate the evolution of the DNA sequence. Let $A(t)$, $G(t)$, $C(t)$ and $T(t)$ be the fraction of the sequence being adenine, guanine, cytosine and thymine at time t. Suppose we know (or can estimate from available data) the probability (or the fraction per unit time) m_{AG}, m_{AC} and m_{AT} that adenine would be replaced by the nucleobase guanine, cytosine and thymine, respectively. Since deletion is not allowed, the rate of change of A would have to be $-\left(m_{AG} + m_{AC} + m_{AT}\right) A \equiv -\Sigma_A A$. Similar rates of change can be estimated for the other three nucleobases. We arrange all these rates of change as a linear system of ODE with coefficient matrix M:

$$\frac{d}{dt} \begin{pmatrix} A \\ G \\ C \\ T \end{pmatrix} = \begin{bmatrix} -\Sigma_A & m_{GA} & m_{CA} & m_{TA} \\ m_{AG} & -\Sigma_G & m_{CG} & m_{TG} \\ m_{AC} & m_{CG} & -\Sigma_C & m_{TC} \\ m_{AT} & m_{TG} & m_{CT} & -\Sigma_T \end{bmatrix} \begin{pmatrix} A \\ G \\ C \\ T \end{pmatrix} \equiv M \begin{pmatrix} A \\ G \\ C \\ T \end{pmatrix},$$

with $\Sigma_A = m_{AG} + m_{AC} + m_{AT}$ to characterize the evolution of the initial DNA sequence with time. Our model of the evolutionary process may be written in vector form

$$\mathbf{y}' = M\mathbf{y}$$

where $\mathbf{y} = (A, G, C, T)^T$. While the elements $\{m_{ij}\}$ of the coefficient matrix M may change with time, it is customary to treat them as time invariant. Their numerical values are key to the evolution of the organism; they are normally estimated from whatever historical replication data of the organism over many generations available for a particular DNA sequence.

4.4.4   *Equal opportunity substitution*

To illustrate the use of differential equations to learn more about genetic mutation, we consider (instead of estimating m_{ij} from available data) the consequences of a highly speculative, one-parameter *equal opportunity* model for the transition matrix M. In this version of the Jukes-Cantor (1969) model, we stipulate that substitution of a nucleobase by any of the other three bases are equally likely with probability $\alpha/3$ where $\alpha > 0$ is a number to be determined, possibly estimated from available data. Available records suggest that α should be a rather small number, ranging from 10^{-8} mutations per site per year for mitochondrial DNA to 0.01 mutations per site per year for the influenza A virus DNA. The corresponding transition matrix M take the form

$$M = \begin{bmatrix} -\alpha & \alpha/3 & \alpha/3 & \alpha/3 \\ \alpha/3 & -\alpha & \alpha/3 & \alpha/3 \\ \alpha/3 & \alpha/3 & -\alpha & \alpha/3 \\ \alpha/3 & \alpha/3 & \alpha/3 & -\alpha \end{bmatrix}. \tag{4.4.1}$$

For such a transition matrix, an ancestral sequence with $\mathbf{y}^o = (1,0,0,0)^T$ would generally have a mixed distribution of nucleobases for $t > 0$ (even after one replication if we model in discrete time).

4.4.4.1 *Non-isolated steady states*

The question of interest is what happens after many generations? Does $\mathbf{y}(t)$ tend to some limiting $\mathbf{y}_\infty$? The origin $\mathbf{y}^{(c)} = \mathbf{0}$ clearly is a critical point for the ODE system and therefore a limiting steady state. But unlike the systems we have already encountered, the coefficient matrix is not invertible given that the rows (and columns) sum up to $\mathbf{0}$ so that the matrix M is not of full rank. It follows that $\lambda_1 = 0$ is an eigenvalue with the associated eigenvector $\mathbf{v}^{(1)} = \frac{1}{2}(1,1,1,1)^T$. In that case, the vector $c\mathbf{v}^{(1)}$ is a steady state for any constant c with the origin corresponding to $c = 0$. Since c can be any real number (even complex number), the critical points associated with different values of c are not isolated as we have been accustomed to seeing from our models of biological phenomena up to this point. The question of interest is to which steady state would a given initial distribution of nucleobases evolve? The answer involves the corresponding complementary solutions of the ODE for the evolution of the DNA sequence.

4.4.4.2 *Stability of critical points*

To find the remaining eigenvalues of the matrix M, we can calculate them by Mathematica or Maple. However, it is not difficult to see that $\lambda_2 = -4\alpha/3$ since

$$| M - \lambda_2 I | = \begin{vmatrix} \alpha/3 & \alpha/3 & \alpha/3 & \alpha/3 \\ \alpha/3 & \alpha/3 & \alpha/3 & \alpha/3 \\ \alpha/3 & \alpha/3 & \alpha/3 & \alpha/3 \\ \alpha/3 & \alpha/3 & \alpha/3 & \alpha/3 \end{vmatrix} = 0.$$

Evidently, $\lambda_2 = -4\alpha/3$ is a multiple eigenvalue of multiplicity 3 since all four rows are the same. The coefficient matrix M being a symmetric real matrix, we know from Theorem 4 that we can find a full set of three eigenvectors. These may be taken to be $\mathbf{v}^{(2)} = \frac{1}{2}\left(1, -1, -1, 1\right)^T$, $\mathbf{v}^{(3)} = \frac{1}{2}\left(1, -1, 1, -1\right)^T$ and $\mathbf{v}^{(4)} = \frac{1}{2}\left(1, 1, -1, -1\right)^T$. The important conclusion from the multiplicity of $\lambda_2 = -4\alpha/3 < 0$ is that there are no other eigenvalues and hence no other steady states aside from the non-isolated states found in the previous subsection. In addition, all the complementary solutions other than the ones associated with $\lambda_1 = 0$ decay to zero as $t \to \infty$.

4.4.4.3   *The unique limiting (steady state) nucleobase distribution*

To determine the limiting distribution $\mathbf{y}(t)$ as $t \to \infty$ for the Equal Opportunity Substitution model given the initial distribution $\mathbf{y}^o$, we apply the superposition principle for the complementary solution and write

$$\mathbf{y}(t) = c_1\mathbf{v}^{(1)} + c_2 e^{-4\alpha t/3}\mathbf{v}^{(2)} + c_3 e^{-4\alpha t/3}\mathbf{v}^{(3)} + c_4 e^{-4\alpha t/3}\mathbf{v}^{(4)}$$

with

$$\mathbf{y}(0) = c_1\mathbf{v}^{(1)} + c_2\mathbf{v}^{(2)} + c_3\mathbf{v}^{(3)} + c_4\mathbf{v}^{(4)}$$
$$= V\mathbf{c} = \mathbf{y}^o$$

where

$$V = \frac{1}{2}\begin{bmatrix} 1 & 1 & 1 & 1 \\ 1 & -1 & -1 & 1 \\ 1 & -1 & 1 & -1 \\ 1 & 1 & -1 & -1 \end{bmatrix}.$$

This linear system can be solved to obtain $\mathbf{c} = V^{-1}\mathbf{y}^o$. As the eigenvectors are orthonormal (mutually orthogonal and of unit magnitude), we have $\mathbf{c} = V^{-1}\mathbf{y}^o = V^T\mathbf{y}^o$ from which we have

$$c_1 = \frac{1}{4}(y_1^o + y_2^o + y_3^o + y_4^o)$$
$$= \frac{1}{4}(A^o + G^o + C^o + T^o)$$

with

$$\lim_{t \to \infty} \mathbf{y}(t) = \frac{1}{4}(A^o + G^o + C^o + T^o)\begin{pmatrix} 1 \\ 1 \\ 1 \\ 1 \end{pmatrix}.$$

Hence, this model of the evolving DNA sequence will tend to one with the same fraction of all four nucleobases whatever the value of α and the initial distribution of the nucleobases may be.

Whether or not the conclusion on the eventuality of the Juke-Cantor equal opportunity model for DNA mutation is reasonable or acceptable, there are other more refined models on the evolution of the phenomenon of DNA mutation. Among these are the Kimura 2-parameter and 3-parameter models with coefficient matrices

$$M_2 = \begin{bmatrix} \alpha & \beta & \gamma & \gamma \\ \beta & \alpha & \gamma & \gamma \\ \gamma & \gamma & \alpha & \beta \\ \gamma & \gamma & \beta & \alpha \end{bmatrix}, \quad M_3 = \begin{bmatrix} \alpha & \beta & \gamma & \delta \\ \beta & \alpha & \delta & \gamma \\ \gamma & \delta & \alpha & \beta \\ \delta & \gamma & \beta & \alpha \end{bmatrix}.$$

These models can also be analyzed similarly.

4.5   Defective Matrices

4.5.1   A simple example

For unsymmetric real matrices with multiple eigenvalues, the situation is more complicated. In most cases, such a matrix is defective and cannot be diagonalized. In that case, the ODE cannot be uncoupled to result in a set of single first order equations each for one unknown only. It turns out that our goal to reduce the problem to solving single first order equations can still be met. To see what this means and how it differs from the nondefective case, let us work out the following simple example:

Example 7

$$\mathbf{y}' = \begin{pmatrix} y_1 \\ y_2 \end{pmatrix}' = \begin{bmatrix} 4 & -2 \\ 2 & 0 \end{bmatrix} \begin{pmatrix} y_1 \\ y_2 \end{pmatrix} = A\mathbf{y}, \qquad \mathbf{y}(0) = \mathbf{y}^o. \tag{4.5.1}$$

The two eigenvalues of the matrix A are the same with $\lambda_1 = \lambda_2 = 2$ and there is only one associated eigenvector $\mathbf{p}^{(1)} = (1,1)^T$ up to a multiplicative factor. Therefore the matrix A cannot be diagonalized as in the nondefective cases and the ODE system cannot be completely de-coupled; otherwise we would have a second eigenvector.

In trying to reduce the problem to solving single first order equations, we form the combination $y_1' - y_2'$ to get

$$(y_1 - y_2)' = 2(y_1 - y_2) = \lambda_1(y_1 - y_2) \tag{4.5.2}$$

which is the expected single first order ODE associated with the (multiple) eigenvalue $\lambda_1 = 2$ for the single unknown $z_2 = y_1 - y_2$. While we will not get a second uncoupled equation for a single other unknown, we can always form a second combination of y_1 and y_2 to get a second ODE that is different from the first and would offer some advantage. One natural choice would be to stay with y_2 and write the ODE y_2' in terms of z_2 and y_2:

$$y_2' = (2y_1 - 2y_2) + 2y_2 = \lambda_1 y_2 + z_2. \tag{4.5.3}$$

In the context of working with two new unknowns that are linear combinations of the original two, we can let the second new combination be $z_1 = y_2$ and write (4.5.2) and (4.5.3) in terms of these new variables to get

$$\mathbf{z}' = J\mathbf{z} \tag{4.5.4}$$

where

$$\mathbf{z} = \begin{pmatrix} z_1 \\ z_2 \end{pmatrix} = \begin{pmatrix} y_2 \\ 2y_1 - 2y_2 \end{pmatrix}, \qquad J = \begin{bmatrix} \lambda_1 & 1 \\ 0 & \lambda_1 \end{bmatrix} = \begin{bmatrix} 2 & 1 \\ 0 & 2 \end{bmatrix}. \tag{4.5.5}$$

It is worth stating explicitly at this point that we have, by design, chosen the new variables z_1 and z_2 so that the resulting new system is in what is known as the *Jordan normal form* (4.5.4). The 2×2 *upper triangular* matrix J in (4.5.5) consisting of the multiple eigenvalue $\lambda_1 = 2$ on the main diagonal and "1" along the *super-diagonal* (the elements above the main diagonal) is called a 2×2 *Jordan block*. The following definition is generally used in the literature of linear algebra:

Definition 8 *An $m \times m$ Jordan block is an $m \times m$ matrix having as its only non-zero elements the same (multiple eigen-) value on its diagonal and a "1" for all the elements along the super-diagonal. (The special case of a 1×1 Jordan block is of the form $[\lambda]$.)*

Short of completely de-coupling the original ODE system (which is not possible for (4.5.1)), the Jordan form of that system is as good as it gets and, more importantly, good enough for our purpose. In scalar form, the system (4.5.4) consists of the following two 1st order linear ODE with constant coefficients:

$$z_2' = \lambda_1 z_2, \qquad z_1' = \lambda_1 z_1 + z_2.$$

The first equation is a single first order separable ODE for z_2 alone and can be solved separately to get

$$z_2 = e^{\lambda_1 t} z_2^0,$$

where z_2^0 is a constant of integration corresponding to the initial value of $z_2(t)$ with $z_2(0) = 2y_1(0) - 2y_2(0) = 2(y_1^0 - y_2^0) \equiv z_2^0$. The second is an inhomogeneous first order linear ODE for z_1 with $z_2(t) = e^{\lambda_1 t} z_2^0$ as a known forcing. It can be solved by the method from Chapter 1 for a single linear first order ODE to get

$$z_1 = e^{\lambda_1 t} \left[z_1^0 + \int_0^t e^{-\lambda_1 \tau} z_2(\tau) d\tau \right] = e^{\lambda_1 t} \left[z_1^0 + z_2^0 t \right].$$

While the new system does not de-couple z_1 from z_2, it does reduce the original problem to solving a sequence of single first order equations. Starting with the last unknown working backwards, the last solution obtained serves as the forcing term in the equation for the next unknown to be solved. That the coupling constant (linking the last solution to the unknown to be solved) being 1 is as simple as it gets.

With z_1 and z_2 completely determined, we can work backwards and solve (4.5.5) for y_1 and y_2 to get

$$y_2 = z_1 = e^{\lambda_1 t} \left[z_1^0 + z_2^0 t \right],$$

$$y_1 = \frac{1}{2} z_2 + z_1 = e^{\lambda_1 t} \left[z_1^0 + z_2^0 \left(\frac{1}{2} + t \right) \right]$$

or

$$\mathbf{y} = \begin{bmatrix} 1 & 1/2 \\ 1 & 0 \end{bmatrix} \begin{pmatrix} z_1 \\ z_2 \end{pmatrix}$$

$$= \begin{bmatrix} 1 & 1/2 \\ 1 & 0 \end{bmatrix} \begin{pmatrix} e^{\lambda_1 t} \left(z_1^0 + z_2^0 t \right) \\ e^{\lambda_1 t} z_2^0 \end{pmatrix} = Q\mathbf{z}.$$

4.5.2   Systematic reduction to Jordan form

To develop a systematic method of reducing the same problem to its Jordan form, we work backwards and let

$$\mathbf{y} = Q\mathbf{z} = Q \begin{pmatrix} z_1 \\ z_2 \end{pmatrix} \quad \text{(or } \mathbf{z} = Q^{-1}\mathbf{y}),$$

for some invertible 2×2 matrix Q. After pre-multiplying the original ODE by Q^{-1} to get

$$\mathbf{z}' = \left(Q^{-1}AQ\right)\mathbf{z},$$

we stipulate the yet unspecified matrix Q to have the property $Q^{-1}AQ = J$, where J is a 2×2 Jordan block. This requires $AQ = QJ$ or

$$\left[A\mathbf{q}^{(1)},\ A\mathbf{q}^{(2)}\right] = \left[\mathbf{q}^{(1)}, \mathbf{q}^{(2)}\right] \begin{bmatrix} \lambda_1 & 1 \\ 0 & \lambda_1 \end{bmatrix}$$

$$= \left[\lambda_1\mathbf{q}^{(1)},\ \lambda_1\mathbf{q}^{(2)} + \mathbf{q}^{(1)}\right].$$

It follows that

$$A\mathbf{q}^{(1)} = \lambda_1\mathbf{q}^{(1)}, \qquad A\mathbf{q}^{(2)} = \lambda_1\mathbf{q}^{(2)} + \mathbf{q}^{(1)}.$$

The first equation determines the one eigen-pair (which we expect to get) for our problem to be $\lambda_1 = 2$ and $\mathbf{q}^{(1)} = (1,1)^T$, up to a multiplicative factor as usual. The second is an inhomogeneous linear algebraic system:

$$\begin{bmatrix} 2 & -2 \\ -2 & 2 \end{bmatrix} \begin{pmatrix} q_1 \\ q_2 \end{pmatrix} = \begin{pmatrix} 1 \\ 1 \end{pmatrix}$$

that does not completely determine all the unknowns. Instead, we obtain a one-parameter family of solution $q_1 = q_2 + 1/2$. Since we only need a nontrivial solution for this problem, we can take $q_2 = 0$ to get $\mathbf{q}^{(2)} = (\frac{1}{2}, 0)^T$ leading to

$$Q = \begin{bmatrix} 1 & 1/2 \\ 1 & 0 \end{bmatrix}, \quad \text{with} \quad Q^{-1} = \begin{bmatrix} 0 & 1 \\ 2 & -2 \end{bmatrix}.$$

With Q found, we can use it to obtain the new unknowns, i.e., the needed combinations of the original unknowns

$$\mathbf{z} = Q^{-1}\mathbf{y},$$

to transform the original ODE into Jordan form.

4.5.3   Matrices with a single eigenvector

The method of reduction is the same for a general $n \times n$ matrix with a single eigenvector (and hence an eigenvalue of multiplicity n). We illustrate this by the following system of three linear homogeneous ODE with constant coefficients:

Example 8

$$\mathbf{y}' = \begin{pmatrix} y_1 \\ y_2 \\ y_3 \end{pmatrix}' = \begin{bmatrix} 1 & 1 & 1 \\ 2 & 1 & -1 \\ -3 & 2 & 4 \end{bmatrix} \begin{pmatrix} y_1 \\ y_2 \\ y_3 \end{pmatrix} = A\mathbf{y}, \qquad \mathbf{y}(0) = \mathbf{y}^o.$$

For this problem, the coefficient matrix A has an eigenvalue of multiplicity 3: $\lambda_1 = \lambda_2 = \lambda_3 = 2$ with only one associated eigenvector $\mathbf{q}^{(1)} = (0, 1, -1)^T$. We again expect possible reduction to a 3×3 Jordan block so that there are three linear independent vectors $\{\mathbf{q}^{(1)}, \mathbf{q}^{(2)}, \mathbf{q}^{(3)}\}$ and a corresponding pseudo modal matrix $Q = [\mathbf{q}^{(1)}, \mathbf{q}^{(2)}, \mathbf{q}^{(3)}]$ for which

$$AQ = \begin{bmatrix} A\mathbf{q}^{(1)}, & A\mathbf{q}^{(2)}, & A\mathbf{q}^{(3)} \end{bmatrix} = \begin{bmatrix} \mathbf{q}^{(1)}, \mathbf{q}^{(2)}, \mathbf{q}^{(3)} \end{bmatrix} \begin{bmatrix} \lambda_1 & 1 & 0 \\ 0 & \lambda_1 & 1 \\ 0 & 0 & \lambda_1 \end{bmatrix}$$

$$= \begin{bmatrix} \lambda_1 \mathbf{q}^{(1)}, & \lambda_1 \mathbf{q}^{(2)} + \mathbf{q}^{(1)}, & \lambda_1 \mathbf{q}^{(3)} + \mathbf{q}^{(2)} \end{bmatrix} = QJ.$$

The three columns of the matrix relation above give three vector equations:

$$(A - \lambda_1 I) \mathbf{q}^{(1)} = 0, \quad (A - \lambda_1 I) \mathbf{q}^{(2)} = \mathbf{q}^{(1)}, \quad (A - \lambda_1 I) \mathbf{q}^{(3)} = \mathbf{q}^{(2)}. \tag{4.5.6}$$

The first equation of (4.5.6) determines the eigenvector

$$\mathbf{q}^{(1)} = (0, 1, -1)^T.$$

The second corresponds to the following system of three equations

$$\begin{bmatrix} -1 & 1 & 1 \\ 2 & -1 & -1 \\ -3 & 2 & 2 \end{bmatrix} \mathbf{q}^{(2)} = \begin{pmatrix} 0 \\ 1 \\ -1 \end{pmatrix}$$

which has a one-parameter family of solutions:

$$q_1^{(2)} = 1, \qquad q_2^{(2)} + q_3^{(2)} = 1.$$

By setting $q_3^{(2)} = 0$, we get

$$\mathbf{q}^{(2)} = (1, 1, 0)^T.$$

Note that setting $q_2^{(2)} = 0$ instead would give

$$\hat{\mathbf{q}}^{(2)} = (1, 0, 1)^T$$

which is not independent from the other two already found since

$$\mathbf{q}^{(1)} - \mathbf{q}^{(2)} + \hat{\mathbf{q}}^{(2)} = \mathbf{0}.$$

Similarly, we get from the last equation of (4.5.6) the linear system

$$\begin{bmatrix} -1 & 1 & 1 \\ 2 & -1 & -1 \\ -3 & 2 & 2 \end{bmatrix} \mathbf{q}^{(3)} = \begin{pmatrix} 1 \\ 1 \\ 0 \end{pmatrix}$$

with another one-parameter family of solution

$$q_1^{(3)} = 2, \qquad q_2^{(3)} + q_3^{(3)} = 3.$$

By taking $q_3^{(3)} = 0$, we get $\mathbf{q}^{(3)} = (2, 3, 0)^T$.

Altogether, the three vectors $\{\mathbf{q}^{(1)}, \mathbf{q}^{(2)}, \mathbf{q}^{(3)}\}$ may be put together as the three columns of the pseudo modal matrix Q:

$$Q = \begin{bmatrix} 0 & 1 & 2 \\ 1 & 1 & 3 \\ -1 & 0 & 0 \end{bmatrix}.$$

With Q in hand, we can set $\mathbf{z} = Q^{-1}\mathbf{y}$ and rewrite the ODE in terms of $\mathbf{z}(t)$ to get

$$\mathbf{z}' = J\mathbf{z} = \begin{bmatrix} 2 & 1 & 0 \\ 0 & 2 & 1 \\ 0 & 0 & 2 \end{bmatrix} \mathbf{z}.$$

The corresponding set of scalar ODE (for the components of $\mathbf{z}$) are

$$z_1' = 2z_1 + z_2, \qquad z_2' = 2z_2 + z_3, \qquad z_3' = 2z_3. \qquad (4.5.7)$$

We solve the three equations in (4.5.7) successively, starting with the last equation for z_3, to get

$$z_3(t) = e^{2t}c_3, \qquad z_2(t) = e^{2t}(c_2 + c_3 t), \qquad z_1(t) = e^{2t}\left(c1 + c_2 t + \frac{1}{2}c_3 t^2\right),$$

or

$$\mathbf{z}(t) = e^{2t} \begin{bmatrix} 1 & t & \frac{1}{2}t^2 \\ 0 & 1 & t \\ 0 & 0 & 1 \end{bmatrix} \mathbf{c}.$$

The constant vector $\mathbf{c}$ is determined by the initial condition $\mathbf{z}(0) = \mathbf{c} = Q^{-1}\mathbf{y}(0) = Q^{-1}\mathbf{y}^0$. In terms of the original vector unknown, we have

$$\mathbf{y}(t) = e^{2t}Q \begin{bmatrix} 1 & t & \frac{1}{2}t^2 \\ 0 & 1 & t \\ 0 & 0 & 1 \end{bmatrix} Q^{-1}\mathbf{y}^0.$$

4.5.4 *Multiple but insufficient eigenvectors*

From the last two examples, we learned how to solve a linear constant coefficient system of ODE with a defective coefficient matrix that has only one eigenvector. The key of the method of solution is to postulate the existence of a similarity transformation $Q^{-1}AQ$ that transforms the coefficient matrix (for the examples considered above) into a Jordan block matrix J. In both cases, the pseudo modal matrix Q was found by implementing the postulate $AQ = QJ$. However, there is no assurance that this works for other matrices, i.e., the needed matrix Q may not exist or the transformation may not be to a single Jordan block. The following example helps to clarify the second possibility and enables us to address the first:

Example 9 $y' = Ay = \begin{bmatrix} 5 & -3 & -2 \\ 8 & -5 & -4 \\ -4 & 3 & 3 \end{bmatrix} y, \qquad y(0) = y^0.$

As in the previous example, the matrix A here also has a single eigenvalue of multiplicity 3: $\lambda_1 = \lambda_2 = \lambda_3 = 1$. But unlike the previous example, we now have two associated eigenvectors since

$$(A - \lambda_1 I)\, q = \begin{bmatrix} 4 & -3 & -2 \\ 8 & -6 & -4 \\ -4 & 3 & 2 \end{bmatrix} q = 0$$

consists of a single condition $4q_1 - 3q_2 - 2q_3 = 0$ to give a two-parameter family of solutions for the three components of $(q_1, q_2, q_3)^T$. For example, we may take $q^{(1)} = (3/4, 1, 0)^T$ and $q^{(2)} = (1/2, 0, 1)^T$ corresponding to setting $q_3 = 0$ and $q_2 = 0$, respectively. If we now use the same recipe and try to find a third vector $q^{(3)}$ to form a linearly independent set from

$$(A - \lambda_1 I)\, q^{(3)} = p$$

by taking p to be $q^{(1)}$ or $q^{(2)}$, we would find no solution as (in the language of linear algebra) $q^{(1)}$ and $q^{(2)}$ are not in the column space of the matrix

$$A - \lambda_1 I = A - I = \begin{bmatrix} 4 & -3 & -2 \\ 8 & -6 & -4 \\ -4 & 3 & 2 \end{bmatrix}.$$

The solution is to take p to be a linear combination of all the available eigenvectors that is in the column space of $A - \lambda_1 I$. For our present example, we take $p = \alpha q^{(1)} + \beta q^{(2)}$ (since we have altogether two eigenvectors) and choose α and β so that p is in the column space of $A - \lambda_1 I$ and solve $(A - \lambda_1 I)\, q^{(3)} = p$ for the third linearly independent vector to form Q. It is straightforward to find that $p = -2q^{(1)} + q^{(2)} = (-1, -2, 1)^T$ is in the column space of $A - I$ with $q^{(3)}$ found to be $(1/4, 0, 0)^T$ from solving

$$\begin{bmatrix} 4 & -3 & -2 \\ 8 & -6 & -4 \\ -4 & 3 & 2 \end{bmatrix} q^{(3)} = \begin{pmatrix} -1 \\ -2 \\ 1 \end{pmatrix}.$$

With

$$Q = \begin{bmatrix} 3/4 & -1 & 1/4 \\ 1 & -2 & 0 \\ 0 & 1 & 0 \end{bmatrix},$$

we get

$$Q^{-1}AQ = J = \begin{bmatrix} J_1 & O \\ O & J_2 \end{bmatrix} = \begin{bmatrix} 1 & 0 & 0 \\ 0 & 1 & 1 \\ 0 & 0 & 1 \end{bmatrix}$$

where J_1 is a 1×1 Jordan block and J_2 is a 2×2 Jordan block while O in the upper and lower corner is an 1×2 and a 2×1 zero matrix, respectively. Note that the diagonal elements of J_1 and J_2 are the same eigenvalue in this case while the dimension $n_k \times n_k$ of the kth block J_k happens to be the same of the block's subscripted index k; these are not always the case.

4.5.5   General defective matrix

We can now state the general result for a linear systems of ODE with constant coefficients. From the developments in this section, it should be clear that we need only to focused on homogeneous systems.

Theorem 5 *For the homogeneous linear system* $\mathbf{y}' = A\mathbf{y}$ *where A is an $n \times n$ constant matrix, there exists an invertible matrix Q such that a change of variable to $\mathbf{z} = Q^{-1}\mathbf{y}$ transforms the linear system to $\mathbf{z}' = J\mathbf{z}$ where*

$$
J = \begin{bmatrix}
J_1 & O & . & . & . & O \\
O & J_2 & O & . & . & O \\
& & \cdots & & & \\
& & \cdots & & & \\
O & O & O & . & . & J_m
\end{bmatrix},
\tag{4.5.8}
$$

where J_k is an $n_k \times n_k$ Jordan block and $\sum_{k=1}^{m} n_k = n$.

Remark 2 *It is customary to place all the 1×1 blocks (for true eigenvectors) at the upper left corner of J and other blocks in the order of their size. Sometimes, the dimension of the kth Jordan block is indicated by a superscript in the form $J_k^{(n_k)}$. As seen from Example 9, there may be more than one Jordan blocks associated with a single multiple eigenvalue. Hence the eigenvalues of several Jordan blocks may be the same in some cases.*

Theorem 5 assures us that we can always find the needed pseudo modal matrix for the desired similarity transformation. The theorem is proved in most advanced linear algebra text and will not be pursued here. From the perspective of a mathematical modeler, it makes it possible to reduce the solution of any linear ODE system with constant coefficients (homogeneous or inhomogeneous) associated with the dynamical system model for the evolution of the phenomenon being investigated to solving a set of scalar linear first order linear ODE sequentially, one at a time (which we know how to do).

Chapter 5

Nonlinear Autonomous Interactions

5.1 Predator-Prey Interaction

5.1.1 *Rabbits and foxes*

The interaction between a population of rabbits and a population of preying foxes is a typical predator-prey relation. However, it is not appropriate to model it by a linear system of the form (4.1.1) since the "loss rate" of the rabbit population cannot be simply proportional to the fox population. The loss rate constant a_{12} must be a function of the size of the rabbit population. If there is no rabbit left, no fox population can reduce it further. For a rabbit habitat with finite resources, we know also from our earlier studies of the growth of a single population that the growth of the rabbit population is usually limited by the carrying capacity of the habitat even in the absence of any prey. A more realistic model would have the linear growth rate of the usual prey growth model be replaced by a logistic type growth rate. In this section, we investigate the consequences of the following modified Lotka-Volterra model for the evolution of the rabbit population $R(t)$ and the fox population $F(t)$ with time t,

$$R' = aR - \beta R^2 - \gamma RF \equiv f(R, F) \tag{5.1.1}$$

$$F' = -cF + \delta RF \equiv h(R, F) = g(R, F) \tag{5.1.2}$$

with $(\ )' = d(\ )/dt$. The five quantities a, β, γ, c and δ in the equations are parameters that help form various gain and loss rates of the two populations. When these five are all positive, the two interacting populations are in a predator-prey relation. The ratio of the parameters a and β, $R_c = a/\beta$, is the *carrying capacity* of the rabbit habitat. The two simultaneous nonlinear differential equations (5.1.1)–(5.1.2) are usually augmented by the two initial conditions at $t = 0$:

$$R(0) = R_0, \qquad F(0) = F_0, \tag{5.1.3}$$

to form an IVP for the evolution of rabbit and fox population with time.

We note in passing that we could in principle also modify the second equation of the Lotka-Volterra model above to allow for a limiting size for the predator population. However, we refrain from such a modification as there is already a natural mechanism for limiting the predator growth. A large population of predators invariably reduces the size of the prey population. When R is sufficiently low so that the unit growth rate δR in (5.1.2) becomes less than the unit loss rate c, the predator population would decline without an imposed limit capacity.

5.1.2 Reduction of order

5.1.2.1 Rabbits and foxes

A first integral of the two ODE for R and F (5.1.1)–(5.1.2) is possible for an autonomous system so that its order is reduced by one. This is accomplished by using the chain rule in differentiation to form

$$\frac{dF}{dR} = \frac{F'}{R'} = \frac{F\left(-c + \delta R\right)}{R\left(a - \beta R - \gamma F\right)}, \tag{5.1.4}$$

which is a first order ODE relating F and R without involving the independent variable t.

For the special case $\beta = 0$, this first order ODE becomes separable and can be solved exactly to give

$$R^c F^a = C_0 e^{\delta R + \gamma F}$$

where C_0 is a constant of integration. It can be used to fit the prescribed initial data taken in the form

$$F_0 = F(t = 0) = F(R(t = 0)) = F(R_0).$$

The final solution for $\beta = 0$ is

$$\left(\frac{R}{R_0}\right)^c \left(\frac{F}{F_0}\right)^a = e^{\delta(R - R_0) + \gamma(F - F_0)}.$$

An explicit solution in terms of elementary functions is not available for $\beta > 0$ unless $\gamma = 0$. (This special case is no longer a predator-prey interaction.) Nevertheless, the reduction of a general second order autonomous ODE system to a single first order ODE often yield additional understanding and insight to the modeled phenomenon. One of the most talked about example of such successes is the motion of the over-the-top pendulum in physics discussed in the next subsection. It is chosen as an illustration for its appeal to readers' intuition.

5.1.2.2 The "over the top" pendulum

The equation of motion for a (frictionless rigid arm) pendulum is

$$m\ell\theta'' + mg\sin(\theta) = 0 \tag{5.1.5}$$

where the positive parameter m, ℓ and g are the pendulum mass, pendulum arm length and the Earth's gravitational constant, respectively. It is a rotary version of Newton's second law of motion equating the rate of change of angular momentum to the torque responsible for the change.

The second order ODE (5.1.5) can be rewritten as a first order system by setting $y_1 = \theta$ and $y_2 = \theta'$ to get

$$\mathbf{y}' \equiv \frac{d\mathbf{y}}{dt} = \frac{d}{dt}\begin{pmatrix} y_1 \\ y_2 \end{pmatrix} = \begin{pmatrix} y_2 \\ -\frac{g}{\ell}\sin(y_1) \end{pmatrix}. \tag{5.1.6}$$

The system is autonomous since m, ℓ, and g do not change with t (under normal circumstances). Augmented by a prescribed set of initial angular position y^0 and angular velocity v^0:

$$\mathbf{y}(0) = \begin{pmatrix} y_1(0) \\ y_2(0) \end{pmatrix} = \begin{pmatrix} y^0 \\ v^0 \end{pmatrix},$$

the resulting IVP determines the evolution of the pendulum with time.

To apply reduction of order to this system, we form

$$\frac{dy_2}{dy_1} = -\frac{g}{\ell}\frac{\sin(y_1)}{y_2}$$

which is separable with an exact solution

$$\frac{1}{2}(y_2)^2 = \frac{g}{\ell}\cos(y_1) + C_0. \tag{5.1.7}$$

Here, C_0 is a constant of integration to be determined by the given initial conditions. The relation (5.1.7) enables us to eliminate y_2 from the first equation of (5.1.6) to get

$$\frac{dy_1}{dt} = y_2 = \pm\sqrt{\frac{2g}{\ell}\cos(y_1) + C_0}. \tag{5.1.8}$$

Equation (5.1.8) is again a first order separable ODE and can be solved to get

$$t - t_0 = \pm\int_{y^0}^{y_1} \frac{dz}{\sqrt{\frac{2g}{\ell}\cos(z) + C_0}} \tag{5.1.9}$$

where we have made use of the initial condition $y_1(t_0) = y^0$ to determine a second constant of integration. As mentioned in Chapter 1, having t as a function of y_1 is perfectly acceptable as a solution for the original problem, especially for the pendulum problem for which we know that the change of y_1 with t is monotone (either increasing or decreasing) in a certain range of t. We note also that the remaining constant C_0 is determined by a second initial condition, a prescribed initial (angular) velocity, $y_2(t_0) - v^0$. This is done by using both initial conditions to require

$$\frac{1}{2}(v^0)^2 = \frac{g}{\ell}\cos(y^0) + C_0 \quad \text{or} \quad C_0 = \frac{1}{2}(v^0)^2 - \frac{g}{\ell}\cos(y^0). \tag{5.1.10}$$

With C_0 known in terms of the initial data, we can use (5.1.9) to determine the future time when the pendulum is at a certain position. For the time interval when this relation is monotone, the relation can be inverted to show the position of the pendulum for the time in that interval.

The over-the-top pendulum appear to be a problem in physics and not much to do with the life sciences. However, if we revisit the problem of red blood cell production and write the model equation for mature adult (for which ε may be set to zero)

$$\frac{dE}{dt} = -\alpha R + \beta M, \qquad \frac{dM}{dt} = -\gamma E$$

where $E = R - R_c$ is the "excess" in red blood cells (above the normal RBC level), the RBC problem begins to bear some resemblance to the linearized pendulum problem. If we now think of $-\gamma E$ being a first approximation of $-\gamma \sin(E)$ for small E, a more appropriate model for the case of moderate to large E would be to return to the more appropriate characterization of

$$\frac{dE}{dt} = -\alpha R + \beta M, \qquad \frac{dM}{dt} = -\gamma \sin(E)$$

since there is a limit to the amount of bone marrow that can be induced by an excessively low level of RBC ($E \ll 0$). For the pathological case of an abnormally low RBC death rate so that $|\alpha R| \ll |\beta M|$, we would have to a good approximation the following model for the RBC production:

$$\frac{dE}{dt} \simeq \beta M, \qquad \frac{dM}{dt} = -\gamma \sin(E) \tag{5.1.11}$$

which is identical to the system (5.1.6) except for system parameters. Any information extracted from the over-the-top pendulum also applies to this nonlinear RBC production model.

5.1.2.3 *Parametric plot and phase portrait*

An important point about the pendulum problem and its solution is how the information about the solution may be presented. Traditionally, we show the evolution of position and velocity of the pendulum with time by plotting the graphs of the solution for $y_1(t) = \theta(t)$ and $y_2(t) = \theta'(t)$ as functions of t. In physics and engineering dynamics, equally informative is a *parametric plot* of y_2 vs. y_1 which can be done using (5.1.7) with C_0 given by (5.1.10) as in Chapter 3 without solving (5.1.6) for $y_1(t)$ and $y_2(t)$. Such a parametric plot is called a *phase portrait* for the problem. We will have occasions to see the usefulness of phase portraits later.

To illustrate, we show below three parametric plots of $y_2 = \theta'(t)$ vs. $y_1 = \theta(t)$ for the over the top pendulum model. The top (non-periodic) graph is for the initial conditions $\{y_1(0) = 1, y_2(0) = 2\}$ and the lower (circular) graph is for $\{y_1(0) = 1, y_2(0) = 0\}$.

Exercise 13 *For the rabbit-fox problem, let $r = R/R_0$, $f = F/F_0$ and $a = c = \gamma = \delta = 1$ and graph the parametric solution $f(r)$ for $R_0 = F_0 = 1$.*

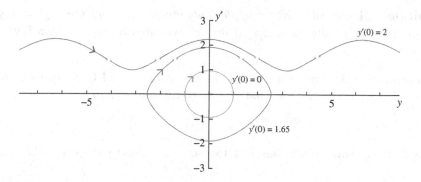

Fig. 5.1   Parametric plots for $x' = y$, $y' = -\sin(x)$ with $y(0) = 1$ and $y'(0) = 0$, 1.65 and 2.

5.1.2.4   *General second order systems*

Given our experience with single first order ODE in Chapter 2, it is not completely surprising that explicit exact solution in terms of elementary and special functions are infrequent for higher order ODE systems. But for autonomous system, the method of reduction of order in previous subsections enables us to reduce a second order system of two first order ODE to a single first order ODE and the techniques discussed in Chapter 2 and in Appendix A can then be brought to bear on the resulting scalar first order ODE.

For the general second order system of two first order autonomous ODE,

$$\frac{dy_1}{dt} = f_1(y_1, y_2), \qquad \frac{dy_2}{dt} = f_2(y_1, y_2), \qquad (5.1.12)$$

we obtain by the chain rule

$$\frac{dy_2}{dy_1} = \frac{dy_2}{dt}\frac{dt}{dy_1} = \frac{dy_2/dt}{dy_1/dt} = \frac{f_2(y_1, y_2)}{f_1(y_1, y_2)} \qquad (5.1.13)$$

where we have made use of (5.1.12) in the last step. The relation (5.1.13) is a scalar first order ODE for y_2 with y_1 as an independent variable. There is nothing special about picking y_1 as the new independent variable; we could have formed a different ODE with y_2 as independent variable:

$$\frac{dy_1}{dy_2} = \frac{f_1(y_1, y_2)}{f_2(y_1, y_2)}. \qquad (5.1.14)$$

5.1.3   *Critical points*

For higher order systems, the same technique still reduces the order of the system by 1. Whether the reduction would lead to an exact solution in terms of elementary or special functions would depend on the particular problem. More often than not, no further progress is possible after reducing the order of the system by one. With knowledge gained from studying single first order ODE, we expect better understanding and more insight to come from critical points and their linear stability analysis. We illustrate this with the rabbit-fox interaction problem.

Definition 9 *A critical point* $\mathbf{y}^c$ *of the autonomous (vector) ODE* $\mathbf{y}' = \mathbf{f}(\mathbf{y})$ *is a constant vector* $\mathbf{y}^c$ *that is a zero of the (vector) growth rate function* $\mathbf{f}(\mathbf{y})$, *i.e.,* $\mathbf{f}(\mathbf{y}^c) = \mathbf{0}$.

To determine the critical points of the rabbit-fox model ODE system, we set $R(t) = R^{(c)}$ and $F(t) = F^{(c)}$ so that (5.1.1)–(5.1.2) simplify to

$$\left[a - \beta R^{(c)} - \gamma F^{(c)} \right] R^{(c)} = 0, \qquad \left(-c + \delta R^{(c)} \right) F^{(c)} = 0. \tag{5.1.15}$$

The two steady state conditions (5.1.15) now give rise to three possible critical points:

$$\left(R^{(1)}, F^{(1)} \right) = (0, 0), \qquad \left(R^{(3)}, F^{(3)} \right) = (a/\beta, 0), \tag{5.1.16}$$

$$\left(R^{(2)}, F^{(2)} \right) = \left(\frac{c}{\delta}, \frac{a}{\gamma} \left(1 - \frac{c/\delta}{a/\beta} \right) \right).$$

Thus, if the initial rabbit and fox populations should happen to coincide with anyone of these three fixed points, the two populations would remain unchanged thereafter. For example, if $R(0) = R_0 = a/\beta$ and $F(0) = F_0 = 0$, then $R(t) = a/\beta$ and $F(t) = 0$ for all $t \geq 0$. If there should be no fox population initially, there would be no fox offsprings and the rabbits are free to grow to its carrying capacity.

The two coupled nonlinear ODE (5.1.1)–(5.1.2) have three critical points. In general, there is no correlation between the order of the ODE system and the number of its critical points. Some nonlinear systems, such as the over the top pendulum, have an infinite number of critical points.

5.1.4 *Linear stability analysis*

Suppose the initial conditions of the rabbit-fox interaction problem are not exactly at a critical point of the model ODE but very close to one of them. To be concrete, suppose $(R_0, F_0) = \varepsilon(\bar{r}, \bar{f})$ with $|\bar{r}| \leq 1, |\bar{f}| \leq 1$ and $0 < |\varepsilon| \ll 1$ so that the initial populations are close to the critical point $(0, 0)$. The solution of the IVP for $R(t)$ and $F(t)$ should remain small of order ε at least for a short while after the starting time $t = 0$. In that case, we may write

$$R(t) = \varepsilon r(t), \qquad F(t) = \varepsilon f(t)$$

with

$$r(0) = \bar{r}, \qquad f(0) = \bar{f}$$

where $f(t)$ is not to be confused with the usual growth rate function $f(y)$. The ODE (5.1.1)–(5.1.2) may then be rewritten in terms of $r(t)$ and $f(t)$:

$$r' = r \left(a - \varepsilon \beta r - \varepsilon \gamma f \right), \qquad f' = f \left(-c + \varepsilon \delta r \right).$$

Since $r(t)$ and $f(t)$ should remain of order 1 for at least a short time, the terms with a multiplicative factor ε should be negligibly small compared to the one without the factor for sufficiently small $|\varepsilon|$. We may then approximate the system of ODE for $r(t)$ and $f(t)$ by

$$r' \simeq ar, \qquad f' \simeq -cf$$

or

$$r(t) \simeq \bar{r}e^{at}, \qquad f(t) \simeq \bar{f}e^{-ct}.$$

With both $a > 0$ and $c > 0$, we have $f(t) \to 0$ as $t \to \infty$. However, the same is not true for $r(t)$. While we should not take $r(t) \to \infty$ literally since the approximation of dropping terms with a factor ε would not be justified after a while, the solution of the simplified problem correctly tells us that $r(t)$ increases with time and hence moves away from the critical point $(0,0)$. As such, $(0,0)$ is seen from a linear stability analysis to be *unstable*.

The stability of the other two critical points can also be determined by a linear stability analysis. Together the stability of the three critical points provides us with the local behavior of the solution near each of these critical points. Two important observations can be made from the results of this one particular problem. In the first instance, the local behavior of the solution near a critical point is clearly determined by the matrix eigenvalue problem for the associated linearized autonomous ODE system. Equally clear from the linear stability analysis is the need for some way to determine what happens if the initial data is not near one of the critical points. In particular, what if (R_0, F_0) is located in the midst of the three? It turns out that much can be learned about the second if we dwell little more on linear autonomous ODE systems but now from the geometrical viewpoint (see Chapter 4), a direction that has taught us much about the growth or decline of a single population.

Exercise 14 *Determine the stability of the critical point $(a/\beta, 0)$.*

Exercise 15 *Determine the stability of the third critical point.*

5.2 Linear Autonomous Systems

5.2.1 *Critical point and its stability*

Consider a linear autonomous systems

$$\mathbf{y}' = A\mathbf{y} + \mathbf{h}, \qquad \mathbf{y}(t_0) = \mathbf{y}^0$$

for an unknown vector $\mathbf{y}(t)$ with an $n \times n$ non-defective constant coefficient matrix A and an inhomogeneous (forcing) term independent of time (or whatever the independent variable t may be). Definition 9 of a critical point for a general autonomous system applies to give the unique critical point

$$\mathbf{y}^c = -A^{-1}\mathbf{h}$$

provided that A is invertible (which we will assume until further notice). If we wish, we can move (not only the initial time to the origin so that we may take $t_0 = 0$ but also) the only critical point to the origin by setting

$$\mathbf{x}(t) = \mathbf{y}(t) - \mathbf{y}^c$$

so that the original ODE becomes

$$\mathbf{x}' = A\mathbf{x}$$

since $\mathbf{y}^c$ does not depend on t. The transformed system in terms of $\mathbf{x}(t)$ has exactly one critical point $\mathbf{x} = \mathbf{0}$ (while the original system has exactly one critical point at $\mathbf{y}(t) = \mathbf{y}^c$).

For this non-defective system, we know from Chapter 4 that A can be diagonalized using its *modal matrix* V (the matrix with the full set of eigenvectors lined up as its columns) to get

$$\mathbf{z}' = \Lambda \mathbf{z}, \qquad \mathbf{z} = V^{-1}\mathbf{x}.$$

The solution of this transformed system is

$$\mathbf{z}(t) = e^{\Lambda t}\mathbf{c}, \qquad e^{\Lambda t} = \begin{bmatrix} e^{\lambda_1 t} & 0 & \dots & 0 \\ 0 & e^{\lambda_1 t} & \dots & 0 \\ & \dots & & \\ 0 & \dots & 0 & e^{\lambda_n t} \end{bmatrix}$$

with

$$\mathbf{x}(t) = V e^{\Lambda t}\mathbf{c}.$$

The following observations are evident from the exact solution for $\mathbf{x}(t)$:

- The only critical point $\mathbf{x}^{(c)} = \mathbf{0}$ is *unstable* if the real part of any of the eigenvalues of A is *positive*. In other words, for any initial value $\mathbf{x}^o \neq \mathbf{0}$, the magnitude of the solution $\mathbf{x}(t)$ generally would grow exponentially without bound if $Re(\lambda_i) > 0$, for some i, $1 \leq i \leq n$.
- $\mathbf{x}^{(c)} = \mathbf{0}$ is *asymptotically stable* if the real parts of all eigenvalues of A are negative. Given any $\mathbf{x}^o \neq \mathbf{0}$, the magnitude of the solution $\mathbf{x}(t)$ decays (exponentially) to zero if $Re(\lambda_i) < 0$ for all $i \leq n$.
- $\mathbf{x}^{(c)} = \mathbf{0}$ is *stable* (but not asymptotically stable) if $Re(\lambda_i) \leq 0$ for all $i \leq n$ and $Re(\lambda_k) = 0$ for at least one k, $0 \leq k \leq n$. The solution components corresponding to pure imaginary eigenvalues are oscillatory in time (sine and cosine functions of t) and the magnitude of the solution $\mathbf{x}(t)$ for any $\mathbf{x}^o \neq \mathbf{0}$ generally neither grow without bound nor decay to zero.
- If $Re(\lambda_i) > 0$ for some but not all indices i, the critical point has already been classified as *unstable* since some component(s) of the solution would grow without bound. But if the initial value $\mathbf{x}^o \neq \mathbf{0}$ is so restricted to eliminate all $z_i(t)$ components with $Re(\lambda_i) > 0$, the corresponding solution would either oscillate around the critical point at the origin or tends to 0 as $t \to \infty$. Eigenvectors restricted by such initial values are said to span a *stable subspace* of the space of solutions of the ODE. Similarly those initial values leading only to $z_i(t)$ components with $Re(\lambda_i) > 0$ are said to span an *unstable subspace*.

- For the remaining case with $Re(\lambda_i) \leq 0$, for $1 \leq i \leq n$ and $Re(\lambda_i) = 0$ for some i (so that the critical point is stable but not asymptotically stable), the asymptotically stable subspace of the solution space is spanned by solution components z_i with eigenvalues $Re(\lambda_i) < 0$.

For the homogeneous linear system $\mathbf{x}' = A\mathbf{x}$ with a non-defective constant coefficient matrix A (and therewith only an isolated critical point at the origin), we have now a simple way to determine its solution behavior: just compute the eigenvalues of A. More information, especially the geometrical and topological type, can then be extracted from the eigen-pairs as we show in the examples below:

5.2.2   Phase portrait

For $n = 2$ or 3 and a fixed value of t, we can think of the components of $\mathbf{x}(t)$ as the Cartesian coordinates of a point in (the two- or three-dimensional) space. As t varies over an interval (a, b), the graph of $\mathbf{x}(t)$ traces out a curve and the components of $\mathbf{x}(t)$ themselves constitute a parametric representation of the curve. While t may be the arclength of the curve or any of a number of other suitable variables, we think of it here as time and the curve being traced out as time increases from some initial time t_0. Like the trajectory of a spacecraft traveling to Mars, the independent variable t does not appear explicitly in the plot of the parametric curve. "Time" is only shown indirectly through the movement of the spacecraft in space. In fact, terms like *trajectories* and *orbits* are often used in discussing solutions of an ODE system to draw upon the analogy for suggestions of properties about the solution not apparent from the mathematical results. Visualizing solutions of ODE systems as space curves, characterizations such as trajectories and orbits are very helpful to our understanding of the solution behavior. We illustrate the benefits of the geometrical interpretation of solutions in this section by way of some second order ODE systems.

5.2.2.1   Parabolic trajectories for a node

Recall that for the constant matrix A in Example 2 of Chapter 4, we have $\lambda_1 = -1$ and $\lambda_2 = -3$ with associated eigenvectors $\mathbf{v}^{(1)} = (1, 1)^T$ and $\mathbf{v}^{(2)} = (1, -1)^T$, respectively. By working with the new variables z_1 and z_2 that diagonalize A (and visualizing them as the new two-dimensional Cartesian frame), we have

$$z_1 = e^{-t} z_1^o, \quad z_2 = e^{-3t} z_2^o, \quad \text{with} \quad z_2 = c \, (z_1)^3, \quad c = \frac{z_2^o}{(z_1^o)^3}.$$

The relation $z_2 = c \, (z_1)^3$ enables us to visualize the trajectory (solution curve) of the given ODE from any initial value (z_1^o, z_2^o) in the Cartesian (z_1, z_2) plane (which can be computed from the prescribed (x_1^o, x_2^o)). The trajectory starts at (z_1^o, z_2^o) and heads toward the origin along the cubic curve connecting these two points as t increases, approaching the origin as $t \to \infty$. Depending on the sign of z_1^o and z_2^o, the cubic curve segment may be in any of the four quadrants but all heading toward the origin as the critical point is asymptotically stable. Also, all trajectories are tangent to the z_1-axis at the origin except the special case of $z_1^o = 0$ corresponding to the z_2-axis. In terms of the Cartesian plane for the original variables $(x_1, x_2) = (x, y)$, the upright cubic curve in the (z_1, z_2) plane becomes distorted but the stability property remains unchanged. The asymptotic stability of the critical point at the origin is indicated by arrows along the trajectories pointing toward the origin as the direction of increasing time. Such a critical point is called a *stable node* or a *sink*. Trajectories for the system given by Example 2 of Chapter 4 are shown in Figure 5.2.

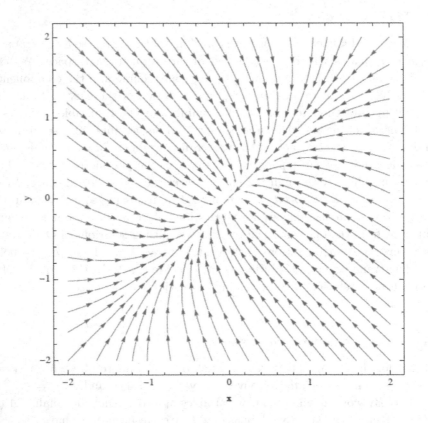

Fig. 5.2   Phase portrait for $x' = -2x + y$, $y' = x - 2y$.

For Example 3 of Chapter 4, we have $\lambda_1 = 2$ and $\lambda_2 = 4$ with associated eigenvectors $\mathbf{v}^{(1)} = (1, 3)^T$ and $\mathbf{v}^{(2)} == (1, 1)^T$, respectively. In terms of the new variables z_1 and z_2 that diagonalize A, we have

$$z_1 = e^{2t} z_1^o, \qquad z_2 = e^{4t} z_2^o$$

with

$$z_2 = c\,(z_1)^2, \qquad c = \frac{z_2^o}{\left(z_1^o\right)^2}, \qquad \text{if} \quad z_1^o \neq 0$$

and

$$z_1 = 0 \qquad \text{if} \quad z_1^o = 0.$$

For $z_1^o = 0$, the solution trajectory traverses along the z_2-axis. For $z_1^o \neq 0$, the relation $z_2 = c\,(z_1)^2$ shows that the trajectory (solution curve) of the given ODE from any initial value (z_1^o, z_2^o) are one half of a parabola in the Cartesian (z_1, z_2)-plane. The trajectory starts at (z_1^o, z_2^o) and runs away from the origin along the parabolic curve connecting (z_1^o, z_2^o) and the origin, approaching ∞ as $t \to \infty$. Depending on the sign of z_1^o and z_2^o, the parabolic segment may again be in any of the four quadrants but all moving away from the origin as the critical point is unstable. Also, all trajectories are tangent to the z_1-axis at the origin except the special case of $z_1^o = 0$ corresponding to the z_2-axis. In terms of the Cartesian plane for the original variables (x_1, x_2), the upright parabola in the (z_1, z_2) plane becomes distorted. The instability of the critical point at the origin (preserved after the change of variables) is generally indicated by arrows along the trajectories pointing away from the origin as the direction of increasing time. Such a critical point is called an *unstable node* or a *source* with local trajectories similar to those in Figure 5.2 but with arrows pointing in opposite direction.

5.2.2.2 *Hyperbolic trajectories of a saddle point*

For Example 4 of Chapter 4, we have $\lambda_1 = 2$ and $\lambda_2 = -3$ with associated eigenvectors $\mathbf{v}^{(1)} = (1,1)^T$ and $\mathbf{v}^{(2)} = (-4,1)^T$, respectively. In terms of the new variables z_1 and z_2 that diagonalize A, we have

$$z_1 = e^{2t}z_1^o, \qquad z_2 = e^{-3t}z_2^o \quad \text{with} \quad (z_2)^2\,(z_1)^3 = c, \qquad c = (z_2^o)^2\,(z_1^o)^3.$$

The relation $(z_2)^2\,(z_1)^3 = c$ shows that the trajectory (solution curve) of the given ODE from any initial value (z_1^o, z_2^o) with $z_1^o z_2^o \neq 0$ is generally a piece of a *hyperbola* in the Cartesian (z_1, z_2) plane (see Figure 5.3).

Suppose (z_1^o, z_2^o) is in the first quadrant, then the trajectory moves in the direction of decreasing z_2 and increasing z_1 with (z_1, z_2) passing through the point of minimum distance from the origin (if it is not already to the right of it at the start) and tending toward $(\infty, 0)$ monotonically. The situation is similar for (z_1^o, z_2^o) in any one of the remaining three quadrants; the trajectories are symmetric with respect to both z_1-axis and z_2-axis. For every trajectory in the first quadrant, there is a mirror image of it in the second and fourth quadrants with respect to the z_2-axis and the z_1-axis, respectively. They in turn have the same image trajectory in the third quadrant with respect to the z_1- and z_2-axis, respectively.

For initial points on the z_1-axis, i.e., $z_2^o = 0$, the corresponding trajectories move along that axis (and for Example 4) away from the origin. For initial points on the

z_2-axis, i.e., $z_1^o = 0$, the corresponding trajectories move along that axis (and for Example 4) toward the origin (but do not get there in finite time). These four straight trajectories are known as *separatrices* for the critical point. The z_2-axis constitutes the stable subspace for the critical point.

The trajectories remain hyperbolas (or straight lines in the case of separatrices) in the Cartesian plane of the original variables x_1 and x_2 but the hyperbolas are now distorted. A typical configuration of the distorted hyperbolas is shown in Figure 5.3. The critical point is called a *saddle point* and the separatrices are the asymptotes of the hyperbolas. The slopes of these asymptotes correspond to the slopes of the eigenvectors.

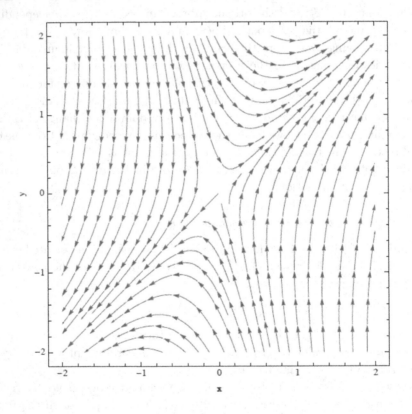

Fig. 5.3 Phase portrait for $x' = x + y$, $y' = 4x - 2y$.

5.2.2.3 *Closed trajectories for a center*

For Example 5, the eigenvalues are $\lambda_1 = \mu + i$ and $\lambda_2 = \mu - i$ and the solution was taken in the real-valued form (4.3.11). For $\mu = 0$, it is straightforward to verify that the trajectory for any initial condition is a circle. It is certainly at the same radial distance from the origin for all time given

$$x_1^2 + x_2^2 = (x_1^o)^2 + (x_2^o)^2 \equiv r_0^2;$$

the constant r_0 evidently is the radius of the circular trajectory. To see how the polar angle changes with time, we note the relation $(\tan \theta)' = (x_2/x_1)'$ with

$$(\tan \theta)' = \frac{\theta'}{\cos^2 \theta}, \qquad \left(\frac{x_2}{x_1}\right)' = \frac{1}{x_1^2}(x_1 x_2' - x_2 x_1') = -\frac{r_0^2}{x_1^2} = -\frac{1}{\cos^2 \theta}$$

so that

$$\theta' = -1 \qquad \text{or} \qquad \theta = -t + \theta_0.$$

Given $\mathbf{x}^o = (x_1^o, x_2^o)^T$ in the (x_1, x_2)-plane, the trajectory determined by our ODE is a close circle of radius r_0 which starts at the angular position θ_0 (with both r_0 and θ_0 determined by the initial conditions x_1^o and x_2^o) and is traversed clockwise in time repeatedly with unit angular speed. Such a critical point is called a *center*; it is *stable but not asymptotically stable* as the trajectory does not head toward the critical point at the origin or toward infinity as $t \to \infty$.

5.2.2.4 *Spiral trajectories for a focus*

For $\mu \neq 0$, the angular motion continues to be given by $\theta = -t + \theta_0$ while the magnitude function $r = \sqrt{x_1^2 + x_2^2}$ is now given by

$$x_1^2 + x_2^2 = r_0^2 e^{2\mu t} = r_0^2 e^{-2\mu(\theta - \theta_0)} \qquad \text{or} \qquad r(\theta) = r_0 e^{-\mu(\theta - \theta_0)}.$$

Starting with the initial point $\mathbf{x}^o = (x_1^o, x_2^o)^T$, the trajectory is evidently a spiral whose radial distance from the origin increases with time if $\mu > 0$ (keeping in mind that $\theta(t)$ is a decreasing function of time starting from θ_0. The critical point at the origin is *unstable* in this case. For $\mu < 0$, the trajectory spiral toward the origin with increasing time; the critical point is therefore *asymptotically stable*.

For a general planar system with two complex conjugate eigenvalues $\mu \pm i\omega$, the critical point is called a *spiral point* (or a *focus*) if $\mu \neq 0$. It is unstable for $\mu > 0$ and asymptotically stable if $\mu < 0$. Typical trajectories are shown in Figure 5.4. The critical point is a *center* if $\mu = 0$ and the corresponding local trajectories are closed ellipses (including circles as a special case). It is possible to re-scale time by setting $\tau = \omega t$ so that the angular velocity is 1 in suitable measurement unit.

5.2.3 *Stability regions*

The examples above show four different types of trajectories for planar linear systems: generalized parabola segments (including the special case of straight lines), generalized hyperbolas, spirals and (closed) ellipses. Upon embellishment by arrows for direction of increasing time, they pretty much cover the possible geometrical configurations for the trajectories of planar ODE systems with non-defective invertible constant coefficient matrix. These geometrical configurations of the solution curves

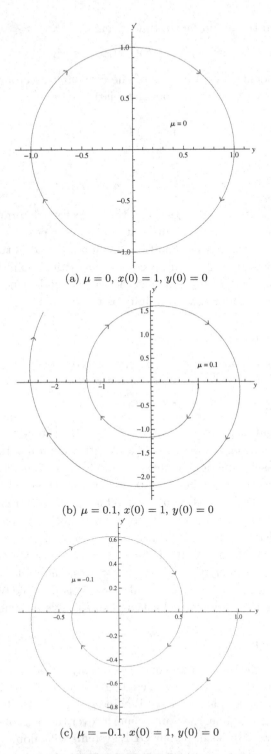

(a) $\mu = 0$, $x(0) = 1$, $y(0) = 0$

(b) $\mu = 0.1$, $x(0) = 1$, $y(0) = 0$

(c) $\mu = -0.1$, $x(0) = 1$, $y(0) = 0$

Fig. 5.4   Phase portrait for $x' = \mu x + y$, $y' = -x + \mu y$.

for a linear planar autonomous system of ODE provide an intuitive and visual reading of the stability of the system's only critical point. In this sense, we think of such an approach to analyzing the stability of the critical point as a geometrical approach.

Mathematically, the behavior of the trajectories around the critical point can in fact be read effectively from the combinations of eigenvalues of the coefficient matrix A. The actual combination of $\{\lambda_1, \lambda_2\}$ for a linear *planar* system in turn depends only on two combinations of the elements of the coefficient matrix $\{a_{ij}\}$: the *trace* of A, denoted by $tr(A) = a_{11} + a_{22} \equiv p$, and the *determinant* of A denoted by $det(A) = a_{11}a_{22} - a_{12}a_{21} \equiv q$. In terms of these two quantities, we have

$$|A - \lambda I| = \lambda^2 - tr(A)\lambda + det(A) = 0.$$

The two solutions of this characteristic equation for the eigenvalues can be given in terms of $tr(A)$ and the *discriminant* $\Delta(A)$:

$$\begin{pmatrix} \lambda_1 \\ \lambda_2 \end{pmatrix} = \frac{1}{2}\left[tr(A) \pm \sqrt{\Delta(A)} \right]$$

with

$$\Delta = [tr(A)]^2 - 4\det(A) \equiv p^2 - 4q.$$

The signs of the real parts of the two eigenvalue, the stability type of the critical point at the origin and geometrical features of the associated trajectory are neatly classified in Figure 5.5 for the different regions in the (p, q)-plane.

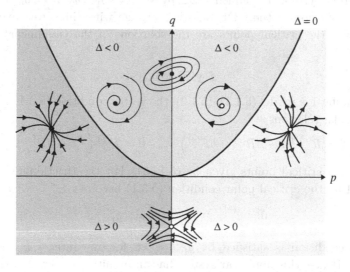

Fig. 5.5 Stability region for autonomous planar systems.

As we have qualified earlier, the discussions in the previous sections are for the most common situations; they do not exhaust all possible trajectories for linear autonomous systems, not even for the planar case. Those excluded from our discussion

so far include: 1) A is not invertible, 2) A is defective, and 3) A has repeated eigenvalues but is not defective. Some examples of these exceptional cases are included
among the exercises.

For cases where A is not invertible, the inhomogeneous term $\mathbf{h}$ in (4.3.2) has to
be in the column space of A in order for the ODE system to have a critical point.
If $\mathbf{h}$ is in the column space of A, then the critical points may not be unique and
possibly may not be isolated.

5.3   Nonlinear Autonomous Systems

5.3.1   *Critical points*

While an explicit exact solution for a higher order ODE system is an infrequent
event, we learned from Chapter 3 that much of what we need to know about the
solution behavior can be learned from a geometrical study of the ODE system. We
expect the same to be true for an autonomous system, possibly in some modified
form, given that we are in higher dimension. The key to the geometrical approach
lies in the *critical points* corresponding to time independent steady state solutions
or equilibrium configurations of the phenomenon modeled by the ODE

$$\mathbf{y}' = \mathbf{f}(\mathbf{y})$$

for an n-vector function $\mathbf{y}(t) = (y_1(t), y_2(t), \ldots, y_n(t))^T$. We denote a *critical point*
for an nth order system of coupled ODE by $\mathbf{y}^{(c)} = (y_1^{(c)}, y_2^{(c)}, \ldots, y_n^{(c)})^T$. Since a
critical point solution of the ODE does not change with time, we have $(\mathbf{y}^{(c)})' =$
$\mathbf{0}$. This means the critical points are the solutions of the (nonlinear) system of
equations

$$\mathbf{f}(\mathbf{y}^{(c)}) = \mathbf{0}. \tag{5.3.1}$$

For the rabbit-fox model (5.1.1)–(5.1.2), the vector condition (5.3.1) corresponds
to the two scalar relations

$$0 = R^{(c)}\left(a - \beta R^{(c)} - bF^{(c)}\right), \qquad 0 = F^{(c)}\left(-c + dR^{(c)}\right)$$

leading to three critical points given by (5.1.16). For the (frictionless) pendulum
problem (5.1.6), the critical point condition (5.3.1) becomes

$$0 = y_2^{(c)}, \qquad 0 = -\frac{g}{\ell}\sin(y_1^{(c)}).$$

The second condition is satisfied by $y_1^{(c)} = k\pi$ for any integer k, i.e., $k = 0,$
$\pm 1, \pm 2, \ldots$. Hence, this nonlinear system has an infinite number of critical points:

$$\mathbf{y}^{(k)} = (k\pi, 0), \qquad (k = 0, \pm 1, \pm 2, \ldots).$$

Once found, a critical point $\mathbf{y}^{(c)}$ of a nonlinear autonomous ODE system $\mathbf{y}' =$
$\mathbf{f}(\mathbf{y})$ may be relocated to the origin by setting $\mathbf{y} = \mathbf{x} + \mathbf{y}^{(c)}$, the vector ODE becomes

$$\mathbf{x}' = \mathbf{f}(\mathbf{x} + \mathbf{y}^{(c)}) \quad \text{with} \quad [\mathbf{f}(\mathbf{x} + \mathbf{y}^{(c)})]_{\mathbf{x}=\mathbf{0}} = \mathbf{f}(\mathbf{y}^{(c)}) = \mathbf{0}.$$

However, unlike a linear system with an invertible constant coefficient matrix, a nonlinear autonomous system may have more than one critical points and we normally need to investigate all critical points. Hence it may not be practical to keep track of the different transformed systems with several different critical points relocated to the origin. For that reason, we typically work with the ODE for $\mathbf{y}$ without relocating any of its critical points to the origin (unless it is there naturally as in the two examples above).

As we have learned from the scalar case, the locations of critical points and their stability provide much of the essential information about the solution of autonomous ODE. Having located some or all of the critical points $\{\mathbf{y}^{(k)}\}$ of the system, we investigate next the stability of each of these critical points. As we do that in the next few sections of this chapter, we need to face the fact that determination of the stability of a critical point of a nonlinear system is no longer simply graphing one or more scalar growth rate functions and reading off whether or not the critical point is stable from these graphs (as we did in Chapter 3). Instead, the task requires some mathematical analysis and computation. One kind of analysis that can be done without extensive background preparation is the linear stability analysis introduced previously for a single autonomous ODE. We see in sections below how it can be extended to handle interacting populations.

5.3.2   *Linear stability analysis*

For a general autonomous ODE system $\mathbf{y}' = \mathbf{f}(\mathbf{y})$, suppose we have found its time-independent steady state solutions from the critical point (vector) condition (5.3.1). To determine the stability of one of these isolated critical points $\mathbf{y}^{(c)}$ in a small neighborhood around it, we assume that all components of $\mathbf{f}(\mathbf{y})$ to be at least twice continuously differentiable in some region of interest in the n dimensional Euclidean space where $\mathbf{y}^{(c)}$ is located. Suppose the initial value $\mathbf{y}^o$ is not exactly equal to $\mathbf{y}^{(c)}$ but close to it so that $\left|\mathbf{y}^o - \mathbf{y}^{(c)}\right| \ll 1$. In that case we can apply Taylor's theorem to $\mathbf{f}(\mathbf{y})$ about $\mathbf{y}^{(c)}$ to obtain

$$\mathbf{f}(\mathbf{y}) = \mathbf{f}\left(\mathbf{y}^{(c)}\right) + F\left(\mathbf{y}^{(c)}\right)\left(\mathbf{y} - \mathbf{y}^{(c)}\right) + O\left(\left|\mathbf{y} - \mathbf{y}^{(c)}\right|^2\right),$$

where $F(\mathbf{u})$ is the Jacobian matrix for $\mathbf{f}(\mathbf{y})$ evaluated at $\mathbf{u}$:

$$F(\mathbf{u}) = \begin{bmatrix} f_{1,1}(\mathbf{u}) \,, & \cdots \,, & f_{1,n}(\mathbf{u}) \\ & \cdots & \\ f_{n,1}(\mathbf{u}) \,, & \cdots \,, & f_{n,n}(\mathbf{u}) \end{bmatrix}. \tag{5.3.2}$$

To get (5.3.2), we expand each component of $\mathbf{f}(\mathbf{y})$ as a Taylor series about the critical point $\mathbf{y}^{(c)}$ (since $\mathbf{f}(\mathbf{y})$ is assumed to be at least twice continuously differentiable):

$$f_k(\mathbf{y}) = f_k(\mathbf{y}^{(c)}) + \left\{ f_{k,1}\left(\mathbf{y}^{(c)}\right)(y_1 - y_1^{(c)}) + \cdots + f_{k,n}\left(\mathbf{y}^{(c)}\right)(y_n - y_n^{(c)}) \right\}$$
$$+ \cdots .$$

Putting each component back to its place in the vector $\mathbf{f}(\mathbf{y})$, we get

$$
\mathbf{f}(\mathbf{y}) = \begin{pmatrix} f_1(\mathbf{y}^{(c)}) \\ \cdot \\ \cdot \\ f_n(\mathbf{y}^{(c)}) \end{pmatrix} + \begin{pmatrix} f_{1,1}(\mathbf{y}^{(c)})z_1 + \cdots + f_{1,n}(\mathbf{y}^{(c)})z_n \\ \cdot \\ \cdot \\ f_{n,1}(\mathbf{y}^{(c)})z_1 + \cdots + f_{n,n}(\mathbf{y}^{(c)})z_n \end{pmatrix} + \cdots .
$$

Given $\mathbf{f}(\mathbf{y}^{(c)}) = \mathbf{0}$, the linearization of the ODE $\mathbf{y}' = \mathbf{f}(\mathbf{y})$ in a small neighborhood of $\mathbf{y}^{(c)}$ takes the form:

$$
\mathbf{z}' = F^{(c)}\mathbf{z}, \qquad F^{(c)} \equiv F(\mathbf{y}^{(c)}) = \left[f_{i,j}(\mathbf{y}^{(c)})\right] \equiv \begin{bmatrix} f_{1,1}^{(c)}, & \cdots, & f_{1,n}^{(c)} \\ & \cdots & \\ f_{n,1}^{(c)}, & \cdots, & f_{n,n}^{(c)} \end{bmatrix}, \qquad (5.3.3)
$$

with $\mathbf{z} = \mathbf{y} - \mathbf{y}^{(c)}$ and $f_{i,j} = \partial f_i/\partial y_j$ being the (i,j) — element of the Jacobian matrix F of $\mathbf{f}(\mathbf{y})$. For a particular critical point $\mathbf{y}^{(c)}$, the ODE in (5.3.3) for $\mathbf{z}$ is a homogeneous linear system. Its coefficient matrix is the Jacobian matrix $F(\mathbf{y})$ evaluated at $\mathbf{y}^{(c)}$ and is therefore a constant matrix. We can then compute its eigenvalues and read off the stability of $\mathbf{y}^{(c)}$ (now as a critical point of the linearized system (5.3.3)) from the signs of the real parts of the eigenvalues. Below are some examples of such computation.

Example 10 $y_1' = 1 - y_1 + \frac{1}{3}y_1y_2,$ $y_2' = y_2 - \frac{1}{3}y_1y_2.$

The condition $\left[y_2^{(c)}\right]' $ is met by either $y_2^{(1)} = 0$ or $y_1^{(2)} = 3$. In the first case, the condition $\left[y_1^{(c)}\right]' = 0$ gives $y_1^{(1)} = 1$ while we get $y_2^{(2)} = 2$ for the second case. The two critical points for this system are therefore $\mathbf{y}^{(1)} = (1,0)^T$ and $\mathbf{y}^{(2)} = (3,2)^T$. To study the stability of the first critical point $\mathbf{y}^{(1)} = (1,0)^T$, we compute the Jacobian matrix and evaluate it at $\mathbf{y}^{(1)}$ to get

$$
F^{(1)} = \begin{bmatrix} -1 + y_2/3 & y_1/3 \\ -y_2/3 & 1 - y_1/3 \end{bmatrix}_{(1,0)} = \begin{bmatrix} -1 & 1/3 \\ 0 & 2/3 \end{bmatrix}
$$

giving the linearized system

$$
\mathbf{z}' = \begin{bmatrix} -1 & 1/3 \\ 0 & 2/3 \end{bmatrix}\mathbf{z} \qquad (5.3.4)
$$

approximating the original nonlinear system for $\mathbf{y}$ adequately in a small neighborhood of $\mathbf{y}^{(1)} = (1,0)^T$.

The eigenvalues of the matrix $F^{(1)}$ are $\lambda_1 = -1$ and $\lambda_2 = 2/3$. Then $\mathbf{y}^{(1)} = (1,0)^T$ as a critical point for the linearized system (5.3.4) is a *saddle point*. The critical point $\mathbf{y}^{(1)} = (1,0)^T$ of the original nonlinear system is said to be *unstable by a linear stability analysis*.

Similarly, for the second critical point $\mathbf{y}^{(2)} = (3, 2)^T$, we have the following Jacobian evaluated at that critical point

$$F^{(2)} = \begin{bmatrix} -1 + y_2/3 & y_1/3 \\ -y_2/3 & 1 - y_1/3 \end{bmatrix}_{(3,2)} = \begin{bmatrix} -1/3 & 1 \\ -2/3 & 0 \end{bmatrix}$$

and the corresponding linearized system:

$$\mathbf{w}' = \begin{bmatrix} -1/3 & 1 \\ -2/3 & 0 \end{bmatrix} \mathbf{w}.$$

The eigenvalues of the matrix $F^{(2)}$ are

$$\begin{pmatrix} \lambda_1 \\ \lambda_2 \end{pmatrix} = \frac{1}{6}(-1 \pm i\sqrt{23}).$$

Hence $\mathbf{y}^{(2)} - (3, 2)^T$, as a critical point of the linearized system, is an asymptotically stable focus (or spiral point). The critical point $\mathbf{y}^{(2)} = (3, 2)^T$ of the original nonlinear system is then said to be *asymptotically stable by a linear stability analysis*.

Example 11 *Consider the equation of motion for the simple frictionless pendulum but now with $g/\ell = 1$ for simplicity:*

$$\mathbf{y}' = \begin{pmatrix} y_2 \\ -\sin(y_1) \end{pmatrix} = \begin{pmatrix} f_1(\mathbf{y}) \\ f_2(\mathbf{y}) \end{pmatrix}. \tag{5.3.5}$$

The critical points for this system have already been found earlier to be $\mathbf{y}^{(k)} = (k\pi, 0)$. From Taylor's formula for f_1 and f_2 truncated after terms linear in $(y_i - y_i^{(k)})$, we have

$$z_1' \simeq f_{1,1}^{(k)} z_1 + f_{1,2}^{(k)} z_2 = z_2,$$
$$z_2' \simeq f_{2,1}^{(k)} z_1 + f_{2,2}^{(k)} z_2 = -\cos(k\pi) z_1$$

with $\mathbf{z} = (y_1 - k\pi, y_2)^T$, or

$$\mathbf{z}' = F^{(k)} \mathbf{z} = \begin{bmatrix} \frac{\partial f_1}{\partial y_1} & \frac{\partial f_1}{\partial y_2} \\ \frac{\partial f_2}{\partial y_1} & \frac{\partial f_2}{\partial y_2} \end{bmatrix}_{(k\pi, 0)} \mathbf{z} = \begin{bmatrix} 0 & 1 \\ -\cos(k\pi) & 0 \end{bmatrix} \mathbf{z}.$$

The two eigenvalues of $F^{(k)}$ are $\pm i$ for k even (including zero) and ± 1 for k odd. Thus, all critical points $(k\pi, 0)^T$ are *stable but not asymptotically stable* for the corresponding linearized system when k is zero or an even integer and are *unstable* when k is odd. The trajectories of the original nonlinear system in the neighborhood of the critical points with k even are expected to be circular (and hence periodic) while those near the critical points with k odd are expected to be hyperbolas.

5.3.3 *Hyperbolic critical points*

Unlike the exact results of Section 2, the conclusions on the stability of the critical points of a nonlinear autonomous system in this section are approximate. The method of linear stability analysis is effective because it works with linearized systems adequate in a sufficiently small neighborhood of each critical point (by omitting nonlinear terms in a Taylor expansion about the critical point) so that the rather complete set of results of the previous section for linear systems apply. The nonlinear terms are expected to be negligibly small in a small neighborhood of the critical point being investigated, at least for a short while, i.e., $|t - t_0| \ll 1$. It is tempting to assume that the conclusion on the stability of a critical point by a linear stability analysis applies to the original nonlinear system. The following two examples show that this assumption is not always appropriate.

Example 12 $x' = -x + y^2, \qquad y' = y^2.$

For its only critical point $(0,0)$, the linearized system is:

$$\begin{pmatrix} x \\ y \end{pmatrix}' = \begin{bmatrix} -1 & 0 \\ 0 & 0 \end{bmatrix} \begin{pmatrix} x \\ y \end{pmatrix}$$

with $\lambda_1 = -1$ and $\lambda_2 = 0$. If the linear stability analysis were truly reliable, we would conclude that the critical point $(0,0)$ is stable (though not asymptotically stable). But the exact solution for any $y^o > 0$ is

$$y(t) = \frac{y^o}{1 - y^o t}$$

which becomes unbounded in finite time $(t_c = 1/y^o)$. Correspondingly, the ODE for $x(t)$ becomes a scalar first order linear ODE

$$x' + x = \left(\frac{y^o}{1 - y^o t} \right)^2$$

with a forcing term singular at $t_c = 1/y^o$. Hence, the critical point $(0,0)$ is unstable (at least for $y^o > 0$).

Example 13 $x' = y + \alpha x \left(x^2 + y^2 \right), \qquad y' = -x + \alpha y \left(x^2 + y^2 \right).$

The only critical point for this system is again at the origin and the corresponding linearized system is

$$x' = y, \qquad y' = -x.$$

Hence the critical point is a stable (but not asymptotically stable) *center* by a linear stability analysis. However, by forming the combination $xx' + yy'$, we get

$$\frac{1}{2}\rho' = \alpha\rho^2 \qquad \text{or} \qquad r' = \alpha r^{3/2}$$

where $\rho = r^2 = x^2 + y^2$. The exact solution for any prescribed $(x^o, y^o) \neq (0,0)$ is

$$\frac{\rho}{\rho^o} = \frac{1}{1 - 2\alpha\rho^o t}, \qquad \rho^o = (x^o)^2 + (y^o)^2.$$

Thus the radial distance of a point on the solution trajectory for $x^o \neq 0$ or $y^o \neq 0$ grows without bound if $\alpha > 0$ and decays to zero if $\alpha < 0$. Only for $\alpha = 0$ (for which the ODE system is linear) is the critical point truly a center. To show the spiral nature of the trajectories, we note the two relations

$$(\tan\theta)' = \sec^2\theta \frac{d\theta}{dt} = \frac{r^2}{x^2}\frac{d\theta}{dt}, \qquad \left(\frac{y}{x}\right)' = \frac{xy' - yx'}{x^2}. \tag{5.3.6}$$

With $x = r\cos\theta$ and $y = r\sin\theta$, we get from (5.3.6)

$$r^2\frac{d\theta}{dt} = xy' - yx' = -r^2 \qquad \text{or} \qquad \theta(t) = -t + \theta^o,$$

where $\tan\theta^o = y^o/x^o$. It follows that the trajectories are clockwise spirals with increasing time uncoiling away from the origin if $\alpha > 0$ or spiraling into the origin if $\alpha < 0$. The critical point at the origin is either unstable (if $\alpha > 0$) or asymptotically stable (if $\alpha < 0$).

Before leaving this example, it should be noted that the use of polar coordinates to transform a given system into a more tractable system is a very useful technique. An important class of methods has been developed for stability analysis based on such transformations.

For at least the two examples above, linear stability is unreliable as the omitted nonlinear terms were decisive in determining the stability of the critical points of the nonlinear system, not the eigenvalues of the relevant Jacobian matrix. On the other hand, the method of linear stability analysis is known to give correct results at least in a sufficiently small neighborhood of a critical point for many problems as confirmed by their exact solution. Hence the method can be an invaluable tool for cases when there is no exact solution if we can establish the conditions for its applicability.

In order to uncover and delimit the range of applicability of linear stability analysis, it is useful to recall the situation for a scalar first order ODE. With $f^{(c)} = f(y^{(c)}) = 0$ for the critical point $y^{(c)}$, we have

$$(y - y^c)' = f(y) = f_{,y}^{(c)}\,(y - y^c) + \frac{1}{2!}f_{,yy}^{(c)}\,(y - y^c)^2 + \;\cdots,$$

where we have made use of the notation $f_{,y}^{(c)} = [df/dy]_{y=y^c}$. Evidently, a linear stability analysis for the scalar problem would not be conclusive if $f_{,y}^{(c)} = 0$ and the first nonvanishing nonlinear terms in the Taylor series would be needed for determining the stability of the critical point. A naive analogue of $f_{,y}^{(c)} = 0$ for a nonlinear autonomous system would be the vanishing of the Jacobian matrix (5.3.2) at the critical point:

$$F^{(c)} = F(\mathbf{y}^c) = \begin{bmatrix} f_{1,1}^{(c)}, \cdots, f_{1,n}^{(c)} \\ \cdots \\ f_{n,1}^{(c)}, \cdots, f_{n,n}^{(c)} \end{bmatrix} = O.$$

But Example 12 showed that linear stability analysis fails to determine the correct stability of the critical point with a nonvanishing Jacobian matrix. So we need to relax the requirement to get an appropriate analogue of $f_{,y}^{(c)} = 0$.

Now, stability for the (isolated) critical point of a linear system is determined by the eigenvalues of the Jacobian matrix $F^{(c)}$ and the condition $f_{,y}^{(c)} = 0$ can be thought of as the vanishing of the (only) eigenvalue for the 1×1 Jacobian matrix of the scalar first order ODE. For Example 12, a linear stability analysis fails when one of the eigenvalues of $F^{(c)}$ vanishing. It would seem a linear stability analysis would be reliable if none of the eigenvalues $\{\lambda_i\}$ should vanish. However, the eigenvalues in Example 13 for which linear stability also failed to meet this requirement (given that none of its eigenvalues vanishes). We are then led to the following definition for determining the applicability of the linear stability analysis to general nonlinear autonomous systems:

Definition 10 *A critical point* $\mathbf{y}^c$ *of the* n^{th} *order autonomous system* $\mathbf{y}' = \mathbf{f}(\mathbf{y})$ *is said to be a* **hyperbolic** *critical point if* $Re(\lambda_k) \neq 0$ *for* $1 \leq k \leq n$. *A critical point is non-hyperbolic if it is not hyperbolic.*

We will also need the following characterization of an *almost linear system*:

Definition 11 *Suppose* $\mathbf{y}^c$ *is an isolated critical point of the nonlinear systems:* $\mathbf{y}' = \mathbf{f}(\mathbf{y})$ *and* $\mathbf{f}(\mathbf{y}) = F^{(c)}(\mathbf{y} - \mathbf{y}^c) + \mathbf{h}(\mathbf{y})$ *with* $\mathbf{h}(\mathbf{y}^c) = \mathbf{0}$. *Let* $\{\lambda_k\}$, $1 \leq k \leq n$, *be the collection of all the eigenvalues of the Jacobian matrix* $F^{(c)}$. *The nonlinear system for* $\mathbf{y}$ *is said to be* **almost linear** *(with respective to* $\mathbf{y}^c$) *if*

$$(i) \ \mathrm{Re}(\lambda_k) \neq 0 \ \ for \ 1 \leq k \leq n, \qquad (ii) \ \lim_{|\mathbf{y}-\mathbf{y}^c| \to 0} \frac{|\mathbf{h}(\mathbf{y})|}{|\mathbf{y} - \mathbf{y}^c|} = 0, \qquad (5.3.7)$$

where $|\mathbf{v}|$ *is the magnitude of the vector* $\mathbf{v}$.

The definition requires $\mathbf{h}(\mathbf{y})$ not to contain any term linear in $\mathbf{y} - \mathbf{y}^c$ in the Taylor expansion for $\mathbf{f}(\mathbf{y})$ with respect to $\mathbf{y}^c$ (when it exists) but allows for the use of a different vector norm than the usual vector magnitude. We can now state without proof the following informal version of the (Hartman-Grobman) sufficiency theorem on the applicability of linear stability analysis:

Theorem 6 *If* $\mathbf{y}^c$ *is an isolated* **hyperbolic** *critical point of an almost linear autonomous system, then the stability of* $\mathbf{y}^c$ *is the same as that found by a linear stability analysis.*

Note that the hyperbolicity is only a sufficient condition and not necessary for the linear stability analysis to be appropriate for the stability of the critical point. This can be seen from Example 11 for sufficiently small $|y_2^o|$. Physically, the pendulum would continue to oscillate about the lower equilibrium points ($\theta = 2k\pi, \theta' = 0$) as long as the initial velocity $\theta'(0) = \omega^o$ is not too high (not high enough to swing the pendulum over the top).

While we need additional mathematical apparatus to characterize structural stability of autonomous systems, it is not difficult to see that the system is structurally unstable if it has a non-hyperbolic critical point. To appreciate the complexity of the issue of structural stability in higher order systems, we note that a system whose critical points are all hyperbolic does not ensure structurally stability (see [17]).

5.4 The Phase Portrait

Even if the stability of a hyperbolic critical point is the same as that for the linearized system, the trajectories of the nonlinear system may be considerably different in appearance except in a small neighborhood of the critical point. However, it can be shown that the slopes at which trajectories approach the critical point as $t \to \infty$ (in the case of asymptotic stability) or as $t \to -\infty$ (in the case of instability) are given correctly by the linearized system. Beyond a small neighborhood of the critical points, the correct shape of the trajectories can be obtained qualitatively with the help of graphical use of the direction fields specified by the ODE [1] and quantitatively by accurate numerical solutions.

The two figures below for the direction field of the same over-the-top pendulum, $(y_2, -\sin(y_1))$, illustrate the power of this approach. For readers familiar with Mathematica, Figure 5.6 is by "VectorPlot" showing the direction of the trajectory passing through different points in the (y_1, y_2)-plane (known as the *phase plane* for the system). The size of and arrow reflecting the (kinetic + potential) energy of the system at that point. Figure 5.7 is by "StreamPlot" showing the trajectories inferred from the direction field. From these figures, we see the unique trajectory generated by a particular set of initial data (y^0, v^0) with the two trajectories in Figure 5.1 of this chapter being two such realizations. We also see from these two figures how the trajectory may change qualitatively as we vary the initial data.

Judicious use of direction fields is often helpful in reducing or eliminating the unnecessary tedium or drudgeries associated with the use of graphical methods. Loci of horizontal and vertical tangent of the trajectories, isoclines and possible asymptotes (that form separatrices) from the ODE often provide enough information to allow us to correctly sketch the global phase portrait efficiently and effectively without the use of any mathematical software.

The more important business of this section however is to discuss some theoretical results that enable us to decide on important features of the trajectories that cannot be decided by the coarse grain direction fields. We mention here only two such features to show the power of theoretical results when they are applicable. Since the illustrating examples are planar systems, the two components of the vector unknown $\mathbf{y}(t)$ may be occasionally denoted by $x(t)$ and $y(t)$, respectively (instead of y_1 and y_2).

5.4.1 *Limit cycles*

Example 14 $x' = y + x - x(x^2 + y^2),$ $y' = -x + y - y(x^2 + y^2).$

The origin is the only critical point for this nonlinear system (should try to show

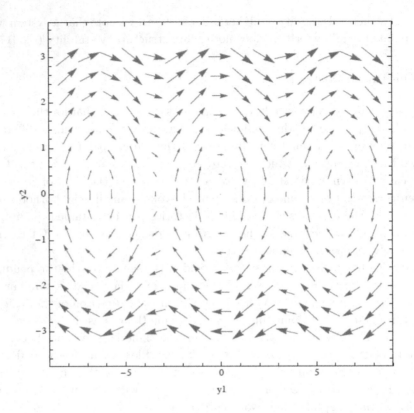

Fig. 5.6 Direction field for $dy_1/dt = y_2$, $dy_2/dt = -\sin(y_1)$ (by "VectorPlot" of Mathematica).

this). The linearized system about the critical point gives two complex conjugate eigenvalues $1 \pm i$ so that the critical point is an unstable spiral point. Since there is no other critical point, it would seem that all trajectories from $(x^o, y^o) \neq (0,0)$ should spiral outward to ∞. However, for $x^2 + y^2 \gg 1$, the components of the growth rate function $\mathbf{f}(\mathbf{y})$ are dominated by terms cubic in the unknown components x and y. Consequently, both x' and y' are negative, at least in the first quadrant, and the trajectories may not increase without bound as suggested by a linear stability analysis.

To get a better understanding of the behavior of the trajectories, we form

$$xx' + yy' = (x^2 + y^2)\left[1 - (x^2 + y^2)\right], \qquad \theta' = \frac{1}{r^2}(xy' - yx') = -1$$

(where the two ODEs have been used to simplify the right-hand side of both expressions). By setting $\rho = x^2 + y^2$, we get

$$\rho' = 2\rho(1 - \rho), \qquad \theta' = -1$$

so that the trajectories in fact spiral outward from any point near the origin for $0 < |\mathbf{y}^o| < 1$. But for $\rho = x^2 + y^2 > 1$, we have $\rho' < 0$ and any trajectory starting

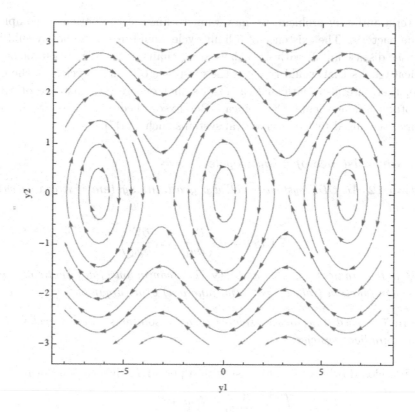

Fig. 5.7   Trajectories of $dy_1/dt = y_2$, $dy_2/dt = -\sin(y_1)$ (by "StreamPlot" of Mathematica).

from any initial point $|\mathbf{y}^o| > 1$ would spiral inward toward the origin. The scalar ODE for $\rho(t)$ alone has a critical point $\rho = 1$. Hence, for any initial conditions $|\mathbf{y}^o| = 1$, the trajectory is the circle $\rho = x^2 + y^2 = 1$ traversed clockwise at unit angular speed. Trajectories starting from initial points inside/outside the unit circle spiral outward/inward toward the circle with smaller and smaller radial increments, tending to the circular trajectory in the limit as $t \to \infty$ (but never get there in finite time). The circular trajectory $x^2 + y^2 = 1$ corresponds to a periodic solution in time and is called a *limit cycle* of the ODE system.

More generally, a limit cycle is a closed trajectory. It is therefore a *periodic solution* of the ODE system for some period T with $\mathbf{y}(t+T) = \mathbf{y}(t)$. It is a dynamic equivalent of an isolated critical point. Once you are on a limit cycle, you stay on it. More precisely, if the initial condition is a point on a limit cycle, the solution would continue to traverse it returning to the same initial point periodically.

We worked and found a limit cycle for Example 14 because the behavior of trajectories near the critical point and those far away from the critical point suggests the possibility of its existence. What if an ODE system is sufficiently complex that there is no obvious hint of a limit cycle and we need to know whether there is one.

For bacteria-antibody problem, we may want to know if the patient is completely rid of the bacteria. The existence of a limit cycle would mean the host would have to carry the disease for life with some days worse than others. We present below two theoretical results that would rule out the existence of a limit cycle. At the other extreme, there are also mathematical results that ensure the existence of a limit cycle; notable among them is the Poincaré-Bendixson theorem. These are discussed in a more specialized text on dynamical systems such as [17].

5.4.2 Non-existence of a limit cycle

Definition 12 *An ODE system is called a* **gradient system** *if it can be written in the form*

$$\mathbf{y}' = -\nabla V(\mathbf{y}) = -\left(\frac{\partial V}{\partial y_1}, \ldots, \frac{\partial V}{\partial y_n}\right)^T,$$

where ∇ is the gradient operator and $V(\mathbf{y})$ is a continuously differentiable, single-valued scalar function (called a potential function) of its arguments $(y_1, \ldots, y_n)$.

Theorem 7 *Closed trajectories (isolated periodic solutions or limit cycles) do not exist for a gradient system.*

Proof If a closed orbit C should exist with a period T, then we can form

$$\int_{t_0}^{t_0+T} \frac{dV}{dt} dt = [V]_{t_0}^{t_0+T} = 0.$$

On the other hand, we also have

$$\int_{t_0}^{t_0+T} \frac{dV}{dt} dt = \int_{t_0}^{t_0+T} \nabla V \cdot \mathbf{y}' dt = -\int_{t_0}^{t_0+T} \mathbf{y}' \cdot \mathbf{y}' dt < 0$$

since $\mathbf{y}' \neq 0$ (as we would have a critical point otherwise). This contradicts the assertion of a periodic solution. Hence, a closed orbit is not possible.   □

Example 15 $\mathbf{y}' = (\sin(y_2),\ y_1 \cos(y_2))^T.$

This is a gradient system for the potential $V = -y_1 \sin(y_2)$ and hence has no limit cycle.

Another simple nonexistence result applies only to planar systems taken in the form

$$\mathbf{y}' = \begin{pmatrix} x \\ y \end{pmatrix}' = \begin{pmatrix} F(x,y) \\ G(x,y) \end{pmatrix} = \mathbf{f}(\mathbf{y}). \tag{5.4.1}$$

Theorem 8 *Suppose F and G are continuously differentiable in both variables in a simply connected region D of the (x,y)-plane and (with $f_{,z} = \partial f/\partial z$) $F_{,x} + G_{,y}$ is of one sign. Then the ODE (5.4.1) has no closed orbit (periodic solution) inside D.*

Proof Suppose a closed trajectory C exists insider D with a period T. Let A be the simply-connected area inside C. Then by two-dimensional divergence theorem, we have

$$\iint_A (F_{,x} + G_{,y})\, dxdy = \oint_C (\mathbf{f}(x,y) \cdot \mathbf{n})\, ds = \oint_C (Fdy - Gdx)$$

$$= \int_{t_0}^{t_0+T} (x'y' - y'x')dt = 0.$$

But the integrand on the left-hand side is of one sign and the integral is therefore non-zero. This contradicts the right-hand side which vanishes. Hence there cannot be a closed trajectory inside D. $\qquad\qquad\square$

Example 16 $\qquad x' = -x + 2y^3 - 2y^4, \qquad y' = -x - y + xy.$

For this system, we have $F_{,x} + G_{,y} = -2 + x$. Hence, the system has no closed trajectories in the left half (x,y)-plane.

5.4.3 *Reversible systems*

Example 13 shows that a linear center may not be a true center for the original nonlinear ODE system. On the other hand, the linear center at the origin of the (x,y)-plane for the nonlinear simple pendulum without damping is in fact a center of the actual equation of motion for initial conditions not too far away from the critical point $\mathbf{y}^c = \mathbf{0}$. Consistent with these examples, Theorem 6 is silent on the true type of critical point when it is a *linear center,* i.e., a center by linear stability analysis. Often we need the true nature of the critical point to complete the global phase portrait as illustrated by the following predator-prey problem:

Example 17 $\qquad x' = x - xy, \quad y' = -y + xy.$

The system has two critical points at $\mathbf{y}^{(1)} = (0,0)^T$ and $\mathbf{y}^{(2)} = (1,1)^T$. It is straightforward to calculate the eigen-pair for each critical point. For $\mathbf{y}^{(1)} = (0,0)^T$, we have $\{\lambda_1 = -1, \mathbf{p}^{(1)} = (1,1)^T\}$ and $\{\lambda_2 = 1, \mathbf{p}^{(2)} = (1,-1)^T\}$. Hence, $\mathbf{y}^{(1)} = (0,0)^T$ is a saddle point and unstable. For $\mathbf{y}^{(2)} = (1,1)^T$, we have $\{\lambda_1 = i, \mathbf{p}^{(1)} = (1,-i)^T\}$ and $\{\lambda_2 = -i, \mathbf{p}^{(2)} = (1,i)^T\}$ so that $\mathbf{y}^{(2)} = (1,1)^T$ is a center. Since $\mathbf{y}^{(2)}$ is a non-hyperbolic critical point, its stability as determined by a linear stability analysis may not be reliable; the critical point may or may not be a center of the original nonlinear system. Without knowing the true behavior of the trajectories near $\mathbf{y}^{(2)} = (1,1)^T$, we will not be able to get the global phase portrait, particularly how the trajectories near one critical point relate to those near the other.

There are several techniques that can assist us in finding out whether a linear center is a true center for the original nonlinear system. One of these would be to find an explicit solution of the ODE system; but analytical solution are rare (though it is possible in this case). A technique that does not depend on an explicit solution applies if the original nonlinear system has a certain kind of symmetry.

For example, in watching a movie on a swinging pendulum without damping, we would not observe anything unusual if the movie should be run in reverse. In fact, without some marker to help identify the order of the motion, we could not tell the difference when the movie is run backward. The motion of the dynamical system characterized by this kind of ODE system is said to have *time-reversal symmetry* and the ODE system itself is said to be a *reversible system*.

Mathematically, a second order (planar) reversible system of ODE of the form (5.4.1) is defined as follows.

Definition 13 *A planar system of the form (5.4.1) is a **reversible system** with respect to the x-axis if the ODE system is invariant after a change from t to −t and y to −y, i.e.,*

$$F(x, -y) = -F(x, y), \qquad G(x, -y) = G(x, y). \qquad (5.4.2)$$

Theorem 9 *If the critical point $\mathbf{y}^c = \mathbf{0}$ is a linear center of a planar nonlinear reversible ODE system with respect to the x-axis, then the critical point is also a true center of the planar nonlinear system (with only closed trajectories in a sufficiently small neighborhood of $\mathbf{y}^c = \mathbf{0}$).*

Proof (sketched) Suppose $\phi(t) = (\varphi(t), \psi(t))^T$ is the solution of the ODE (5.4.1) with the initial condition $\mathbf{y}(t_1) = (x_1, 0)^T$ at some time $t_1 \geq t_0$ with $x_1 < 0$. And suppose $\phi(t_2) = (x_2, 0)^T$ at some later time $t_2 > t_1$ with $x_2 > 0$. (While this is anticipated given that $\mathbf{y}^c = \mathbf{0}$ is a linear center, it needs to be proved in a formal proof which we will not pursue here.) Now let $\tau = -t$ and $\mathbf{z}(\tau) = (\tilde{x}(\tau), \tilde{y}(\tau))^T = (x(t), -y(t))^T$. Then $\mathbf{z}(t)$ satisfies the same ODE as $\mathbf{y}(t)$:

$$\frac{d\mathbf{z}}{d\tau} = \left(\frac{dz_1}{d\tau}, \frac{dz_2}{d\tau}\right)^T = \left(-\frac{dx}{dt}, \frac{dy}{dt}\right)^T = (-F(x,y), \ G(x,y))^T$$
$$= (F(x, -y), \ G(x, -y))^T = (F(\tilde{x}, \tilde{y}), G(\tilde{x}, \tilde{y}))^T = \mathbf{f}(\mathbf{z}).$$

Hence, starting at the same initial point $(x_1, 0)$, we have

$$\mathbf{z}(\tau) = \phi(\tau) = (\varphi(\tau), \psi(\tau))^T \quad \text{or} \quad \mathbf{z}(\tau) = (\varphi(t), -\psi(t))^T.$$

In other words, the trajectory $\mathbf{z}(\tau)$ is a reflection of $\mathbf{y}(t)$ about the x-axis traversed in the opposite direction and with its end point $(x_2, 0)^T$ coinciding with the end point of $\mathbf{y}(t)$. Once $\mathbf{y}(t)$ reaches $(x_2, 0)^T$, the image curve is its continuation steering the trajectory back to starting point at $(x_1, 0)^T$ and thereby forming a close trajectory. Uniqueness theorem for the IVP for (5.4.1) rules out other possible trajectory. □

It is easy to verify that the nonlinear equation of motion for the simple pendulum is a reversible system. The theorem assures us that trajectories near the critical point $\mathbf{y}^c = \mathbf{0}$ are closed even when $\mathbf{y}^c = \mathbf{0}$ is a non-hyperbolic critical point. On the other hand, the theorem makes no guarantee for initial points far away from the origin. As we know, with a sufficient large velocity, the pendulum may be swung over the top and continues to whirl around in the same direction (and therefore not periodic).

There is nothing special about the x-axis in the above result. A similar statement can be made about *time-reversal symmetry about the y-axis* or any other direction. Note that the predator-prey system of Example 17 is reversible with respect to neither the x-axis nor the y-axis. It is however a reversible system with respect to some direction. The use of Theorem 9 to show that $(1,1)^T$ is a true center will be left as an exercise.

5.4.4 *Multiple eigenvalues*

Up to now, we have avoided defective coefficient matrices, i.e., matrices with multiple eigenvalues and an incomplete set of eigenvectors, for the linearized problem. The following example suggests that the critical point is likely to be unstable:

Example 18 $y' = \begin{bmatrix} 0 & 1 \\ 0 & 0 \end{bmatrix} y.$

With $\lambda_1 = \lambda_2 = 0$, the exact solution of this system is

$$y_2(t) = y_2^o, \qquad y_1(t) = y_1^o + y_2^o t.$$

Even with $\lambda_k = 0$, $k = 1, 2$, the component $y_1(t)$ grows linearly with time so that the critical point is generally unstable. The space of solutions for this ODE system does have a stable subspace characterized by (the line) $\{y_2 = 0\}$. Relative to any initial value y^o on the y_1-axis, the critical point at the origin is stable (but not asymptotically stable).

There is another kind of complexity associated with this example. The origin is not the only critical point of this linear system. In fact, all points on the y_1-axis, $(y_1^o, 0)$, are non-isolated critical points (all stable but not asymptotically stable).

Example 19 $y' = \begin{bmatrix} \mu & 1 \\ 0 & \mu \end{bmatrix} y, y(0) = y^o.$

With $\lambda_1 = \lambda_2 = \mu$, the exact solution of this system is

$$y_2(t) = y_2^o e^{\mu t}, \qquad y_1(t) = (y_1^o + y_2^o t) e^{\mu t}.$$

The critical point $y^c = 0$ is clearly unstable if $\mu > 0$. The space of solutions for this case does not have a stable subspace as in the previous example. On the other hand, $y^c = 0$ is asymptotically stable if $\mu < 0$. For the special case of $\mu = 0$, the critical point was found to be also unstable but tends to infinity at a much slower rate (linearly instead of exponentially). While the stability of the critical point changes as μ passes through the critical value $\mu_c = 0$, there is no change in the number or location of the critical point, the system being linear.

The Jacobian matrix at a critical point of a nonlinear autonomous systems may involve multiple eigenvalues. While the stability of the linearized system may be delineated after reduction to Jordan form, the applicability of the results to the original nonlinear system again requires further investigation if the critical point is non-hyperbolic.

5.5   Basic Bifurcation Types

A scientist studying a physical system (which may be mechanical, biological, economical or social in nature) or an engineer designing such a system must be concerned whether the system under consideration is reliable in performing its assigned task. Correspondingly, the vector ODE that models the system should be examined for its structural stability as in the scalar case in Chapter 3. For the scalar ODE $y' = f(y; \mu)$, structural instability is manifested in the transition of a hyperbolic critical point $y^c(\mu)$ of the ODE to a non-hyperbolic critical point as the parameter μ reaches a critical value μ_c, called the *bifurcation point*, with the property $f_{,y}(y^c; \mu_c) = 0$. For an isolated bifurcation point μ_c, the critical point $y^c(\mu)$ becomes hyperbolic again after passing through μ_c.

   In Chapter 3, the three basic types of bifurcation for a scalar ODE share the common feature that two or more critical points coalesce at bifurcation. A non-hyperbolic critical point was taken there to be a critical point of the ODE that is also a stationary point of the growth rate function, i.e., $f_{,y}^{(c)} = f_{,y}(y^c; \mu_c) = 0$, or equivalently the eigenvalue of the 1×1 matrix $f_{,y}^{(c)}$ vanishes. It is evident from the examples discussed in previous sections of this chapter that the analogue for coupled ODE systems is not the Jacobian matrix being a zero matrix or all eigenvalues vanishing. Instead, the examples suggested that we define (as we did earlier in Definition 10) a hyperbolic critical point for an ODE system to be one with all eigenvalue of its Jacobian matrix at the critical point to have non-vanishing real parts; it is a non-hyperbolic critical point otherwise. We show by some examples in the next subsections that an ODE system experiences structural instability when a critical point is non-hyperbolic by this definition.

5.5.1   *The bacteria-antibody problem*

Recall the ODE system from Example 10 modeling the antibody-bacteria dynamics, taken in the form

$$a' = \mu - a + \frac{1}{3}ab, \qquad b' = b - \frac{1}{3}ab. \qquad (5.5.1)$$

For $\mu = 1$ as in Chapter 4, the two critical points are $\mathbf{y}^{(1)} = (1, 0)^T$ and $\mathbf{y}^{(2)} = (3, 2)^T$ with the first being an unstable saddle point and the second an asymptotically stable spiral point. The eigenvectors of the saddle point $\mathbf{y}^{(1)}$ are $(1, 0)^T$ and $(1, 5)^T$, giving the direction of an asymptotically stable separatrix and an unstable one respectively. Direction field of the ODE system suggests that the latter (unstable) separatrix heads toward the spiral point and spiraling into $\mathbf{y}^{(2)}$ clockwise. Since $a(t)$ and $b(t)$ are both nonnegative, all trajectories starting in the first quadrant with some finite bacteria population spiral into the asymptotically stable spiral point $\mathbf{y}^{(2)} = (3,0)^T$. (Whether the biological host survives the level of steady state bacterial population depends on the constitution of the host, which is not a part of the model.) Only when $b(0) = 0$ would the system evolves toward $\mathbf{y}^{(1)} = (1,0)^T$ along the a-axis with $b(t) = 0$ for all $t > 0$. If there is no bacteria initially, there will be no bacteria thereafter (since there is no infusion of bacteria in this model).

Intuitively, it is expected that a higher level of antibody synthesis rate would lead to disproportionately more antibodies and hence a reduction of the eventual steady state bacteria population. If so there would be some threshold value μ_c beyond which there should be a switch in the stability type of the two critical points (and/or a change in the number of critical points). In that case, the ODE system would be structurally unstable at μ_c and hence a bifurcation to take place at that point. We undertake a bifurcation analysis below to determine the existence and location of one or more bifurcation points for the model (5.5.1) to illustrate the general approach to bifurcation problems for vector ODE.

For a general μ, the two critical points for the system (5.5.1) are $\mathbf{y}^{(1)} = (\mu,0)^T$ and $\mathbf{y}^{(2)} = (3, 3-\mu)^T$ reducing to $(1,0)^T$ and $(3,2)^T$ for $\mu = 1$. For $0 < \mu < 3$, it is a matter of repeating the solution process for the $\mu = 1$ case to arrive at the same qualitative conclusion as summarized above for that case. (We will not be concerned with the case $\mu \le 0$ in the context of the antibody-bacteria problem.) For $\mu = 3$, we have $\mathbf{y}^{(1)} = (3,0)^T = \mathbf{y}^{(2)}$ so that the two distinct critical points for $0 < \mu < 3$ and for $\mu > 3$ coalesce and we have only one critical point for the special value $\mu_c = 3$. The change from 2 to 1 to 2 critical points as the system transitions through the critical value μ_c is similar to the situation for a transcritical bifurcation for the scalar ODE case. It remains to investigate stability type of the critical points on the two sides of μ_c to establish the nature of the bifurcation.

For the stability of the critical points for trajectories nearby, we need the Jacobian for the ODE system:

$$F(\mathbf{y}^c) = \begin{bmatrix} -1+\frac{1}{3}b & \frac{1}{3}a \\ -\frac{1}{3}b & 1-\frac{1}{3}a \end{bmatrix}_{\mathbf{y}^c}.$$

With $\mathbf{y}^{(1)}(3) = \mathbf{y}^{(2)}(3) = (3,0)^T$ and

$$F(3,0) = \begin{bmatrix} -1 & 1 \\ 0 & 0 \end{bmatrix},$$

we see that the critical point is non-hyperbolic (since $\lambda_2 = 0$) at $\mu_c = 3$. Hence the critical value $\mu_c = 3$ also satisfies one of the characteristics of a bifurcation point.

We know from previous discussion the stability of the two critical points for $\mu < \mu_c$. For $\mu > \mu_c = 3$, we have

$$F(\mu,0) = \begin{bmatrix} -1 & \frac{1}{3}\mu \\ 0 & 1-\frac{1}{3}\mu \end{bmatrix}, \qquad F(3,3-\mu) = \begin{bmatrix} -\frac{1}{3}\mu & 1 \\ -1+\frac{1}{3}\mu & 0 \end{bmatrix}$$

so that the eigenvalues for $\mathbf{y}^{(1)} = (\mu, 0)^T$ are

$$\lambda_1^{(1)} = -1 < 0, \qquad \lambda_2^{(1)} = 1 - \frac{1}{3}\mu < 0$$

and those for $\mathbf{y}^{(2)} = (3, 3 - \mu)^T$ are

$$\begin{pmatrix} \lambda_1^{(2)} \\ \lambda_2^{(2)} \end{pmatrix} = \frac{1}{6}\left(-\mu \pm \sqrt{\mu^2 + 12\mu - 36}\right)$$

with $\lambda_2^{(2)} < 0 < \lambda_1^{(2)}$. By the Hartman-Grobman theorem, $\mathbf{y}^{(1)} = (\mu, 0)^T$ is an asymptotically stable node and $\mathbf{y}^{(2)} = (3, 3 - \mu)^T$ is a saddle point. Hence there is a switch in stability type for the two critical points as we transition through the bifurcation point $\mu^{(c)} = 3$. At bifurcation, the two critical points coalesce and the single critical point at $(3, 0)^T$ becomes non-hyperbolic. Plotting the two components $a^{(1)} = \mu$ and $a^{(2)} = 3$ (of the two critical points $\mathbf{y}^{(1)} = (\mu, 0)^T$ and $\mathbf{y}^{(2)} = (3, 3 - \mu)^T$) vs. μ including their stability type gives a bifurcation diagram qualitatively identical to those for a transcritical bifurcation for the scalar ODE case. A graph of $b^{(1)} = 0$ and $b^{(2)} = 3 - \mu$ (of the two critical points $\mathbf{y}^{(1)} = (\mu, 0)^T$ and $\mathbf{y}^{(2)} = (3, 3 - \mu)^T$) vs. μ including their stability type also gives a similar bifurcation diagram. We designate such bifurcation for a system of ODE also a transcritical bifurcation.

5.5.2 *The three basic types of bifurcation*

The example in the previous subsection outlined the process for investigating bifurcation in autonomous ODE systems. The process consists of the following steps:

- Determine the critical points in terms of the bifurcation parameter(s).
- Examine possible critical bifurcation parameter values for which critical points coalesce.
- For each potential bifurcation point μ_c, compare the number and stability type of critical points before and after transition through μ_c with their counterpart as appeared in the three basic bifurcation types for scalar ODE to learn the likely type of bifurcation.
- To firm up the bifurcation type, perform a linear stability analysis and find the eigenvalues (and the corresponding eigenvectors) for each critical point on either side of the potential bifurcation point μ_c.
- Draw appropriate bifurcation diagrams to show that they are qualitatively the same as the corresponding diagram for the scalar case. Unless it is obviously unnecessary (e.g., y^c is proportional to x^c as it is the case for the example below), a bifurcation diagram should be plotted for each component of the vector unknown.

Example 20 $x' = y - 2x, \qquad y' = \mu - y + x^2.$

By implementing the steps for bifurcation analysis in higher dimensions, the ODE system above is found to have a saddle-node bifurcation at $\mu_c = 1$. There is no critical point for $\mu > \mu_c = 1$. For $\mu < \mu_c$, the two critical points are

$$\begin{pmatrix} x^{(1)} \\ x^{(2)} \end{pmatrix} = 1 \pm \sqrt{1 - \mu}, \qquad y^{(k)} = 2x^{(k)}, \qquad (k = 1, 2).$$

The eigenvalues for them are

$$2 \begin{pmatrix} \lambda_1^{(1)} \\ \lambda_2^{(1)} \end{pmatrix} = -3 \pm \sqrt{9 + 8\sqrt{1 - \mu}}, \qquad 2 \begin{pmatrix} \lambda_1^{(2)} \\ \lambda_2^{(2)} \end{pmatrix} = -3 \pm \sqrt{9 - 8\sqrt{1 - \mu}},$$

respectively. The critical point $\mathbf{y}^{(1)} = (x^{(1)}, 2x^{(1)})^T$ with $x^{(1)} = 1 + \sqrt{1 - \mu}$, is a saddle point and unstable. The other, $\mathbf{y}^{(2)} = (x^{(2)}, 2x^{(2)})^T$ with $x^{(1)} = 1 - \sqrt{1 - \mu}$, is an asymptotically stable node if $8\sqrt{1 - \mu} \leq 9$ and an asymptotically stable spiral point otherwise.

Example 21 $x' = \mu x + y - x^3, \qquad y' = x - y.$

By the steps for bifurcation analysis, the ODE system above is found to have a pitchfork bifurcation at $\mu_c = -1$. For $\mu < \mu_c = -1$, the only critical point at the origin, $\mathbf{y}^{(1)} = (0, 0)^T$, is an asymptotically stable node given the following two eigenvalues of the relevant Jacobian matrix:

$$2 \begin{pmatrix} \lambda_1 \\ \lambda_2 \end{pmatrix} = -(1 - \mu) \pm \sqrt{(1 - \mu)^2 + 4(1 + \mu)}$$

with $4(1 + \mu) < 4(1 + \mu_c) = 0$.

For $\mu > \mu_c = -1$, the critical point at the origin is now a saddle point and is therefore unstable. In addition, there are two other critical points at $\mathbf{y}^{(2)} = (x^{(2)}, y^{(2)}) = \sqrt{1 + \mu}(1, 1)$ and $\mathbf{y}^{(3)} = -\sqrt{1 + \mu}(1, 1)$.

Exercise 16 *Show that both* $\mathbf{y}^{(2)}$ *and* $\mathbf{y}^{(3)}$ *are asymptotically stable node for* $\mu > \mu_c = -1$.

Bifurcation analysis for other more complex ODE systems can be investigated by the general approach outlined at the start of this section as long as the system is almost linear and the critical points of the system is hyperbolic except possibly at an isolated bifurcation point.

5.5.3   *Hopf bifurcation*

5.5.3.1   *Supercritical Hopf bifurcation*

For the three basic types of bifurcation, the critical point in question is non-hyperbolic at bifurcation with some of its critical point coalescing at the bifurcation point $\mu = \mu_c$. The following example shows something quite different may occur when the critical point is a center at bifurcation.

Example 22 $x' = \mu x + y - x(x^2 + y^2), \qquad y' = -x + \mu y - y(x^2 + y^2).$

The origin is the only critical point of this ODE system. A linear stability analysis shows that $(\lambda_1, \lambda_2)^T = \mu \pm i$ and the critical point is an asymptotically stable spiral point if $\mu < 0$, an unstable spiral point if $\mu > 0$ and a linear center if $\mu = 0$. However, the use of polar coordinates and the usual two combinations $xx' + yy'$ and $xy' - xy'$ yield the following two uncoupled scalar ODE for θ and $r = \sqrt{x^2 + y^2}$:

$$r' = r(\mu - r^2), \qquad \theta' = -1.$$

With $\theta(t) = -t + \theta^o$, all trajectories of the system are traversed clockwise. But now for $\mu > 0$ and $r^2 > \mu$, we have $r' < 0$ so that the radial distance from the origin of a point on any trajectory decreases with time. On the other hand, we have $r' > 0$ for $0 < r^2 < \mu$ so that the radial distance from origin of the (x, y)-plane of a point on the trajectory increases with time.

If the initial condition is such that $r(0) = \sqrt{\mu}$, then the solution of the original nonlinear system is periodic traversing the same circle of radius $\sqrt{\mu}$ repeatedly as time increases. The single trajectory defined parametrically by $r(t) = \sqrt{\mu}$ and $\theta(t) = -t + \theta^o$ is a periodic solution approached by trajectories starting from any point on either side of the circular orbit. In short, it is a limit cycle for the ODE system.

As shown by the observations above, the solution of the ODE system changes qualitatively as μ passes through the critical value $\mu = \mu_c = 0$. With the eigenvalues of the Jacobian matrix for $\mathbf{y}^c = 0$ given by $(\lambda_1, \lambda_2)^T = \mu \pm i$, the only critical point of the ODE system becomes non-hyperbolic for the critical value $\mu_c = 0$ (returns to hyperbolic again for $\mu < 0$). However, the critical point at the origin changes abruptly from an asymptotically stable spiral point to an unstable spiral point as μ passes from negative to positive. Such structural instability entitles $\mu_c = 0$ to be designated as a *bifurcation point*. The additional occurrence of an asymptotically stable circular limit cycle of radius $r^{(1)} = 0$ and $r^{(2)} = \sqrt{\mu}$ (and no other new critical points) distinguishes this type of bifurcation from the other three basic types. It is known as a *Hopf bifurcation*. Plotting the two $r^{(k)}$ vs. μ produces a bifurcation diagram distinguishably different from the previously known types of bifurcation diagram.

In general, Hopf bifurcation involves a change of stability of a spiral point through a (non-hyperbolic) linear center as shown by the example above. The bifurcation in that example is known as a *supercritical* Hopf bifurcation. This kind of Hopf bifurcation typically goes from an asymptotically stable spiral point to an unstable one (through a linear center at bifurcation) with an asymptotically stable limit cycle generated at bifurcation. The sequence of events can be reversed; there is no particular reason why we must go from negative to positive μ values and not the other way around.

5.5.3.2 *Subcritical Hopf bifurcation*

In contrast, the following example exhibits a different and more problematic type of Hopf bifurcation, known as *subcritical* Hopf bifurcation.

Example 23 $x' = \mu x + y + x(\rho - \rho^2),$ $y' = -x + \mu y + y(\rho - \rho^2),$ with $\rho = x^2 + y^2 = r^2$.

Again, the origin is the only critical point of this ODE system. A linear stability analysis shows that $(\lambda_1, \lambda_2)^T = \mu \pm i$ and the critical point is an asymptotically stable spiral point if $\mu < 0$, an unstable spiral point if $\mu > 0$ and a linear center if $\mu = 0$. The use of polar coordinates and forming the usual two combinations $xx' + yy'$ and $xy' - xy'$ yield the following two uncoupled scalar ODE for $r(t)$ and $\theta(t)$ (keeping in mind $\rho = r^2$):

$$r' = r(\mu + r^2 - r^4), \qquad \theta' = -1.$$

With $\theta(t) = -t + \theta^o$, all trajectories of the system are traversed clockwise. The condition $r' = 0$ yields two other critical points at most for the scalar ODE for $r(t)$ in addition to $r^{(1)} = 0$ keeping in mind that the radial coordinate cannot be negative:

$$\binom{r^{(2)}}{r^{(3)}} = \sqrt{\frac{1}{2}\left(1 \pm \sqrt{1 + 4\mu}\right)}. \tag{5.5.2}$$

For $\mu < -1/4$, there is no other critical point than an asymptotically stable spiral point at the origin since $1 + 4\mu < 0$. For $-1/4 < \mu < 0$, the two expressions in (5.5.2) are real. In that case, $r(t) = r^{(3)} < r^{(2)}$ is an unstable limit cycle for the original problem while $r(t) = r^{(2)}$ is an asymptotically stable limit cycle. For $0 < \mu < \infty$, there is only an unstable spiral point at the origin and an asymptotically stable limit cycle $r(t) = r^{(2)}$ and no others.

In contrast to the problem in the previous subsection, the present problem has an asymptotically stable critical point at the origin as well as one asymptotically stable *limit cycle* at $r(t) = r^{(3)}$ and an unstable one, $r(t) = r^{(2)}$, in between. As the bifurcation parameter increases through the bifurcation point $\mu_c = 0$ (for which the critical point becomes a non-hyperbolic linear center), the unstable limit cycle shrinks and eventually engulfs the asymptotically stable critical point at the origin, rendering it unstable. The nearby trajectories are all destined to head for the asymptotically stable limit cycle $r(t) = r^{(3)}$. This so-called *subcritical Hopf bifurcation* is more dramatic and requires the attention of the designer or engineer servicing the physical system the ODE system characterizes. For the supercritical case, while a small perturbation that pushes μ from a slightly negative value to slightly positive would make the critical point at the origin unstable, trajectories from initial configurations near the origin would approach an asymptotically stable circular limit cycle of radius $\sqrt{\mu}$ generated at bifurcation. For a small change in μ, the limit cycle constitutes a time dependent steady state of small amplitude periodic motion. Though it may be a small nuisance, the change is normally tolerable. For the subcritical example, any small change in μ from a small negative value to a small positive value would result in a drastic change from an at rest configuration to a large amplitude of periodic motion of $r(t) = r^{(3)} \simeq 1$.

In summary, we have added in Hopf bifurcation a new and distinct type to our group of ODE bifurcations. Unlike the other three types of the previous section, Hopf bifurcations involve complex eigenvalues in the relevant linear stability analysis and the occurrence or disappearance of limit cycles. Neither were involved in the other three types. While there are many other kinds of bifurcation, some of which can be found in [27], the four we have studied constitute the most frequently encountered types. We will be contend with their acquaintance and move on to other topics and issues involving boundary value problems in ODE.

Chapter 6

HIV Dynamics and Drug Treatments

6.1 The Human Immunodeficiency Virus (HIV)

6.1.1 *HIV is a retrovirus*

Historically, conventional wisdom, known as the Central Dogma in biology, perceived that genetic information flows from *DNA* to *RNA* to proteins, and proteins are responsible for the functions of all life forms including bacteria and virus. The discovery of reverse transcriptase (RT) that transcribes DNA from RNA led to the identification of a distinct class of viruses, now called retroviruses, that does not follow the prescription of the Central Dogma. Their independent contributions in this area helped David Baltimore and Howard Temin win a share of the 1975 Nobel Prize for Physiology or Medicine.

The *human immunodeficiency virus* (*HIV* for short) is a retrovirus. It contains a genome composed of two copies of single stranded RNA housed in a cone-shaped core surrounded by a membrane envelope. On the surface of the envelop are the protein molecules gp120 and the transmembrane protein gp41. When a virus particle tries to enter and infect a human (CD4+T) cell, gp120 serves as the attachment site for a receptor of the host cell. A transfer RNA located near the 5' end of each RNA serves as an initiation site for reverse transcription. Viral enzymes housed in the core include reverse transcriptase, protease, and integrase. These specific details on the composition of a HIV virion as shown in the cartoon in Figure 6.1 will be important to our discussion of different drug treatments of AIDS, the clinical ramification of HIV.

6.1.2 *The immune system*

In the human body, viruses, bacteria and other pathogens do not replicate unabated but are opposed by immune responses. The backbones of the immune system are the B cells and T cells, collectively referred to as lymphocytes. There are 10^{12} lymphocytes (B cells and T cells) in adults.

B cells carry antibody molecules on their surface. For any foreign molecule entering the human body, there would be an antibody that would bind to it like a

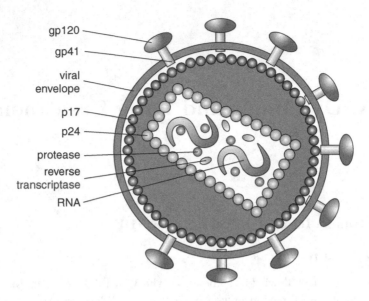

gp120
gp41
viral
envelope
p17
p24
protease
reverse
transcriptase
RNA

Fig. 6.1   HIV structure.

lock, neutralizing it or marking it for elimination by other cells (e.g., macrophage) and blood enzymes of the immune system.

Different T cells are distinguished by the type of receptors on their surface. CD8+T cells are killer cells, CD4+T cells are helper cells, etc. CD4+T cells are particularly important as they activate different components of the immune system such as proliferating CD8+T Cells and macrophage (through faster division) and energizing them when foreign molecules are detected.

What distinguishes HIV from other kinds of virus is that its main activity is to impair CD4+T cells (through the 8 phase life cycle described below), thereby suppressing the immune system in human body and leaving the individual defenseless against other infectious diseases. A more complete discussion of the human immune system can be found on the internet (e.g., HHS.gov).

6.1.3   The virus life cycle

All viruses need to infect cells for their reproduction. HIV virions reproduce and proliferate through the following process:

(1) HIV particles (or HIV virions) attach themselves to a host (CD4+T) cell.
(2) A virion or its genetic material enters the cell.
(3) The genetic material of the virus RNA is uncoated.
(4) Viral DNA is produced by the uncoated HIV genetic material and manipulates the host cell.
(5) The viral genome is multiplied by the machinery of the host cell.

(6) Further viral proteins are produced which will be used for completing new virions.

(7) The new virions (virus particles) are assembled and attached to (the inside wall of the) host cell membrane.

(8) The new HIV particles are released from the host cell to invade another (normal) CD4+T cell (before or when the host cell bursts from excessive virions).

The cycle repeats by each of the virions released by the destroyed host cell. For a more extensive description of the HIV life cycle, visit the NIH website on HIV (e.g., https://aidsinfo.nih.gov/education-materials/fact-sheets/19/73/the-hiv-life-cycle). A very nice animation of HIV virus entering the human CD4+T cell can be found on the YouTube https://www.youtube.com/watch?v=RO8MP3wMvqg.

6.1.4   Three phases of HIV infection

6.1.4.1   Phase I

During the first few weeks after inoculation with the (HIV) virus, patients usually develop high virus load, i.e., an abundance of virions in the blood, and show some flu-like symptoms. The virus is distributed to many different organs of the body. CD4+T cells fall transiently and then return to almost normal levels. At the end of this first phase, the virus load falls again and the clinical symptoms disappear.

6.1.4.2   Phase II

During the next (second) phase of HIV disease, the patients generally have very little clinical symptoms. The average length of the asymptomatic phase ranges from a few months to many years. Virions continue to replicate and proliferate, and CD4+T cells fall relentlessly.

6.1.4.3   Phase III

The final (third) phase of HIV disease is characterized by the development of (rather severe) AIDS (Acquired Immune Deficiency Syndrome). CD4+T have fallen to below 200 per $\mu\ell$ (micro liter = 10^{-6} liter). Opportunistic infections begin to appear. The impaired immune system can no longer fight off infections that normally would not be a problem. Patients die from opportunistic infections.

Many HIV infected patients have strong immune responses against the virus and do not enter into the third phase of HIV infection. However, their immune system may not be strong enough to eliminate the virus altogether as we shall see from our mathematical models on the HIV dynamics.

6.1.5   Treatment of HIV

6.1.5.1   Reverse transcriptase

One important agent in making more copies of HIV virions is the reverse transcriptase (RT). It translates the viral RNA into DNA. There are now drugs that bind HIV encoded RT enzymes and prevent the transcription of viral RNA into DNA. This effectively stops further synthesis of HIV virions and therefore constitutes a treatment.

Two complications arise from such a treatment: (1) It is not always perfectively effective, and (2) The virus may mutate and become insensitive to the drug.

6.1.5.2   Protease

Another important agent in making more virions is the virus enzyme protease. Its function is to cleave the gag gene encoded protein into four subunits that coat the genetic material of the virus. In 1995, David Ho treated a small group of HIV patients with a new drug that inhibits the protease enzymes of the HIV virus from this normal function.

When the gag-encoded inner core protein (named Gag) is not cleaved, there is not a coat for the virus' genetic material. So the drug effectively prevents an infected cell from producing the correct form of the Gag protein. The resulting virions do not function properly and cannot infect new cells. The drug therefore constitutes a different kind of treatment for HIV infection by rendering the infected CD4+T cell incapable of producing new active virus particles.

We again have the same two complications: (1) It is not always perfectively effective, and (2) The virus may mutate and become insensitive to the drug.

6.1.5.3   Cocktail treatments

There are other kinds of drug treatments. An example is the use of Chemokines to block the chemokine receptors which are needed for the virus to enter the host cell. In general, specific drugs may be grouped by their underlying mechanism of inhibition into the following five categories:

- Nucleoside reverse transcriptase inhibitors (NRTIs)
- Non-Nucleoside reverse transcriptase inhibitors (NNRTIs)
- Protease Inhibitors (PIs)
- Entry/Fusion Inhibitors
- Integrase Inhibitors.

As indicated in the previous description of single drug treatments, none of them is perfectly effective in eliminating all virus at the maximum dosage that can be tolerated by a patient.

In 1995, a combination treatment known as the "AIDS Cocktail" was introduced to patients infected with HIV/AIDS. This type of therapy is also often referred to as *highly active anti-retroviral therapy* (HAART). It has also been called *combination anti-retroviral therapy* (cART), or simply *anti-retroviral therapy* (ART). Whatever it is called, dramatic improvements have been observed among patients receiving some form of AIDS Cocktail treatment. More specifically, patients who received combination therapy have reported decreased viral loads and increased CD4+T cell counts. The life expectancies of HIV patients have become much closer to general mortality rates since the introduction of anti-retroviral therapy; it is felt that the combined treatment not only has enabled an HIV-infected person to live longer, but also improved the patient's overall quality of life with an increase in CD4+T cell counts closer to the level of a healthy individual.

Consequently, the current recommended treatment of HIV is a combination of three or more medications from at least two of the five classes of medications listed above. The general belief based on the cumulated clinical evidence is that by using medications from more than one class, it is more likely to succeed in eliminating the virus. Evidently, a multi-drug regiment can battle HIV at different stages of it replication. We culminate the short exposition of HIV dynamics in this chapter with a discussion of the theoretical foundation for "AIDS Cocktail" treatments. In particular, we show by mathematical modeling and analysis 1) why a cocktail treatment using a combination of several drugs is (disproportionately) more successful in eradicating (or at least managing) the HIV virions in a patient than any single drug (all known to be not perfectly effective), and 2) how we may quantify the effectiveness of the constituent drugs in a cocktail recipe needed for full recovery.

6.1.6 *The principal direction of HIV research*

There are three principal directions of research on HIV:

- Epidemiology – the transmission dynamics of HIV
- Viral Dynamics – interaction of HIV and the immune system
- Intervention – drug treatments, etc.

We will give an introduction to the last two topics to show the working of the mathematical issues of evolution, equilibrium, stability and bifurcation in a biomedical problem of intense interest. Given the clinical importance of HIV and AIDS, a little more space will be devoted to the mathematical aspect of a cocktail drug treatment of AIDS, the clinical aspects of HIV that is not normally discussed in conjunction with HIV dynamics (see [29]).

6.2 HIV Dynamics

6.2.1 *Model formulation*

6.2.1.1 *A three-component model*

As an introduction to modeling the evolution of an HIV population in human, we consider here a relatively simple three unknown model formulated by Herz *et al.* and its modification (see [29]). For this model, we let

    $x(t)$ = population (in unit of biomass) of uninfected CD4+T cells

    $y(t)$ = population (in unit of biomass) of infected CD4+T cells

    $v(t)$ = population (in unit of biomass) of HIV virions in plasma

and model the interaction of HIV virion population and CD4+T cell population by the three (rate of change) equations

$$x' = s - \mu x - kxv, \qquad y' = kxv - ay, \qquad v' = cy - \gamma v - kxv. \tag{6.2.1}$$

They are simple accounting of the activity components that contribute to the rate of change of each population. The first equation is simply a statement of the rate of increase of (or gain in) the healthy CD4+T cell population being a sum of i) the (constant) rate s of such cells being produced in the body, ii) the rate of reduction (or negative gain rate) through natural death proportional to the size $x(t)$ of that cell population with a rate constant μ, and iii) the loss (negative gain) rate of healthy cells from being infected by HIV also proportional to size $x(t)$ but with a "rate constant" kv proportional to the amount of virus present (since there would be no loss in the absence of either population). The accounting of the other two rate equations can be interpreted similarly.

    In the original formulation of Herz *et al.*, there are only the first two terms in the last equation:

$$x' = s - \mu x - kxv, \qquad y' = kxv - ay, \qquad v' = cy - \gamma v. \tag{6.2.2}$$

The very last term, $-kxv$, in (6.2.1) was added later in [29] to that equation to account for the loss of virions (for invading other cells) once they have entered a healthy cell. While this was thought to be a conceptually significant difference, calculations for a standard set of parameter values [18]

Typical Parameter Values [18]

$s = 0.272/\text{day}/\text{mm}^3$

$\mu = 0.00136/\text{day}/\text{mm}^3$

$k = 0.00027/\text{day}/(\text{virion}/\text{mm}^3)$

$a = 0.33/\text{day}/\text{mm}^3$

$c = 50 \text{ virion}/\text{CD4}/\text{day}$

$\gamma = 2.0/\text{day}$

show about a 1% difference in the resulting nontrivial steady state configuration. Hence, the extra term does not appear to have a significant effect on the HIV dynamics. While it might lead to a qualitatively different outcome for some other combinations of parameter values, we will use the simpler original three-component model (6.2.2) in the subsequent developments but will point out the different outcomes by the modified model (6.2.1) whenever appropriate.

6.2.1.2 Simplification at virus saturation

The specific examples of nonlinear autonomous ODE discussed in the previous chapters have been for two state variables allowing us to look at solution trajectories in the phase plane. It is therefore of some interest to begin our investigation of HIV models by studying related two-component models. We may obtain such a simpler model from the three-component model (6.2.2) when the virus population is near the carrying capacity of the environment so that $dv/dt \approx 0$. In that case, the third ODE becomes approximately

$$v \approx cy/\gamma. \tag{6.2.3}$$

We can use this to eliminate v from the remaining two ODE to get

$$x' = s - \mu x - k_0 yx, \qquad y' = y(k_0 x - a), \tag{6.2.4}$$

where $k_0 = kc/\gamma$.

6.2.1.3 Simplification at early stage of infection

In contrast to the scenario of a virus population near saturation described above, we consider here the dynamics of the virus population shortly after the patient first infected by the HIV virus. At that early stage, the virion population is small and the size of the CD4+T cell population is not significantly reduced by virus infection. In that case, the healthy cell population $x(t)$ is approximately its normal steady state size s/μ (which follows from the steady state prior to the infection with $v(t) = 0$ and $x' = s - \mu x = 0$) at least for a short time $t > 0$.

To examine what the immune system may do to defend against the growth of the small amount of HIV virions at this early stage of infection, we set $x(t) = s/\mu$, so that the last two ODE of the three-component model become

$$y' = \frac{ks}{\mu} v - ay, \qquad v' = cy - \left(\gamma + \frac{ks}{\mu}\right) v \tag{6.2.5}$$

resulting in a different two-component model.

6.2.2 Steady states and bifurcation for virus saturation

6.2.2.1 Critical points

In steady state with $x' = y' = 0$, the two-component system (6.2.4) becomes

$$s - (\mu + k_0 y)\, x = 0, \qquad y(k_0 x - a) = 0.$$

The second equation requires either $y_1 = 0$ or $x_2 = a/k_0$. In each case, the other unknown is calculated from the first equation to give the following two critical points:

$$P_1 = (x_1, y_1) = \left(\frac{s}{\mu}, 0\right), \qquad P_2 = (x_2, y_2) = \left(\frac{a}{k_0}, \frac{\mu}{a}\left(\frac{s}{\mu} - \frac{a}{k_0}\right)\right). \qquad (6.2.6)$$

Note that

- P_1 is always on the positive x-axis corresponding to an individual free of HIV virions.
- P_2 can be anywhere in the right half (x, y)-plane (however it being in the 4th quadrant if $sk_0 < a\mu$ is not biologically realistic).

6.2.2.2 *Bifurcation*

Transition point between an acceptable and (biologically) unacceptable P_2 is $s/\mu = a/k_0$ (for which $P_2 = P_1$). A bifurcation is expected at any combination of parameter values that satisfy $s/\mu = a/k_0$. If we take s as the bifurcation parameter, then the ODE (dynamical) system is expected to have a transcritical bifurcation at the bifurcation point

$$s_c = \frac{\mu a}{k_0}, \qquad \text{with} \quad k_0 = \frac{kc}{\gamma}.$$

The system's two critical points coalesce to one as s reaches the bifurcation point s_c and returns to two critical points after passing through s_c.

6.2.2.3 *Stability*

To investigate the stability of the two critical points, we perform the following linear stability analysis with the help of the Jacobian matrix of this two-component model

$$J_2 = \begin{bmatrix} -\mu - k_0 y & -k_0 x \\ k_0 y & k_0 x - a \end{bmatrix}. \qquad (6.2.7)$$

For P_1, the Jacobian is specialized to

$$J_2\left(\frac{s}{\mu}, 0\right) = \begin{bmatrix} -\mu & -k_0 s/\mu \\ 0 & -a + k_0 s/\mu \end{bmatrix} \qquad (6.2.8)$$

which has as its eigenvalues

$$\lambda_1 = -\mu, \qquad \lambda_2 = \frac{-\mu a + k_0 s}{\mu}$$

so that P_1 is an asymptotically stable node if $s < s_c = a\mu/k_0$ and is a saddle point (and unstable) if $s > s_c = a\mu/k_0$.

For P_2, the Jacobian is specialized to

$$J_2\left(\frac{a}{k_0}, \frac{s}{a} - \frac{\mu}{k_0}\right) = \begin{bmatrix} -k_0 s/a & -a \\ -\mu + k_0 s/a & 0 \end{bmatrix} \qquad (6.2.9)$$

which has as its eigenvalues

$$\begin{pmatrix} \lambda_1 \\ \lambda_2 \end{pmatrix} = \frac{k_0 s}{2a} \left[-1 \pm \sqrt{1 - \frac{4\mu a}{k_0 s^2} \left(\frac{s}{\mu} - \frac{a}{k_0} \right)} \right].$$

Hence, P_2 is asymptotically stable if $s > s_c = a\mu/k_0$ and is a saddle point (and unstable) if $s < s_c = a\mu/k_0$.

Summary 1 *For $s < s_c = a\mu/k_0$, P_2 is biologically unacceptable and is unstable while the other (always meaningful) critical point P_1 is asymptotically stable. For $s > s_c$, P_2 is biologically acceptable and is asymptotically stable while the other critical point P_1 is unstable.*

6.2.2.4 *No limit cycle*

For completeness, we should investigate the possible presence of a limit cycle not discernible from the form of the ODE system (6.2.4). Inasmuch as we do not expect (or wish for) a limit cycle, we begin with the negative criteria for this system. It is easy to see that system is not a gradient system given $F_{,y} = -k_0 x$ and $G_{,x} = k_0 y$ so that the test $F_{,y} = G_{,x}$ fails. With

$$F_{,x} + G_{,y} = k_0 \left(x - y \right) - (a + \mu)$$

the Bendixson negative criterion also does not apply in the first quadrant in the (x, y)-plane. However, with $h(x, y) = 1/y$

$$(hF)_{,x} + (hG)_{,y} = - \left(\frac{\mu}{y} + k_0 \right)$$

which is negative for all (x, y) in the first quadrant. Hence Dulac extension of the Bendixson criterion applies so that a periodic solution is impossible in the first quadrant of the (x, y)-plane [30].

6.2.2.5 *Reproductive ratio*

For $s < s_c = a\mu/k_0$ or

$$R = \frac{s/\mu}{a/k_0} < 1,$$

the HIV virion population eventually extinct and the patient recovers. On the other hand, if $R > 1$, the un-infected CD4+T cell tends to a steady state population of $a/k_0 < s/\mu$ (= HIV free CD4+T cell population). Whether the patient can survive other diseases with this reduced level of CD4+T cell population would depend on the actual numerical value of a/k_0 and exposure to other viruses which impose further demands on the immune system.

The quantity

$$R = \frac{ksc}{\mu a \gamma} \equiv \frac{s/\mu}{a/k_0},$$

is known as the *reproductive ratio* for the evolving HIV virion population. Evidently, it is a critical parameter for the virus proliferation.

Remark 3 *Had we used the modified three-component model (6.2.1) instead of the original model of Herz et al., (6.2.2), the outcome would have been qualitatively the same. The only change in the expressions for the critical points, the eigenvalues and the reproductive ratio would be to replace c in k_0 by $(c - a)$ so that k_0 and R become*

$$k_a = \frac{k(c-a)}{\gamma}, \qquad R_a = \frac{s/\mu}{a/k_a},$$

respectively. With $a > 0$, we have $k_a < k_0$ so that the reproductive ratio R_a would be smaller than R had we worked with the modified model. This is not surprising as less virions are available to infect health CD4+T cells in the modified model.

6.2.3   Steady states and bifurcation for a small virus population

6.2.3.1   The early stage dynamics

The two ODE (6.2.5) for the early stage viral dynamics are linear. The exact solution of this linear ODE system with constant coefficients is

$$\begin{pmatrix} y \\ v \end{pmatrix} = c_1 \mathbf{z}^{(1)} e^{\lambda_1 t} + c_2 \mathbf{z}^{(2)} e^{\lambda_2 t}$$

where $\{\lambda_1, z^{(1)}\}$ and $\{\lambda_2, z^{(2)}\}$ are the two eigen-pairs (eigenvalue and eigenvector) of the coefficient matrix A,

$$A = \begin{bmatrix} ks/\mu & -a \\ c & \gamma + ks/\mu \end{bmatrix}$$

of the linear ODE.

6.2.3.2   Reproductive ratio

We can again express the eigenvalues of A in terms of the reproductive ratio R to get

$$\begin{pmatrix} \lambda_1 \\ \lambda_2 \end{pmatrix} = \frac{\sigma}{2}\left[-1 \pm \sqrt{1 + \frac{4a\gamma}{\sigma^2}(R-1)}\right], \qquad \sigma = a + \gamma + \frac{ks}{\mu} > 0$$

with

$$R = \frac{ksc}{\mu a \gamma} \equiv \frac{s/\mu}{a/k_0}.$$

For the virus to multiply (and its population size to grow), the reproductive ratio R will have to be > 1 again. A sufficiently strong immune system at the initial introduction of the HIV virus does offer the possibility of eliminating the virus.

6.2.3.3 Critical point analysis

For the early stage dynamics two-component system (6.2.5), there is only one critical point at the origin in the finite (y, v)-plane. (This is not surprising given that the system is linear.) The Jacobian is just the coefficient matrix A of the linear ODE system. The stability of the only critical point depends on the eigenvalues of A. They have already been found to change qualitatively with the magnitude of the reproductive ratio. The critical point is asymptotically stable if the reproductive ratio R is < 1 and unstable if $R > 1$. In the first case, the patient should recover and be HIV free eventually. For the latter case, the virus proliferate to the point where a lot of CD4+T cells are infected and $x(t)$ is no longer near its HIV free steady state level. The low virion level approximation is no longer applicable.

Remark 4 *Had we used the modified three-component model (6.2.1) instead of the original model of Herz et al., (6.2.2), the outcome would have been qualitatively the same. The only change in the expressions for the eigenvalues and the reproductive ratio would be to replace c in k_0 by $(c - a)$ so that k_0 and R become k_a and R_a, respectively.*

$$k_a = \frac{k(c - a)}{\gamma}, \qquad R_a = \frac{s/\mu}{a/k_a}$$

with the reproductive ratio R_a smaller than R. As such, the initial healthy state is less likely to be unstable.

6.2.4 The three-component model

6.2.4.1 Critical points

Having gained considerable insight to the viral dynamics of HIV from the two special cases investigated in the previous subsections, we now return to the original three-component model (6.2.2) and examine its consequences.

We begin with finding the critical points of the system by setting $x' = y' = v' = 0$ to get

$$s - \mu x - kxv = 0, \qquad kxv - ay = 0, \qquad cy - \gamma v = 0.$$

Again, we can use the last equation to eliminate v from the first two equations to get

$$\frac{s}{\mu} - x\left(1 - \frac{kc}{\mu\gamma}y\right) = 0, \qquad \left(\frac{kc}{\gamma}x - a\right)y = 0, \qquad (6.2.10)$$

in addition to $v = cy/\gamma$. The second of the two equations in (6.2.10) requires

$$y = 0 \quad \text{or} \quad x = a\gamma/kc \equiv a/k_0, \qquad k_0 = kc/\gamma.$$

From these two options, we get the following two critical points for this three-component system:

$$P_1 = \begin{pmatrix} x_1 \\ y_1 \\ v_1 \end{pmatrix} = \begin{pmatrix} s/\mu \\ 0 \\ 0 \end{pmatrix}, \qquad P_2 = \begin{pmatrix} x_2 \\ y_2 \\ v_2 \end{pmatrix} = \begin{pmatrix} a/k_0 \\ (sk_0 - a\mu)/ak_0 \\ c(sk_0 - a\mu)/ak_0\gamma \end{pmatrix}. \qquad (6.2.11)$$

6.2.4.2   Transcritical bifurcation

Note that P_1 corresponds to a virus free state and is always biologically meaningful but P_2 is not when $sk_0 - a\mu < 0$ or

$$R = \frac{s/\mu}{a/k_0} < 1. \tag{6.2.12}$$

The two critical points coalesce when R reaches the critical value 1 of the reproductive ratio (similar to critical role it plays in the previous sections) emerging with two critical points on either side. All indications again point to a transcritical bifurcation at the bifurcation point $R_c = 1$. We complete the confirmation of this inference by a linear stability analysis in the next subsection.

We note parenthetically that had we used the model (6.2.1) instead of (6.2.2), the only change in the results above would be that c is replaced by $(c - a)$ wherever c appears in the final results.

6.2.4.3   Stability

The Jacobian matrix of (6.2.2) for the critical point $P_1 = (s/\mu, 0, 0)$ is

$$J_3 = \begin{bmatrix} -\mu - kv & 0 & -kx \\ kv & -a & kx \\ 0 & c & -\gamma \end{bmatrix}. \tag{6.2.13}$$

When evaluated at $P_1 = (s/\mu, 0, 0)$, J_3 becomes

$$J_3\left(s/\mu, 0, 0\right) = \begin{bmatrix} -\mu & 0 & -ks/\mu \\ 0 & -a & ks/\mu \\ 0 & c & -\gamma \end{bmatrix}. \tag{6.2.14}$$

The three eigenvalues for this matrix are

$$\lambda_1 = -\mu, \quad \begin{pmatrix} \lambda_2 \\ \lambda_3 \end{pmatrix} = \frac{1}{2}\left[-(a+\gamma) \pm \sqrt{(a+\gamma)^2 - 4k_0\gamma\left(\frac{a}{k_0} - \frac{s}{\mu}\right)}\right]. \tag{6.2.15}$$

It follows from the expressions for the three eigenvalues that

- P_1 is asymptotically stable if $s/\mu < a/k_0$ ($R < 1$) in which case P_2 is not biologically meaningful (see (6.2.11) and (6.2.12))
- P_1 is unstable if $s/\mu > a/k_0$ ($R > 1$) in which case P_2 is biologically meaningful (see (6.2.11)).

In the first case, all virus particles will be eliminated eventually and the patient recovers. In the second case, virus population rise to a steady level. More importantly, the healthy CD4+T cells population is reduced from s/μ to a/k_0 ($< s/\mu$) in steady state. Whether that is acceptable depends on what may be required to protect the patient from other diseases present in the environment.

The requirement that P_2 be in the positive octant is precisely the requirement that the reproductive rate R be > 1! For the set of parameter values used by Phillips in his 1996 Science article (see [18, 29] and references therein), P_1 is unstable and we need to investigate the stability of P_2.

The Jacobian matrix for the critical point P_2 is also obtained from (6.2.13) to be

$$[J_3]_{P_2} = \begin{bmatrix} -sk_0/a & 0 & -a\gamma/c \\ -\mu + sk_0/a & -a & a\gamma/c \\ 0 & c & -\gamma \end{bmatrix}, \tag{6.2.16}$$

with $k_0 = kc/\gamma$. The characteristic polynomial for the eigenvalues is the cubic equation

$$\lambda^3 + c_2\lambda^2 + c_1\lambda + c_0 = 0 \tag{6.2.17}$$

where

$$c_2 = \frac{sk_0}{a} + a + \gamma, \qquad c_1 = \frac{sk_0}{a}(a+\gamma), \qquad c_0 = a\gamma\left(\frac{sk_0}{a} - \mu\right). \tag{6.2.18}$$

For Phillips' set of parameter values, the three roots all have a negative real part and P_2 is therefore asymptotically stable. More generally, the three coefficients $\{c_0, c_1, c_2\}$ are all positive for $sk_0 > a\mu$. Hence the Routh-Hurwitz necessary and sufficient conditions [16, 20] for all three roots of the cubic equation to have a negative real parts are satisfied so that P_2 is always asymptotically stable whenever the reproductive ratio $R > 1$ with the virus proliferates toward the steady state population $(sk_0/a - \mu)/(k_0\gamma/c)$.

In that case, there will always be some virus present in the steady state configuration. More importantly, the steady state normal CD4+T cell population is reduced again to a/k_0 the same as that found in (6.2.6) for the approximate two-component model (6.2.4). (This agreement shows the benefit of working with a simpler model to get considerable insight to the HIV dynamics as well as to establish the robustness of the model.) Whether the reduced CD4+T cell population is adequate to deal with the other infectious diseases encountered by the patient would depend on the environment surrounding the patient. More likely than not, some other virus or bacteria would become fatal or further lower the immunity of the body to allow the HIV population to do further damage to the patient's health.

Remark 5 *Had we used the modified three-component model (6.2.1) instead of the original model of Herz et al., (6.2.2), the outcome would again be qualitatively the same. The only change in the expressions for the critical points, the eigenvalues and the reproductive ratio would be to replace c in k_0 by $(c-a)$ so that k_0 and R become k_a and R_a, respectively,*

$$k_a = \frac{k(c-a)}{\gamma}, \qquad R_a = \frac{s/\mu}{a/k_a}.$$

With $a > 0$, we have $k_a < k_0$ so that the reproductive ratio R_a for the modified model (6.2.1) would be smaller than R. The healthy (cured) state would be more likely to be asymptotically stable. This is not surprising as less virions are available to infect health CD4+T cells in the modified model.

6.3 Drug Treatments

6.3.1 *The activities of HIV*

In order to discuss drug treatments, we recall the gp120 proteins on the surface of the HIV virion envelop. One or more of these would bind to a CD4+T cell receptor. After a while, the two membranes fuse and the core of the virion (its RNA genetic material included) enters the CD4+T cell interior. Once inside, the reverse transcriptase units of the virus begin to copy the viral RNA, produce DNA molecules and insert them into the genome of the host cell. After getting the virion's DNA, the host cell is manipulated to make RNA copies of the HIV viral genome and other components of HIV virions. Thousands of new virions are assembled and in due time exit the cell, taking with them the host cell membrane and destroying the host CD4+T cell in the process.

6.3.1.1 *Blocking reverse transcriptase*

One important agent in making more copies of HIV virions is the reverse transcriptase (RT). It translates the viral RNA into DNA. There are now drug such as AZT that binds HIV encoded RT enzymes and prevent the translation of viral RNA into DNA. This effectively stops further synthesis of HIV virions by the disabled RT enzymes and therefore constitutes a treatment. The effect of this type of treatment is mathematically equivalent to reducing the rate of infection of normal CD4+T cell by a factor $1 - r$ where r is a measure of the effectiveness of the drug treatment. In that case, the ODE of the three-component model for the evolution of HIV population are modified to read

$$x' = s \ - \ \mu x \ - \ (1 \ - \ r)kxv, \qquad y' = (1 \ - \ r)kxv \ - \ ay,$$
$$v' = cy \ - \ \gamma v \ - \ (1 \ - \ r)kxv.$$

As indicated in an earlier section, the treatment is not perfectively effective so that we have $r < 1$ at the maximum dosage that can be tolerated by the patient.

6.3.1.2 *Protease inhibitor*

Another important player in making more virions is the virus enzyme protease. Its function is to cleave the gag gene encoded protein into four subunits that coat the genetic material of the virus. There are drugs that inhibit the protease enzymes of the HIV virus from its normal function. When the gag-encoded inner core protein (named Gag) is not cleaved, there is not a coat for the virus' genetic material. Such drugs effectively prevent an infected cell from producing the correct form of the Gag protein. The resulting virions do not function properly and cannot infect new cells. We have then a different kind of treatment for HIV infection. Again, the treatment is effective only q fraction of the time as administered by the physician. As such, the effect of the use of protease inhibitor treatment is mathematically equivalent to reducing the rate of producing HIV virions capable of infecting normal CD4+T cell so that the third ODE of the three-component model for the evolution of HIV is modified to reflect the fractional success of such treatment:

$$x' = s - \mu x - kxv, \qquad y' = kxv - ay,$$
$$v' = (1-q)cy - \gamma v - kxv, \qquad u' = qcy - \gamma u.$$

The treatment does produce a new population of HIV infected CD4+T cells $u(t)$ incapable of producing new HIV virions. The model assumes that $u(t)$ programmed cell death is at the same death rate γu as the normal HIV particles. In any case, the population size $u(t)$ of infected CD4+T cells that are ineffective in producing virions has no bearing on the patient's recovery.

6.3.2 *Combining RT and protease inhibitor*

6.3.2.1 *The model*

Since cure is not assured for any single drug treatment, a treatment plan that combines both types of inhibitors has been used. This combined treatment may be modeled mathematically by the following set of four ODE:

$$x' = s - \mu x - (1-r)kxv, \qquad y' = (1-r)kxv - ay, \qquad (6.3.1)$$
$$v' = (1-q)cy - \gamma v - (1-r)kxv, \qquad u' = qcy - \gamma u.$$

The first 3 equations are simply those of the original three-component model (6.2.2) now modified to include the effects of the two types of drug treatments. The last equation is to account for the drug-disabled HIV infected CD4+T cells (of biomass $u(t)$) resulting from the work of drug inhibited protease enzymes. This last equation is uncoupled from the other three in that the first three can be solved for x, y and v and the result for $y(t)$ can then be used in the last equation to determine $u(t)$.

6.3.2.2 *Critical points*

Similar to the original three-component model without drug treatments, the first three equations can again be shown to have two critical points: $P_1 = (s/\mu, 0, 0)$ and

$$P_1 = \begin{pmatrix} s/\mu \\ 0 \\ 0 \end{pmatrix}, \qquad P_2 = \begin{pmatrix} a/k_{qr} \\ (sk_{qr} - a\mu)/ak_{qr} \\ c_q(sk_{qr} - a\mu)/ak_{qr}\gamma \end{pmatrix} \qquad (6.3.2)$$

where

$$c_q = c(1-q), \qquad k_{qr} = \frac{kc}{\gamma}(1-q)(1-r) = k_0(1-q)(1-r). \qquad (6.3.3)$$

These results are similar to those for the case of no drug treatment. The only changes consist of replacing c and k_0 by c_q and k_{qr}, respectively.

Similar to the case without drug treatments, P_1 is always biologically meaningful but P_2 is not when $sk_{qr} - a\mu < 0$ or

$$R_{qr} = \frac{s/\mu}{a/k_{qr}} = \frac{s/\mu}{a/k_0}(1-q)(1-r) = R(1-q)(1-r) < 1. \qquad (6.3.4)$$

6.3.2.3 Transcritical bifurcation

The two critical points coalesce when R_{qr} reaches the critical value 1 emerging with two critical points on either side. All indications again point to a transcritical bifurcation at the bifurcation point $R_{qr} = 1$. Given

$$R_{qr} = R(1-q)(1-r) \leq R$$

with strict inequality unless $q = r = 0$, P_2 is less likely to be biologically realistic (and will be shown to be more likely to be unstable) even if just one or the other drug is somewhat effective.

6.3.2.4 Stability

The Critical Point P_1 The Jacobian matrix of (6.3.1) for the critical point P_1 is essentially given by (6.2.14) but with k and c replaced by $k_{qr} = k(1-q)(1-r)$ and $c_q = c(1-q)$, respectively. Consequently, the three eigenvalues are given by

$$\lambda_1 = -\mu, \qquad \binom{\lambda_2}{\lambda_3} = \frac{1}{2}\left[-(a+\gamma) \pm \sqrt{(a+\gamma)^2 - 4k_{qr}\gamma\left(\frac{a}{k_{qr}} - \frac{s}{\mu}\right)} \right].$$

It follows from the expressions for the three eigenvalues that

- P_1 is asymptotically stable if $s/\mu < a/k_{qr}$ (corresponding to $R_{qr} < 1$) in which case P_2 is not biologically meaningful.
- P_1 is unstable if $s/\mu > a/k_{qr}$ (corresponding to $R_{qr} > 1$) in which case P_2 is biologically meaningful.

In the first case, all virus particles will be eliminated eventually and the patient recovers. In the second case, virus population rise to a steady level. More importantly, the healthy CD4+T cells population is reduced from s/μ to $a/k_{qr} = a/\{k_0(1-q)(1-r)\}$ $(< s/\mu)$ in steady state. Whether that is adequate for protecting the patient against other infectious disease depends on the patient's surrounding environment. The benefit of at least one potentially effective drug is a larger population of healthy CD4+T cells to constitute a stronger immune system against opportunistic diseases. Note that the requirement that P_2 be in the positive octant is precisely the requirement that the reproductive rate R_{qr} be > 1!

The Critical Point P_2 The Jacobian matrix for the critical point P_2 is also obtained from (6.2.13) by replacing k and c by $k_{qr} = k(1-q)(1-r)$ and $c_q = c(1-q)$, respectively. Consequently, the characteristic polynomial for the eigenvalues is again given by a cubic equation similar to (6.2.17) but with k_0 replaced by k_{qr} in the three coefficients of the polynomial.

In general, the three coefficients of the cubic polynomial for eigenvalues are all positive for $sk_{qr} > a\mu$. Hence the Routh-Hurwitz necessary and sufficient conditions for all three roots to have a negative real part are satisfied so that P_2 is asymptotically stable whenever the reproductive ratio $R_{qr} > 1$ with the virus proliferates toward the steady state population $(sk_{qr}/a - \mu)/(k_{qr}\gamma/c_q)$.

In that case, there will always be some virus present in the steady state configuration. More importantly, the steady state normal CD4+T cell population is reduced to $a/k_{qr} > a/k_0$ (and $\gg a/k_0$ if q and r are close to 1). Whether the drug-modified CD4+T cell population is adequate to deal with the other infectious diseases encountered by the patient would depend on the environment surrounding the patient. With p and q positive, it is less likely now for some other virus or bacteria to do further damage to the patient's health.

Remark 6 *Had we used the modified three-component model (6.2.1) instead of the original model of Herz et al., (6.2.2), the outcome would again be qualitatively the same. The only change in the expressions for the critical points, the eigenvalues and the reproductive ratio would be to replace c_q by $(c_q - a)$ so that k_{qr} and R_{qr} become $\bar{k}_{qr}$ and $\bar{R}_{qr}$, respectively,*

$$\bar{k}_{qr} = \frac{k(1-r)(c_q - a)}{\gamma}, \qquad \bar{R}_{qr} = \frac{s/\mu}{a/\bar{k}_{qr}}. \tag{6.3.5}$$

With $a > 0$, we have $\bar{k}_{qr} < k_{qr}$ so that the reproductive ratio $\bar{R}_{qr}$ for the modified model (6.2.1) would be smaller than R_{qr}. The health (cured) state would be more likely to be asymptotically stable. This is not surprising as less virions are available to infect health CD4+T cells in the modified model.

6.3.3 *Effectiveness of AIDS cocktail treatments*

Recall that $P_1 = (s/\mu, 0, 0)$ is asymptotically stable if and only if

$$R_{qr} < 1 \tag{6.3.6}$$

where

$$R_{qr} = \frac{s/\mu}{a/k_{qr}} = R(1-q)(1-r) \le R.$$

Consequently, we conclude that with one or both drugs effective

- the virus is less "viral" and less proliferative
- the patient is more likely to be cured, and
- the cocktail AIDS treatment is superior to either drug treatment alone!

Upon writing the *condition for cure* (6.3.6) as

$$(1-q)(1-r) < \frac{a/k_0}{s/\mu}. \tag{6.3.7}$$

We see that

Conclusion 1 *The patient should recover eventually if either drug is perfectly effective so that $r = 1$ or $q = 1$ (and we do not need the other drug).*

For a fixed $r < 1$, we may write (6.3.7) as an expression for the effectiveness index q for the other drug:

$$q > 1 - \frac{1}{1-r} \frac{a/k_0}{s/\mu} \equiv q_{cr}. \tag{6.3.8}$$

In that case, we can take (6.3.8) as a condition on a threshold value for q for cure:

Conclusion 2 *For an RT blocker drug that is not perfectly effective so that $0 < r < 1$, the second drug needs only to have an effectiveness index q greater than q_{cr} for the patient to become healthy again eventually (with $q < 1$ and $r < 1$).*

Since the expression R_{qr} is symmetric in q and r, we have immediately also a similar condition for a fixed $q < 1$:

Conclusion 3 *For a protease inhibiting drug that is not perfectly effective so that $0 < q < 1$, the second drug needs only to have an effectiveness index r greater than r_{cr},*

$$r_{cr} = 1 - \frac{1}{1-q} \frac{a/k_0}{s/\mu}, \tag{6.3.9}$$

for the patient to become virus-free eventually.

6.3.4   *On the modified three-component model*

Note that the symmetry between the two drugs would be loss had we used the modified three-component model (6.2.1). For that model, (6.3.6) is replaced by $\bar{R}_{qr} < 1$ (see (6.3.5)) or

$$(1-r)\left[c(1-q) - a\right] < \frac{a\mu\gamma}{sk}. \tag{6.3.10}$$

For a fixed $r < 1$, the condition (6.3.10) may be written as a condition on the effectiveness of the protease inhibiting drug for virus-free recovery

$$q > 1 - \left[\frac{a/k_0}{(1-r)s/\mu} + \frac{a}{c} \right] = 1 - \frac{a/k_0}{(1-r)s/\mu}\left[1 + \frac{a}{c}\frac{(1-r)s/\mu}{a/k_0} \right] \equiv \tilde{q}_{cr}. \tag{6.3.11}$$

For a fixed $q < 1$ on the other hand, the condition (6.3.10) may be written as a condition on the effectiveness of the RT blocking drug for virus-free recovery

$$r > 1 - \frac{a/k_0}{(1 - a/c_q)(1-q)s/\mu} \equiv \tilde{r}_{cr}. \tag{6.3.12}$$

Given a fixed effectiveness measure q of a protease inhibiting drug, the new effective cure index $\tilde{r}_{cr}$ for the RT blocking drug in the modified model is smaller given a smaller denominator in the second term in (6.3.12) compared to the corresponding denominator in (6.3.9). In other words, the RT blocking drug can have a lower level of effectiveness and still renders the patient virus-free. Depending on the magnitude of a/c_q, the reduction may or may not be significant.

The situation is similar for a fixed level of effectiveness r for the RT blocking drug. The level of effectiveness for the protease inhibiting drug is lower for the modified model given the additional positive term inside the brackets of the expression for $\tilde{q}_{cr}$ in (6.3.11) compared to the corresponding expression for q_{cr} in (6.3.8). The difference again may or may not be significant depending on the magnitude of a/c.

Unlike (6.3.7), the new cure condition for the modified model is not symmetric in r and q since there is no corresponding symmetry in (6.3.10). With $a/c_q < 1$, we may write

$$\frac{1}{1 - a/c_q} = 1 + \frac{a}{c_q} + O\left(\frac{a^2}{c_q^2}\right)$$

and

$$r > 1 - \frac{a/k_0}{(1-q)s/\mu}\left\{1 + \frac{a}{c}\frac{1}{1-q} + O\left(\frac{a^2}{c_q^2}\right)\right\} = \tilde{r}_{cr}.$$

The inequality

$$\frac{1}{1-q} > 1 > \frac{(1-r)s/\mu}{a/k_0},$$

holds generally for a stable P_1 (with $s/\mu < a/k_0$ for the case without drug treatment).

The asymmetry between the two types of drugs in the modified model may have some clinical significance in the administration of a cocktail treatment. Given an imperfect protease inhibiting drug with effectiveness q_1, the effectiveness r_1 of the companion RT blocking drug in a cocktail AIDS treatment required for eventual cure is considerably lower than the analogous requirement on the effectiveness q_2 of a protease inhibitor for a given RT blocker with effectiveness $r_2 = q_1$. In this sense, there may be qualitative and quantitative different predictions from the two three-component models even if the differences are small for the special case of the Phillips type patients.

Chapter 7

Index Theory, Bistability and Feedback

7.1 Poincaré Index for Planar Systems

Our discussion in Chapter 5 on nonlinear autonomous systems is far from complete. For example, we have to rely on a linear stability analysis for such systems to determine the stability and local behavior of trajectories starting from a point near the critical point. This is much more limited than what we can get for a single equation by the graphical method where we know what happen to trajectories starting from locations near an unstable critical point. Furthermore, the finding of a center by linear stability analysis is not conclusive. While there are advanced methods such as the Lyapunov stability analysis that could do more to determine stability, we limit further discussion to the index theory, bistability and feedback to expand the scope of our general effort of critical point analysis.

Index theory for planar systems provides another approach to determine the existence of a limit cycle and delineates when a linear center is a true center. A planar autonomous system,

$$x' = f(x, y), \quad y' = g(x, y),$$

specifies at every point of the (x, y)-plane (the phase plane) a direction vector. Along a closed curve C in the phase plane, not necessarily a trajectory, we have at every point on this curve one such vector specifying the direction (of increasing time) of the trajectory passing through that point. As we move from a point to a neighboring point, say counter-clockwise, the direction vector generally changes. By continuing moving in the same direction along the closed curve, we eventually return to the (original) starting point with the vector pointing in the same direction. Assuming that there is no critical point of the ODE system on C, the angle φ made by the direction vector and the x-axis changes continuously through this process by an integer multiple of 2π by the time it returns to the starting position.

Definition 14 *The Poincaré Index I_C of a simple closed curve C (that does not intersect itself and has no critical point on the curve) is the number of times the vector field on C rotates counter-clockwise after traveling around C once in a counter-clockwise direction.*

As indicated at the end of the paragraph before the definition above, I_C is always an integer. In addition, the following properties of I_C can be shown from the definition:

(1) The index of a periodic orbit C is $+1$ (since the direction vector is always tangent to the trajectory and rotates around once to get back to the starting point).

(2) A loop C (not necessarily a trajectory) around a saddle point has index -1 (as can be seen from Figure 7.1).

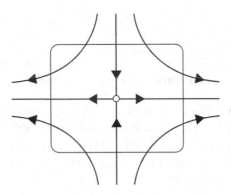

Fig. 7.1   Circuit around a saddle point.

Exercise 17 *Show that the Poincaré index I_C of a closed curve surrounding only one critical point of a planar linear autonomous system is $+1$ if the critical point is not a saddle point.*

(3) If a closed curve C_1 can be morphed (deformed continuously) into another closed curve C_2 without crossing a critical point of the ODE system, then their Poincaré indices are the same. (Indices of two close curves can only differ by an integer. Hence two curves differ ever so slightly cannot differ by so much and must therefore be the same.)

(4) The Poincaré index of a close curve that encloses no critical point must be zero. (By the property above, we can morph the given closed curve to as small a loop as we wish with the new loop having the same Poincaré index. But for a very small loop, the director vector is nearly the same at every point on the loop and hence the Poincaré index is (as close to zero as we wish and hence) zero.

(5) When all direction vectors around a given closed curve reverse their direction (equivalent to reversing time by changing from t to $-t$), the index of the close curve does not change.

Theorem 10 *The index of a closed curve surrounding a number of critical points of the ODE system is the sum of the indices of the curves around each critical point.*

Proof (Exercise) □

The following example shows how index theory can help to settle the question of existence or non-existence of a limit cycle

Example 24 $x' = \frac{1}{2}xy^2 + x^2 - 1 = f(x, y), \qquad y' = -2y + yx^2 = g(x, y).$

The system is not a gradient system since $f_{,y} = xy \neq g_{,x} = -2xy$. For the Dulac (also known as the Bendixson) criterion, we have

$$f_{,x} + g_{,y} = \frac{1}{2}y^2 + 2x - 2 + x^2 = \left(\frac{y}{\sqrt{2}}\right)^2 + (x+1)^2 - 3.$$

Hence the system does not have a limit cycle lying completely inside or completely outside the circle (in terms of the variables $y/\sqrt{2}$ and x)

$$\left(\frac{y}{\sqrt{2}}\right)^2 + (x+1)^2 = 3 \qquad (7.1.1)$$

centered at the point $(-1, 0)$.

To eliminate the remaining possibility of one or more limit cycles straddling the two complementary regions (inside and outside the aforementioned circle), we note the following two properties of the given system:

- The system has only four critical points at $(\pm 1, 0)$ and $(-\sqrt{2}, \pm\sqrt[4]{2})$ with no critical point along $x = \sqrt{2}$ (since $x' = (y^2/\sqrt{2}) + 1 > 0$ for all y). Only $(1, 0)$ is outside the circle (7.1.1),
- From the Jacobian matrix

$$J = \begin{bmatrix} \frac{1}{2}y^2 + 2x & xy \\ 2xy & -2 + x^2 \end{bmatrix},$$

  we see that the three critical points $(-\sqrt{2}, \pm\sqrt[4]{2})$ and $(1, 0)$ are saddle points and the remaining critical point $(-1, 0)$ is an asymptotically stable node.
- A periodic solution must encircle only $(-1, 0)$ by Theorem 10 and the first property of the system above.
- Any initial point $(x_0, 0)$ with $x_0 > -1$ generates a trajectory moving away from $(1, 0)$ along the x-axis that tends to $(-1, 0)$.
- Any initial point $(x_0, 0)$ with $x_0 < -1$ also generates a trajectory along the x-axis that tends to $(-1, 0)$ along the x-axis.

It follows that no closed trajectory of the given autonomous system (straddling the two complementary regions or not) can enclosed $(-1, 0)$. Such a closed trajectory would have to cross the x-axis where there is always another trajectory along $y = 0$ and trajectories of the same autonomous system do not cross.

If there should be a limit cycle (or periodic orbit) straddling the two complementary regions, then it can only encircle one of the two saddle points $(-\sqrt{2}, \pm\sqrt[4]{2})$. But this is not possible since the index of a limit cycle (or periodic orbit) must be $+1$ by property (1) after the definition of the Poincaré index while the index of a saddle point is -1. Hence there is no limit cycle for this ODE system.

7.2   Bistability — A Dimerized Reaction

Critical points and their stability have been found useful in the study of interacting biochemical substances. In the models involving nonlinear interactions investigated so far, there are usually several critical points but typically only one of them is biologically meaningful (for non-negativity of size or concentration) and asymptotically stable. One of the important phenomena of interest in biology is the presence of more than one asymptotically stable critical points giving rise to uncertainty in the steady state behavior of the interacting populations. The obvious question, both interesting and important, is to which stable critical point do the populations evolve eventually? If the populations are in one stable configuration, what does it take to move them to another stable configuration?

In this section, we discuss a simple example of two interacting populations for which there are two asymptotically stable critical points. Another example of bistability in competing populations with ecological implications will be analyzed in the following section.

7.2.1   *Interaction involving a dimer*

The first mathematical model of two interacting populations with more than one asymptotically stable critical points is a simple enzyme-substrate biochemical interaction. For this simple model, we have two substrate and (substrate-enzyme complex) concentrations $S(t)$ and $C(t)$ interacting according to the following nonlinear planar autonomous system:

$$S' = -\alpha S + \beta C \tag{7.2.1}$$

$$C' = -\gamma C + \frac{\mu S^2}{1 + S^2}. \tag{7.2.2}$$

The two equations follow from the generally accepted Michaelis and Menten's Model of enzyme kinetics but specialized to the case of abundant enzymes (relative to the substrates) so that we may take the enzyme concentration E to be approximately $E(0) = E_0$ at least for a short while (see [24]):

$$S' = -k_{se}E_0 S + k_{rb}C. \tag{7.2.3}$$

$$C' = -(k_{cp} + k_{rb})C + k_{se}E_0 S^*. \tag{7.2.4}$$

In the last term of (7.2.4), we introduce notion of a dimer which is formed by two copies of the substrate molecule with the contribution of S^* to the synthesis of the substrate-enzyme complex taken to be equivalent to $S^2/(1 + S^2)$. In any case, the system (7.2.3)–(7.2.4) is of the form (7.2.1)–(7.2.2) motivating our interest in the latter.

7.2.2 Fixed points and bifurcation

For the critical points $(S^{(k)}, C^{(k)})$ of the system (7.2.1)–(7.2.2), we have

$$C^{(k)} = \frac{\alpha}{\beta} S^{(k)}, \qquad C^{(k)} = \frac{\mu}{\gamma} \frac{\left[S^{(k)}\right]^2}{1 + \left[S^{(k)}\right]^2}.$$

These two relations generally give rise to three fixed points,

$$S^{(1)} = 0, \qquad \begin{pmatrix} S^{(2)} \\ S^{(3)} \end{pmatrix} = \sigma \mp \sqrt{\sigma^2 - 1}, \tag{7.2.5}$$

$$C^{(k)} = \frac{\alpha}{\beta} S^{(k)}, \tag{7.2.6}$$

with

$$2\sigma = \frac{\mu\beta}{\alpha\gamma} > 0. \tag{7.2.7}$$

The following observations suggest the possibility of a bifurcation at the bifurcation point

$$\sigma_c = 1.$$

- There is only one real critical point $(S^{(1)}, C^{(1)}) = (0,0)$ if $0 < \sigma < 1$.
- There is a second critical point $(S^{(2)}, C^{(2)}) = (\sigma, \mu/2\gamma)$ if $\sigma = \sigma_c = 1$.
- There are three critical points given by (7.2.5) and (7.2.6) if $\sigma > 1$.

With the number of fixed points changing from 3 to 1 as σ decreases from $\sigma > 1$ through the bifurcation point to $\sigma < 1$, the bifurcation might appear to be of the pitchfork type. However, a plot of a bifurcation diagram (even before determining the stability of these fixed points) would show that the bifurcation is *saddle-node* with $(S^{(2)}, C^{(2)})$ and $(S^{(3)}, C^{(3)})$ for $\sigma > 1$ coalescing at $\sigma = 1$ and disappearing for smaller values of σ. All the while the fixed point $(S^{(1)}, C^{(1)}) = (0,0)$ is unaffected by the change in σ (in both its location and (as we shall see) its stability).

7.2.3 Linear stability analysis

For the purpose of illustrating bistability, we are interested principally in the range $\sigma > 1$ for which we have three distinct fixed points. For the stability of these fixed points, we need the Jacobian matrix for the problem:

$$J_k = \begin{bmatrix} -\alpha & \beta \\ \dfrac{2\mu S^{(k)}}{\left(1 + \left[S^{(k)}\right]^2\right)^2} & -\gamma \end{bmatrix}.$$

7.2.3.1 Stability of $(S^{(1)}, C^{(1)})$

For the critical point $(0,0)$, the Jacobian becomes

$$J_1 = \begin{bmatrix} -\alpha & \beta \\ 0 & -\gamma \end{bmatrix}.$$

The corresponding eigenvalues are

$$\lambda_1^{(1)} = -\alpha < 0, \qquad \lambda_2^{(1)} = -\gamma < 0.$$

Proposition 3 *The critical point $(S^{(1)}, C^{(1)}) = (0,0)$ is asymptotically stable for any positive value of σ*

Since there is no change in the stability of $(S^{(1)}, C^{(1)}) = (0,0)$ for all σ, the critical point is not involved in any bifurcation that may be in the reaction.

7.2.3.2 Stability of $(S^{(k)}, C^{(k)})$ for $k > 1$

For the other two critical points with $S^{(k)} = \sigma \mp \sqrt{\sigma^2 - 1}$, the Jacobian matrix is

$$J_k = \begin{bmatrix} -\alpha & \beta \\ \frac{\mu/2\sigma^2}{\sigma \mp \sqrt{\sigma^2-1}} & -\gamma \end{bmatrix} = \begin{bmatrix} -\alpha & \beta \\ \frac{\mu}{2\sigma^2}\left(\sigma \pm \sqrt{\sigma^2-1}\right) & -\gamma \end{bmatrix} \quad (k = 2, 3). \quad (7.2.8)$$

For $\sigma > 1$, the two "larger" eigenvalues $\lambda_1^{(k)}$ of the Jacobian matrix (7.2.8) for the critical points $(S^{(k)}, C^{(k)})$, $k = 2$ and 3, are

$$\begin{pmatrix} \lambda_1^{(2)} \\ \lambda_1^{(3)} \end{pmatrix} = \frac{1}{2}\left\{ (-\alpha - \gamma) + \sqrt{(\alpha - \gamma)^2 + \frac{2\beta\mu}{\sigma^2}\left(\sigma \pm \sqrt{\sigma^2-1}\right)} \right\}$$

while the two "smaller" eigenvalues $\lambda_2^{(k)}$ are

$$\begin{pmatrix} \lambda_2^{(2)} \\ \lambda_2^{(3)} \end{pmatrix} = \frac{1}{2}\left\{ (-\alpha - \gamma) - \sqrt{(\alpha - \gamma)^2 + \frac{2\beta\mu}{\sigma^2}\left(\sigma \pm \sqrt{\sigma^2-1}\right)} \right\}.$$

For the critical point $(S^{(2)}, C^{(2)}) = (\sigma - \sqrt{\sigma^2-1})(1, \alpha/\beta)$, we have with $2\beta\mu/\sigma = 4\alpha\gamma$

$$\lambda_1^{(2)} > \frac{1}{2}\left\{ (-\alpha - \gamma) + \sqrt{(\alpha - \gamma)^2 + \frac{2\beta\mu}{\sigma}} \right\}$$

$$= \frac{1}{2}\left\{ (-\alpha - \gamma) + \sqrt{(\alpha - \gamma)^2 + 4\alpha\gamma} \right\} = 0$$

and consequently the following result:

Proposition 4 *The critical point $(S^{(2)}, C^{(2)})$ is unstable for $\sigma > 1$.*

For the critical point $(S^{(3)}, C^{(3)})$, we have

$$-\frac{1}{2}(\alpha + \gamma) < \lambda_1^{(3)} \tag{7.2.9}$$

$$< \frac{1}{2}\left\{(-\alpha - \gamma) + \sqrt{(\alpha - \gamma)^2 + \frac{2\beta\mu}{\sigma}}\right\} = 0.$$

Similarly, we have

$$\lambda_2^{(3)} < \frac{1}{2}\left\{(-\alpha - \gamma) - \sqrt{(\alpha - \gamma)^2}\right\}$$

$$< \begin{cases} -\alpha & (\alpha < \gamma) \\ -\gamma & (\gamma < \alpha) \end{cases} < 0.$$

Altogether, we have proved the following proposition:

Proposition 5 *The critical point $(S^{(3)}, C^{(3)})$ is asymptotically stable for $\sigma > 1$ by a linear stability analysis.*

In summary, we have for the range $\sigma > 1$ (and for the biologically realistic range of values of α and γ) two *asymptotically stable* critical points at $(S^{(1)}, C^{(1)})$ and $(S^{(3)}, C^{(3)})$. This establishes the existence of bistability for this reaction. For an initial condition (S_0, C_0) sufficiently close to the linearly stable critical point $(S^{(1)}, C^{(1)}) = (0, 0)$, it seems reasonable to expect the corresponding concentrations to evolve toward $(S^{(1)}, C^{(1)})$. If (S_0, C_0) is sufficiently close to the linearly stable critical point $(S^{(3)}, C^{(3)})$, the subsequent concentrations are expected to evolve toward $(S^{(3)}, C^{(3)})$. But for (S_0, C_0) not close to either linearly stable critical points, toward which would the solution of the autonomous system evolve? For an answer to this important question, we need more sophisticated mathematical tool for determining the global portrait of the solution trajectories to be found in more advanced texts in dynamical systems such as [17].

With the two distinct critical points $(S^{(2)}, C^{(2)})$ and $(S^{(3)}, C^{(3)})$ exhibiting different stability types for $\sigma > 1$, coalescing to same fixed point for $\sigma = 1$ and transitioning to no fixed point for smaller σ, the reaction has a saddle-node bifurcation with $\sigma_c = 1$ being the bifurcation point.

7.3   Bistability — Two Competing Populations

Many other biological processes involve the phenomenon of bistability. While the mathematics involved is similar to that used in the simple example in the previous section, the description of most of these processes would require considerable background information in biology and biochemistry. For example, cytoplasmic calcium concentration, denoted by $\left[Ca^{2+}\right]_i$, is of great importance for the life and death of cells and bistability occurs in conjunction with its concentration in many cells, including bullfrog sympathetic ganglion neurons and pancreatic beta cells (see [5] and references therein). Another group of rather complex biological processes exhibiting bistability is the area of morphogenesis in developmental biology. One simpler one is the role of cytoplasmic calcium concentration ($[Ca^{2+}]_i$ again) in the folding of a sheet of cells in the shaping of tissues and organs. The effects of $[Ca^{2+}]_i$ on the Newtonian dynamics of the shape changes of a cell sheet lead to the possibility of bistability on the sheet shape in a certain range of system parameter values (see [22] and references therein). Instead of embarking on a description of the modeling and analysis of these or other rather complex biological processes at the cellular level, we formulate and analyze in this section a model of two competing populations which is only a natural extension of those we have become totally familiar in the earlier chapters.

7.3.1 Leopards and hyenas

Leopards and hyenas both feed on smaller animals such impalas, gazelles, dikdiks and other members of the antelope family (and other small animals cohabiting in the same habitat). If it is the only predator for these smaller animals, each (leopards or hyenas) population would multiply subject to some size limiting growth rate which we will take to be the logistic growth rate. When both co-exist in the same habitat, as in the East Africa game reserves (e.g., Serengeti National Park), they would have to compete with each other for the same preys. While they do not prey on each other (except for occasional fights for the same kills), the presence of each inhibits the growth of the other. Assuming a sizable population of preys, the growth of the two *competing* populations is adequately modeled by the following two first order nonlinear differential equations:

$$L' = aL - \alpha L^2 - bLH \equiv g(L,H), \qquad (7.3.1)$$
$$H' = cH - \gamma H^2 - dLH \equiv h(L,H).$$

As usual, we limit

- α, γ, b and d to be all positive;
- a and c to be generally in the interval $(0,1)$, i.e., $0 < a,c < 1$ (allowing $0 < a,c < 2$ in extreme cases).

Note that without the two interaction terms ($b = d = 0$), the leopards and hyenas simply grow according to their own logistic growth rate. The terms proportional to $-LH$ constitute the negative effect (inhibition) on the growth of the competing populations.

Exercise 18 *Analyze the linear stability of critical points for the special case of* $\alpha = \gamma = 0$ *and* $b = d$.

7.3.2 Critical points

At a critical point, we have $L' = H' = 0$ so that L and H are both constant in time. In that case, the system (7.3.1) becomes

$$g(L, H) = L(a - \alpha L - bH) = 0, \tag{7.3.2}$$
$$h(L, H) = H(c - \gamma H - dL) = 0.$$

The four solutions of (7.3.2) are

$$(L^{(1)}, H^{(1)}) = (0,0), \quad (L^{(2)}, H^{(2)}) = \left(\frac{a}{\alpha}, 0\right), \tag{7.3.3}$$

$$(L^{(3)}, H^{(3)}) = \left(0, \frac{c}{\gamma}\right),$$

$$(L^{(4)}, H^{(4)}) = \left(\frac{bc - a\gamma}{db - \alpha\gamma}, \frac{da - \alpha c}{db - \alpha\gamma}\right).$$

Since $a > 0$ and $c > 0$, the critical points $(L^{(2)}, H^{(2)})$ and $(L^{(3)}, H^{(3)})$ do *not* degenerate to coincide with $(L^{(1)}, H^{(1)}) = (0,0)$ or with each other. (Otherwise, there may be bifurcation in the model.) The possible coalescing of $(L^{(4)}, H^{(4)})$ and one or more other critical points will be investigated after we determine the stability of some of the critical points.

7.3.3 Linear stability analysis

To determine the stability of the critical points, we need the Jacobian matrix for the system (7.3.1):

$$J_k = \begin{bmatrix} a - 2\alpha L^{(k)} - bH^{(k)} & -bL^{(k)} \\ -dH^{(k)} & c - 2\gamma H^{(k)} - dL^{(k)} \end{bmatrix}.$$

7.3.3.1 Stability of $(L^{(1)}, H^{(1)})$

For the critical point $(L^{(1)}, H^{(1)}) = (0,0)$, the Jacobian becomes

$$J_1 = \begin{bmatrix} a & 0 \\ 0 & c \end{bmatrix}.$$

The two eigenvalues are

$$\lambda_1^{(1)} = a > 0, \quad \lambda_2^{(1)} = c > 0.$$

Proposition 6 *The critical point $(L^{(1)}, H^{(1)}) = (0,0)$ is unstable independent of the values of all the other parameters.*

The fact that $(L^{(1)}, H^{(1)}) = (0,0)$ do not change its stability for all parameter values (given the restrictions $0 < a, c < 1$) shows that this critical point is not involved in any bifurcation of the evolution of the two competing populations.

Stability of $(L^{(2)}, H^{(2)})$ For the critical point $(L^{(2)}, H^{(2)}) = (a/\alpha, 0)$, the Jacobian becomes

$$J_2 = \begin{bmatrix} -a & -ba/\alpha \\ 0 & c - da/\alpha \end{bmatrix}.$$

The two eigenvalues are

$$\lambda_1^{(2)} = -a < 0, \qquad \lambda_2^{(2)} = c - \frac{da}{\alpha} = -\frac{da - \alpha c}{\alpha}.$$

Proposition 7 *The critical point $(L^{(2)}, H^{(2)}) = (a/\alpha, 0)$ is*
   i) unstable if $a/\alpha < c/d$,
   ii) asymptotically stable if $c/d < a/\alpha$.

The fact that $(L^{(2)}, H^{(2)}) = (a/\alpha, 0)$ switches its stability as the parameter $\mu = c\alpha - ad$ changes sign suggests the possibility of a bifurcation of the evolution of the two competing populations at $\mu = 0$. This possibility will be investigated after we have determined the stability of the remaining critical points.

7.3.3.2   Stability of $(L^{(3)}, H^{(3)})$

The situation for the critical point $(L^{(3)}, H^{(3)}) = (0, c/\gamma)$ is analogous to that of $(L^{(2)}, H^{(2)})$ with the role of b, c, and γ interchanged with that of d, a and α, respectively. In particular, the Jacobian for $(L^{(3)}, H^{(3)})$ becomes

$$J_3 = \begin{bmatrix} a - bc/\gamma & 0 \\ -dc/\gamma & -c \end{bmatrix}.$$

The two eigenvalues are

$$\lambda_1^{(3)} = a - \frac{bc}{\gamma} = -\frac{bc - \gamma a}{\gamma}, \qquad \lambda_2^{(3)} = -c < 1$$

Proposition 8 *The critical point $(L^{(3)}, H^{(3)}) = (0, c/\gamma)$ is*
   i) unstable if $c/\gamma < a/b$,
   ii) asymptotically stable if $a/b < c/\gamma$.

Similar to the situation encountered in the stability of $(L^{(2)}, H^{(2)})$, the fact that $(L^{(3)}, H^{(3)}) = (0, c/\gamma)$ switches its stability as the parameter $\zeta = a\gamma - bc$ changes sign suggests the possibility of a bifurcation at $\zeta = 0$. This possibility will be discussed after we have determined the stability of $(L^{(4)}, H^{(4)})$.

7.3.4   Existence of bistability

If $c/d < a/\alpha$ and $a/b < c/\gamma$, both $(L^{(2)}, H^{(2)}) = (a/\alpha, 0)$ and $(L^{(3)}, H^{(3)}) = (0, c/\gamma)$ are asymptotically stable. It follows that bistability exists for our competing populations of leopards and hyenas. Having established the existence of this bistability, there remains two questions of interest:

1) Is $(L^{(4)}, H^{(4)})$ stable or unstable when $(L^{(2)}, H^{(2)})$ and $(L^{(3)}, H^{(3)})$ are asymptotically stable?

2) Are there other bistability configurations for our model?

We address the first question in the next subsection assuming the conditions

$$\frac{c}{d} < \frac{a}{\alpha} \quad \text{and} \quad \frac{a}{b} < \frac{c}{\gamma} \tag{7.3.4}$$

to ensure the asymptotic stability of $(L^{(2)}, H^{(2)})$ and $(L^{(3)}, H^{(3)})$.

7.3.4.1 Stability of $(L^{(4)}, H^{(4)})$

For the fixed point $(L^{(4)}, H^{(4)}) = (bc-a\gamma, da-\alpha c)/(db - \alpha\gamma)$, the Jacobian becomes

$$J_4 = \begin{bmatrix} -\alpha L^{(4)} & -bL^{(4)} \\ -dH^{(4)} & -\gamma H^{(4)} \end{bmatrix},$$

after using the expressions for $L^{(4)}$ and $H^{(4)}$ to simplify the elements of J_4. The two eigenvalues are

$$\begin{pmatrix} \lambda_1^{(4)} \\ \lambda_2^{(4)} \end{pmatrix} = \frac{1}{2} \left\{ \left(-\alpha L^{(4)} - \gamma H^{(4)} \right) \pm \sqrt{(\alpha L^{(4)} - \gamma H^{(4)})^2 + 4bdL^{(4)}H^{(4)}} \right\}.$$

We are interested here in the special case when the inequalities in (7.3.4) hold so that both $(L^{(2)}, H^{(2)})$ and $(L^{(3)}, H^{(3)})$ are asymptotically stable. Note that from (7.3.4) follows the inequality $(c/d)(a/b) < (a/\alpha)(c/\gamma)$ or

$$\alpha\gamma < bd \tag{7.3.5}$$

given $ac > 0$. Important consequences of (7.3.5) include

- $L^{(4)} > 0$ and $H^{(4)} > 0$ so that the fixed point $(L^{(4)}, H^{(4)})$ is biologically realizable.
- $4bdL^{(4)}H^{(4)} > 4\alpha\gamma L^{(4)}H^{(4)}$ so that

$$\sqrt{(\alpha L^{(4)} - \gamma H^{(4)})^2 + 4bdL^{(4)}H^{(4)}} > \sqrt{(\alpha L^{(4)} + \gamma H^{(4)})^2}. \tag{7.3.6}$$

Proposition 9 *If either inequality of (7.3.4) holds along with (7.3.5), the fixed point $(L^{(4)}, H^{(4)})$ is unstable.*

Proof It follows from (7.3.6) that

$$\lambda_1^{(4)} > \frac{1}{2}\left\{ \left(-\alpha L^{(4)} - \gamma H^{(4)} \right) + \sqrt{(\alpha L^{(4)} + \gamma H^{(4)})^2} \right\} = 0. \qquad \square$$

Corollary 2 *If both inequalities of (7.3.4) hold, the fixed point $(L^{(4)}, H^{(4)})$ is unstable.*

Proof The two inequalities in (7.3.4) imply (7.3.5) so that Proposition 9 applies.
$$\square$$

It is of some interest to investigate the stability of $(L^{(4)}, H^{(4)})$ for the inequalities

$$\frac{c}{d} > \frac{a}{\alpha} \quad \text{and} \quad \frac{a}{b} > \frac{c}{\gamma}, \tag{7.3.7}$$

instead of (7.3.4). From (7.3.7), we have $(c/d)(a/b) > (a/\alpha)(c/\gamma)$ so that

$$\alpha\gamma > bd. \tag{7.3.8}$$

As consequences of (7.3.7) and (7.3.8), we have

- $L^{(4)} > 0$ and $H^{(4)} > 0$ so that the fixed point $(L^{(4)}, H^{(4)})$ is biologically realizable, and
- $\sqrt{(\alpha L^{(4)} - \gamma H^{(4)})^2 + 4bdL^{(4)}H^{(4)}} < \sqrt{(\alpha L^{(4)} + \gamma H^{(4)})^2}$

and therewith

$$\lambda_1^{(4)} < \frac{1}{2}\left\{ \left(-\alpha L^{(4)} - \gamma H^{(4)}\right) + \sqrt{(\alpha L^{(4)} + \gamma H^{(4)})^2} \right\} = 0$$

and

$$\lambda_2^{(4)} < \frac{1}{2}\left\{ \left(-\alpha L^{(4)} - \gamma H^{(4)}\right) - \sqrt{(\alpha L^{(4)} - \gamma H^{(4)})^2} \right\}$$

$$= -\min\left\{ \alpha L^{(4)}, \gamma H^{(4)} \right\} < 0.$$

Proposition 10 *If either of the inequalities in (7.3.7) holds along with (7.3.8), then the critical point $(L^{(4)}, H^{(4)})$ is asymptotically stable (by linear stability analysis).*

Proof The proposition follows from $\lambda_1^{(4)} < 0$ and $\lambda_{12}^{(4)} < 0$.
$$\square$$

Corollary 3 *If both inequalities in (7.3.7) hold, then the critical point $(L^{(4)}, H^{(4)})$ is asymptotically stable.*

Proof The two inequalities in (7.3.7) imply (7.3.8) so that Proposition 10 applies.
$$\square$$

7.3.5 Other bistability configurations?

7.3.5.1 *No bistability with unstable $(L^{(2)}, H^{(2)})$ and $(L^{(3)}, H^{(3)})$*

If instead of (7.3.4), we have the opposite inequalities (7.3.7), then $(L^{(2)}, H^{(2)})$ and $(L^{(3)}, H^{(3)})$ are unstable. Since $(L^{(1)}, H^{(1)})$ is also unstable independent of the relative size of the four products bc, $a\gamma$, da and αc (not to mention $\alpha\gamma$ and bd), we have immediately the following proposition:

Proposition 11 *Bistability does not exist if both inequalities in (7.3.7) hold.*

Proof The three fixed points $(L^{(1)}, H^{(1)})$, $(L^{(2)}, H^{(2)})$ and $(L^{(3)}, H^{(3)})$ are all unstable. Only $(L^{(4)}, H^{(4)})$ is asymptotically stable for this case by Proposition 10.

$\square$

7.3.5.2 *Stable $(L^{(2)}, H^{(2)})$ and unstable $(L^{(3)}, H^{(3)})$*

Suppose only the first inequality of (7.3.4) holds but not the second so that $(L^{(2)}, H^{(2)})$ is asymptotically stable but $(L^{(3)}, H^{(3)})$ is unstable. Whether there is bistability now depends on the stability of

$$(L^{(4)}, H^{(4)}) = \left(\frac{bc - a\gamma}{db - \alpha\gamma}, \frac{da - \alpha c}{db - \alpha\gamma} \right). \tag{7.3.9}$$

Now, the product $(c/d)(a/b)$ may be less than or greater than $(a/\alpha)(c/\gamma)$ (or db may be less or greater than $\alpha\gamma$ given the inequality $a/b > c/\gamma$ may more than compensate for (7.3.4), i.e.,

$$i)\ db - \alpha\gamma > 0, \quad \text{or} \quad ii)\ db - \alpha\gamma < 0.$$

(i) $db - \alpha\gamma > 0$: Together with $c/d < a/\alpha$, we have from Proposition 9 that $(L^{(4)}, H^{(4)})$ is unstable. This leaves $(L^{(2)}, H^{(2)})$ as the only asymptotically stable critical point and hence no bistability in this case.

Furthermore, with $a/b > c/\gamma$ so that

$$L^{(4)} = \frac{bc - a\gamma}{db - \alpha\gamma} < 0$$

(while $H^{(4)}$ can be seen to be positive), the corresponding critical point is not biologically realizable as we cannot have a negative leopard population.

(ii) $db - \alpha\gamma < 0$: Together with $a/b > c/\gamma$, we have from Proposition 10 that $(L^{(4)}, H^{(4)})$ is asymptotically stable. This critical point and the asymptotically stable $(L^{(2)}, H^{(2)})$ constitute bistability for this case (while the other two remaining critical points are unstable).

However, with $da > \alpha c$ so that

$$H^{(4)} = \frac{da - \alpha c}{db - \alpha\gamma} < 0$$

(while $L^{(4)}$ can be seen to be positive), $(L^{(4)}, H^{(4)})$ is not biologically realizable as we cannot have a negative hyenas population.

We summarize the results above in the following proposition:

Proposition 12 *If $c/d < a/\alpha$ and $a/v > c/\gamma$ so that $(L^{(2)}, H^{(2)})$ is asymptotically stable and $(L^{(3)}, H^{(3)})$ is unstable, bistability occurs only if $db < \alpha\gamma$ with a biologically unrealizable $(L^{(4)}, H^{(4)})$ being asymptotically stable in that case.*

7.3.5.3 Stable $(L^{(3)}, H^{(3)})$ and unstable $(L^{(2)}, H^{(2)})$

Now suppose only the second inequality (but not the first) of (7.3.4) holds so that $(L^{(3)}, H^{(3)})$ is asymptotically stable but $(L^{(2)}, H^{(2)})$ is unstable. Whether there is bistability depends again on the stability of $(L^{(4)}, H^{(4)})$. The situation is a mirror image of the last subsection. Without repeating a similar argument, we simply state below a similar set of results regarding bistability for this case.

(i) $db - \alpha\gamma > 0$:   Together with $a/b < c/\gamma$, we have from Proposition 9 that $(L^{(4)}, H^{(4)})$ is unstable. This leaves $(L^{(3)}, H^{(3)})$ as the only asymptotically stable critical point and the model exhibits no bistability in this case.

In addition, with $c/d > a/\alpha$ so that

$$H^{(4)} = \frac{da - \alpha c}{db - \alpha\gamma} < 0$$

(while $L^{(4)}$ can be seen to be positive), the corresponding critical point is not biologically realizable as we cannot have a negative leopard population.

(ii) $db - \alpha\gamma < 0$:   Together with $c/d > a/\alpha$, we have from Proposition 10 that $(L^{(4)}, H^{(4)})$ is asymptotically stable. This critical point and the asymptotically stable $(L^{(3)}, H^{(3)})$ constitute bistability for this case (while the other two remaining critical points are unstable).

However, with $a/b < c/\gamma$ so that

$$L^{(4)} = \frac{bc - a\gamma}{db - \alpha\gamma} < 0$$

(while $H^{(4)}$ can be seen to be positive), $(L^{(4)}, H^{(4)})$ is not biologically realizable as we cannot have a negative leopard population.

We summarize the results above in the following proposition:

Proposition 13 *If $a/b < c/\gamma$ and $c/d > a/\alpha$ so that $(L^{(3)}, H^{(3)})$ is asymptotically stable and $(L^{(2)}, H^{(2)})$ is unstable, bistability occurs only if $db < \alpha\gamma$ with a biologically unrealizable $(L^{(4)}, H^{(4)})$ being asymptotically stable in that case.*

7.3.6 Multiple bifurcation parameters

With $\mu = c\alpha - ad$, we have the following observations based on linear stability analysis:

- If $\mu < 0$, we have from Proposition 9 that the fixed point $(L^{(2)}, H^{(2)}) = (a/\alpha, 0)$ is asymptotically stable and $(L^{(4)}, H^{(4)}) = (bc - a\gamma, da - \alpha c)/(db - \alpha\gamma)$ is unstable (assuming for the latter $\Delta = db - \alpha\gamma > 0$ to keep $H^{(4)} > 0$ biologically realistic).
- If $\mu > 0$, we have from Proposition 10 that $(L^{(2)}, H^{(2)})$ is unstable and $(L^{(4)}, H^{(4)})$ is asymptotically stable (assuming for the latter $\Delta = db - \alpha\gamma < 0$ to keep $H^{(4)} > 0$ biologically realistic).
- If $\mu = 0$, we have $H^{(4)} = 0$ and $L^{(4)} = a/\alpha$ so that $(L^{(4)}, H^{(4)})$ degenerates to coincide with $(L^{(2)}, H^{(2)})$.

Together, they show that the growth of the competing populations has a transcritical bifurcation at $\mu = c\alpha - ad = 0$.

Similarly, the stability of $(L^{(3)}, H^{(3)})$ and $(L^{(4)}, H^{(4)})$ switches as $\zeta = a\gamma - bc$ changes sign and $(L^{(4)}, H^{(4)})$ coincides with $(L^{(3)}, H^{(3)}) = (0, c/\gamma)$ at $\zeta = 0$. Hence, the competing population growth has another transcritical bifurcation at $\zeta = a\gamma - bc = 0$.

It is left to the reader to investigate the consequences of $\mu = 0$ and $\zeta = 0$ simultaneously.

7.4 Cell Lineages and Feedback

7.4.1 *Multistage cell lineages*

Many biological and biochemical processes are involved in the development and regeneration of tissues and organs of an organism, not necessarily the same for different organs or organisms. One of these involves *cell lineages*, a multi-stage cell division process from first cleavage of a *stem cell* to the ultimate differentiation into (terminal) cells of tissues and organs after a number of subsequent divisions. When a stem cell divides, each offspring (progenitor) cell has the potential either to remain a stem cell or become a transit-amplifying cell (TA) having more limited division potential. A TA progenitor cell may divide to become more specialized with successive divisions, finally giving rise to a progenitor that is reproductively sterile, terminally differentiated cell (TD) with a specialized function, such as a muscle cell, a red blood cell, or a brain cell. Offspring stem cells on the other hand can continue to divide and may be differentiated into any specialized cell type with a subsequent division. As such, stem cells are often characterized as pluripotent (or multipotent) while TA cells (with more limited functions) are said to be obligopotent. Terminally differentiated (TD) cells are unipotent since they are committed to a specific function and can only die (apoptosis) without further division.

Typically, when a stem cell (at stage 0) divides, either progeny (offspring cell) may again be a stem cell with probability p_0 or may become a *transit-amplifying cell* (TA) with probability $1 - p_0$. For an unbranched lineage, the TA progenitor cell at stage 1 divides with each offspring to remain the same TA with probability p_1 or may become a different (more specialized) TA cell with probability $1 - p_1$. The process repeats with subsequent divisions until stage n that results in a terminally differentiated (TD) that does not divide but dies with a death rate constant δ. A schematic diagram of this unbranched cell lineage is shown in Figure 7.2.

When the number of cells is large and they divide asynchronously, we may model the sequence of cell divisions of such an unbranched cell lineage by a set of ODE [11, 15]. For the purpose of this discussion, it suffices to consider the case $n = 2$ so that we have one population each of stem cells, TA cells and TD cells.

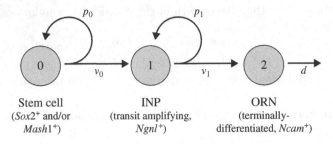

Stem cell | INP | ORN
(*Sox2*⁺ and/or | (transit amplifying, | (terminally-
*Mash*1⁺) | *Ngnl*⁺) | differentiated, *Ncam*⁺)

Fig. 7.2   Cell lineage and feedback.

The ODE model for this simplified cell lineage consists of the following three ODE

$$x_0' = (2p_0 - 1)v_0 x_0$$
$$x_1' = (2p_1 - 1)v_1 x_1 + 2(1 - p_0)v_0 x_0 \qquad (7.4.1)$$
$$x_2' = -\delta x_2 + 2(1 - p_1)v_1 x_1$$

supplemented by the three initial conditions

$$x_0(0) = X_0, \qquad x_1(0) = x_2(0) = 0. \qquad (7.4.2)$$

It is straightforward to obtain the following solution of this linear IVP:

$$x_0(t) = X_0 e^{(2p_0 - 1)v_0 t} \qquad (7.4.3)$$

$$x_1(t) = X_1 \left[e^{(2p_0 - 1)v_0 t} - e^{(2p_1 - 1)v_1 t} \right] \qquad (7.4.4)$$

$$x_2(t) = X_2 \left[\frac{e^{(2p_0 - 1)v_0 t} - e^{-\delta t}}{(2p_0 - 1)v_0 + \delta} - \frac{e^{(2p_1 - 1)v_1 t} - e^{-\delta t}}{(2p_1 - 1)v_1 + \delta} \right] \qquad (7.4.5)$$

where

$$X_1 = \frac{2(1 - p_0)v_0 X_0}{(2p_0 - 1)v_0 - (2p_1 - 1)v_1}, \qquad X_2 = 2(1 - p_1)v_1 X_1. \qquad (7.4.6)$$

The following observations provide important insight to this model:

- $x_2(t)$ becomes unbounded if either $p_0 > 0.5$ or $p_1 > 0.5$ which is biologically unrealistic.
- If $p_0 < 0.5$ and $p_1 < 0.5$, all three cell types eventually run out. If $\delta > 0$ is sufficiently small, there may be a TD cells population to sustain the biological organism for a period (and hence defining its life span which may be very short if δ is not sufficiently small).
- For the special case $p_0 = 0.5$ and $p_1 < 0.5$, the solution becomes

$$x_0(t) = X_0, \qquad x_1(t) = X_1 \left[1 - e^{-(1 - 2p_1)v_1 t} \right] \to X_1,$$

$$x_2(t) = X_2 \left[\frac{1 - e^{-\delta t}}{\delta} - \frac{e^{-(1 - 2p_1)v_1 t} - e^{-\delta t}}{\delta - (1 - 2p_1)v_1} \right] \to \frac{X_2}{\delta}$$

with

$$X_1 = \frac{v_0 X_0}{(1 - 2p_1)v_1} > 0, \qquad X_2 = \frac{2(1 - p_1)v_0}{(1 - 2p_1)} X_0 > 0$$

so that all three cell populations tend to a steady state size.

The three types of outcomes described above persist for the general case of $n > 2$ (see [11]). As such, the model involves a steady state that is unacceptably sensitive to system parameters with a biologically realistic TD steady state possible only for $p_0 = 0.5$ and $p_k < 0.5$ for all $k > 0$ and only when the initial stem cell population is sufficiently large.

7.4.2 *Performance objectives*

In addition to the size of a TD population needed for a normal organism, there are other performance objectives that need to be met by cell lineages. For example, if an injured organ is to be repaired or regenerated, we know that it is completed at a rate much faster than the normal time for development of the organ from an embryo. The ODE model (7.4.1) does not differentiate regeneration from normal development since the growth rates $\{v_k\}$ are specified constants that do not change with functions. Some kind of "feedback" mechanism must be at work to regulate growth to achieve normal development and to induce regeneration after injury.

Experimental evidence from genetic studies since the 1990s has amply confirm the presence of feedback in growth and development of tissues such as skin, liver, bone, brain, blood cells, retina and hair (see [11] for a more in-depth discussion on such biological evidence). We mention here only the example of mammalian *olfactory epithelium* (OE), the neural tissue that senses odor and transmits olfactory information to the brain. The OE is a continually self-renewing tissue and is capable of rapid regeneration. A wealth of experimental data on OE lineage and the molecules that regulate it. For example, the growth and differentiation factor 11 (GDF11) was shown to be produced specifically by cells of the neural lineage of the mouse OE and to provide feedback to inhibit the production of *olfactory receptor neurons* (ORN) from its TA (known as *immediate neuronal precursor* or INP) in that system. This and other experimental results make the OE an attractive system for investigating the relationship between lineages and growth control [11,15].

7.4.3 *Feedback control of output*

For the purpose of limiting growth to "normal" size, some kind of negative feedback mechanism seems appropriate. To see how an appropriate negative feedback actually accomplish nature's performance objective in a way that is both surprisingly well and in an unexpected way, we further simplify the ODE model (7.4.1) to two (instead of three) cell types, eliminating the transit-amplifying cells so that a stem cell offsprings can either be another stem cell or a terminally differentiated cell. The two cell populations at time t are denoted by $x_S(t)$ (stem cells) and $x_D(t)$ (terminally differentiated cells). Their growths are governed by the two ODE

$$x'_s = \{2p(x_S, x_D) - 1\} v x_S \tag{7.4.7}$$
$$x'_D = 2\{1 - p(x_S, x_D)\} v x_S - \delta x_D \tag{7.4.8}$$

starting with the initial conditions

$$x_S(0) = X_S, \qquad x_D(0) = X_D. \tag{7.4.9}$$

(In addition to the simplification indicated above, a new feature in these two ODE models is that the probability p of an offspring cell to remain a stem cell now depends on the two cell population sizes.)

In the general form $p(x_S, x_D)$, the feedback involved is *positive* (activation) if

$$p_{,S} = \frac{\partial p}{\partial x_S} > 0, \qquad p_{,D} = \frac{\partial p}{\partial x_D} > 0,$$

negative (inhibition) if

$$p_{,S} = \frac{\partial p}{\partial x_S} < 0, \qquad p_{,D} = \frac{\partial p}{\partial x_D} < 0,$$

or of a mixed type, e.g.,

$$p_{,S} = \frac{\partial p}{\partial x_S} > 0, \qquad p_{,D} = \frac{\partial p}{\partial x_D} < 0.$$

Remark 7 *Before we analyze this new system with feedback, we note that a feedback on rate of growth constant v does not help with the very restrictive requirement of $p = 0.5$ for the existence of a finite steady state TD population as long as $p(x_S, x_D)$ remains a constant.*

Remark 8 *In the absence of feedback and for $p = 0.5$ to ensure a steady state, the (x_S, x_D) ODE system reduces*

$$x_S' = 0, \qquad x_D' = v x_S - \delta x_D \tag{7.4.10}$$

so that, with $X_D = 0$, we have

$$x_S(t) = X_S, \qquad x_D(t) = \frac{v}{\delta} X_S \left(1 - e^{-\delta t}\right).$$

The final steady state is reached with a half life of $1/\delta$ which is typically a very long time (for normal development).

7.4.4   A finite steady state

To illustrate the general approach to extracting relevant information from the model, we consider here the simple but relevant case of a negative feedback from x_D on the probability of a progenitor to remain a stem cell in the form (of a simple *Hill's function* type)

$$p(x_S, x_D) = \frac{p_0}{1 + c x_D}. \tag{7.4.11}$$

A schematic diagram of a more realistic three-population model is shown in Figure 7.3.

For this feedback, the ODE system (7.4.7)–(7.4.8) becomes

$$x_S' = \left\{ \frac{2 p_0}{1 + c x_D} - 1 \right\} v x_S \tag{7.4.12}$$

$$x_D' = 2 \left\{ 1 - \frac{p_0}{1 + c x_D} \right\} v x_S - \delta x_D. \tag{7.4.13}$$

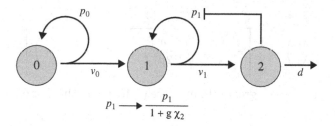

Fig. 7.3   Negative feedback regulation of the INP replication probability.

Whereas the system without feedback (corresponding to $c = 0$) reaches a steady state $\{X_S, vX_S/\delta\}$ only for the very special probability $p = p_0 = 0.5$, the steady state

$$\bar{X}_D = \frac{2p_0 - 1}{c}, \qquad \bar{X}_S = \frac{\delta}{v}\bar{X}_D = \frac{\delta}{vc}(2p_0 - 1) \tag{7.4.14}$$

always exists for the system with feedback (7.4.12)–(7.4.13). Furthermore the steady state is independent of the initial stem cell population size which is not the case for the only steady state possible (for $p = p_0 = 0.5$). The two steady state population sizes $\{\bar{X}_S, \bar{X}_D = v\bar{X}_S/\delta\}$ are both positive as long as

$$p_0 > 0.5 \tag{7.4.15}$$

and now the actual steady state size dictated only by c and p_0 (not the initial stem cell population X_S which is typically small).

Proposition 14 *For all $p_0 > 0.5$, a unique biologically meaningful steady state for terminally differentiated cell population (7.4.14) exists for our two-stage model (7.4.12)–(7.4.13) and is independent of the initial stem cell population.*

7.4.5   Linear stability

Even if it is biologically meaningful, the unique steady state of our model is relevant to the growth process only if it is stable with respect to perturbations. For a linear stability analysis, we compute the Jacobian of our model ODE system to get

$$J\left(\bar{X}_S, \bar{X}_D\right) = \begin{bmatrix} v\left(\frac{2p_0}{1+cX_D} - 1\right) & -\frac{2vcp_0 x_S}{(1+cX_D)^2} \\ 2v\left(1 - \frac{p_0}{1+cX_D}\right) & -\delta + \frac{2vcp_0 x_S}{(1+cX_D)^2} \end{bmatrix}_{(\bar{X}_S, \bar{X}_D)}$$

$$= \begin{bmatrix} 0 & -\delta\left(1 - \frac{1}{2p_0}\right) \\ v & -\frac{\delta}{2p_0} \end{bmatrix}.$$

The two eigenvalues of this Jacobian matrix (evaluated at the unique steady state for the problem) are the two roots of the characteristic equation

$$\lambda^2 + \frac{\delta}{2p_0}\lambda + v\delta\left(1 - \frac{1}{2p_0}\right) = 0$$

or

$$\begin{pmatrix} \lambda_1 \\ \lambda_2 \end{pmatrix} = \frac{1}{2} \left[-\frac{\delta}{2p_0} \pm \sqrt{\left(\frac{\delta}{2p_0}\right)^2 - 4v\delta \left(1 - \frac{1}{2p_0}\right)} \right].$$

Since $2p_0$ is required to be greater than 1 (see (7.4.15)), both λ_1 and λ_2 are negative. It follows that

Proposition 15 *With $2p_0 > 1$ the unique steady state of the ODE system (7.4.12)–(7.4.13) is asymptotically stable by a linear stability analysis.*

Altogether, we have demonstrated that for a two-stage cell lineage model with a negative feedback of the Hill's function type on the probability p that a progenitor cell remains a stem cell, the same probability in the absence of any terminally differentiated cells initially no longer has to be restricted to be $p_0 = 0.5$ precisely for the cell populations to reach a steady state. The outcome is not so surprising since a negative feedback typically acts to inhibit the normal process which is growth in our case. That the inhibition is accomplished through a reduction of stem cell reproduction and thereby decreasing the source progenitor cells for either kind downstream may not be so intuitively obvious. That the steady state differentiated cell population size does not depend on the initial stem cell population size (which is typically small at an early stage of development) is a bonus, especially when it is regulated by the strength of the feedback (through the parameter c).

Similar beneficial outcomes for normal growth and developments have been shown in [11, 15] and elsewhere to result from

- more complex lineages with one or more transit-amplifying progenitor cells prior to terminating in some target tissue or organ cells
- more complex combination models of negative feedbacks.

7.4.6 Fast regeneration time

While the simple negative feedback model (7.4.12)–(7.4.13) have already produced the desired outcome of limited growth (instead of unbounded growth without feedback), it has in store for us a completely unexpected payoff in the area of regenerative growth so important in medical sciences. Suppose that long after an organism has reached its steady state and continues in a maintenance mode, an injury causes the loss of an entire terminally differentiated cell population (that constitutes an organ or a certain type of tissues). The organism naturally tries to regenerate the same body of TD cells to reconstitute the organ or tissue. Clearly it would be preferred that this be accomplished in a time scale considerably faster than that for the development of the original TD cell population (and nature has shown us that this is in fact possible). It turns out that the simple feedback model (7.4.12)–(7.4.13) shows how this may be accomplished by the same feedback mechanism.

By the time an injury to the established TD cell population takes place, say at time $t = t_i$, the system has been in a steady state with a body of corresponding stem cells $\bar{X}_S$ (considerably larger than the original stem cell population responsible for the constitution of the lost TD cells $\bar{X}_D$ prior to injury. The development just prior to injury is to a good approximation characterized by the system

$$x'_S = \left\{ \frac{2p_0}{1 + c\bar{X}_D} - 1 \right\} vx_S$$

$$x'_D = 2\left\{ 1 - \frac{p_0}{1 + c\bar{X}_D} \right\} vx_S - \delta x_D.$$

Right after injury with a complete loss of the terminally differentiated cells, the system dynamics changes to

$$x'_S = \{ 2p_0 - 1 \} vx_S$$

$$x'_D = 2\{ 1 - p_0 \} vx_S - \delta x_D$$

with the initial conditions

$$x_S(t_i) = \bar{X}_S = \frac{\delta}{vc}(2p_0 - 1), \qquad x_D(t_i) = 0. \tag{7.4.16}$$

The change in the system dynamics is rather drastic for two reasons:

(1) The growth rate constant for stem cells increases very substantially from $2p_0 / (1 + c\bar{X}_D) - 1$ to $2p_0 - 1$.
(2) The stem cell population at the start of the regenerative process $\bar{X}_S$ is substantially larger than that at the start of the natural development.

Together, they would lead to a burst of the stem cell generation building a huge population of stem cell in a short time, known to be more than one hundred time faster than that of normal development of the same TD cell population. This large body of new $x_S(t)$ accumulated over a short period of time after t_i enables the system to replenish the lost TD cells in a much shorter time period than the time required for normal development.

The qualitative observations above pertaining to negative feedback on stem cell renewal probably p_0 have been substantiated quantitatively by numerical simulations, i.e., solving numerically the various IVP for a number of different ODE models of different level of complexity. The results of a typical simulations are reproduced in Figure 7.4 (see [11] for more details, other simulations and experimental confirmation of theoretical developments). An inset shows the response at early times in greater detail. Note that progenitor load is now quite low, and regeneration is characterized by a burst of immediate neuronal precursor (INP) proliferation, followed by a wave of ORN production, with improvement in regeneration speed over what would occur in the absence of feedback shown by dashed line.

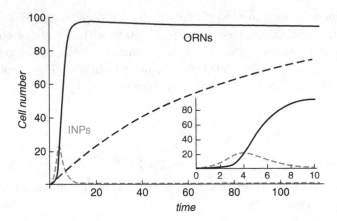

Fig. 7.4   Simulated return to steady state of system after removal of ORN.

Part 3

Optimization

Chapter 8

The Economics of Growth

8.1 Maximum Sustained Yield

8.1.1 *Fishing effort and maximum sustainable yield*

8.1.1.1 *The yield-effort curve*

Consider again a small fish farm with the natural growth of its fish population $y(t)$ (after normalization by the carrying capacity Y_c) adequately characterized by the logistic growth model $y' = dy/dt = y(1 - y) \equiv f_0(y)$ with time measured in units of the linear growth rate constant a^{-1}. Fish is harvested at a rate h that depends on the fishing effort E (corresponding to the equipment and labor committed by the owner/management of the fish farm) and the current size of the fish population. For simplicity, we take this dependence to be of the form $h = qEy$. The growth rate of fish population being harvested is then determined by the first order ODE (see (3.4.5))

$$y' = f_0(y) - qEy = y(1 - qE - y) \equiv f_E(y) \tag{8.1.1}$$

with $E = 0$ corresponding to the natural growth rate (without harvest) denoted by $f_0(y)$. The two critical points of this ODE are

$$y^{(1)} = 0 \quad \text{and} \quad y^{(2)} = 1 - qE. \tag{8.1.2}$$

These two critical points can also be located by the graphical method.

Exercise 19 *Plot natural growth rate curve $f_0(y)$ and the harvest curve $h(y) = qEy$ (for a fixed E) and locate the two critical points and determine their stability from the intersection of these two graphs.*

The bifurcation of these critical points has been analyzed previously and can also be determined by the graphical method. Here we are interested in the critical point $y^{(2)}$ which provides the fishery management a steady fish population for a *sustainable yield* at the rate

$$h_{SY}(E) = qEy^{(2)} = qE(1 - qE). \tag{8.1.3}$$

The graph of the sustainable yield $h_{SY}(E)$ as a function of E is another upside down parabola vanishing at $E = 0$ and $E_c = 1/q$. Not surprisingly, there would be no harvest without effort; but too much effort (larger than the bifurcation point $1/q$) would result in overfishing and drive the fish population to extinction.

Exercise 20 *Plot $y^{(2)}$ and $h_{SY}(E)$ as functions of E.*

8.1.1.2 Maximum sustainable yield

One important observation at this point is that the nontrivial fixed point $y^{(2)} = 1 - qE$ and correspondingly the sustainable yield $h_{SY} = qE(1 - qE)$ varies with the fishing effort E. With the effort parameter E at our disposal, we may choose it to maximize h_{SY}. It is a simple calculus problem to show that the *maximum sustainable yield*, denoted by h_{MSY}, for our particular growth model is

$$h_{MSY} = \frac{1}{4},$$

attained at

$$E_{MSY} = \frac{1}{2q}.$$

Note that E_{MSY} is different from E_c and the corresponding sustainable (steady state) fish population is

$$y_{MSY} = \frac{1}{2}.$$

Note that y_{MSY} and h_{MSY} do not depend on E; they are the intrinsic properties of the natural growth rate function. Had the harvest rate been a constant and served as the control instrument and the bifurcation parameter, then the maximum sustainable yield would be 1/4 again and the maximum sustainable (steady state) population would still be 1/2. However the control instrument (now the harvest rate itself) responsible for the maximum sustainable yield would be $h_{MSY} = h_c = (= 1/4)$ found to be the bifurcation point for the bifurcation parameter h.

Exercise 21 *With E as the control instrument, show that the bifurcation point would be $E_c = 1/q$ instead.*

Early literature on fishery focused on achieving maximum sustainable yield. Similar to the case of specifying the harvest rate h, the two critical points of the problem coalesce at the bifurcation point $E_c = 1/q$ and the one remaining critical point is semi-stable. The highly productive Peruvian anchovy fishery was estimated to be at maximum sustainable yield in the early 1970s. The government did not heed the warning from scientists and allowed nearly normal fishing effort when hit by the arrival of an El Nino. The fishery never recovered from that "small" perturbation of the maximum sustainable yield.

Exercise 22 *Use the h_{SY} vs. E graph to show that fishing with $E > E_c$ reduces the equilibrium fish population and the h_{SYM}.*

Our discussion here is to work our way toward a scientifically based fishing strategy by looking into the various economic issues that contribute to a sensible management policy for fishery.

8.1.2 *Depensation and other growth rates*

8.1.2.1 *Depensation growth with harvest*

Instead of the logistic (natural) growth, most ecologists would consider the depensation growth rate to be a more appropriate characterization of the growth rate of a normal fish population. After normalization, we take the natural (unharvested) growth rate of the fish to be $f_0(y) = y(2-y)(y-1)$ so that the actual growth rate is governed by

$$y' = f_0(y) - qEy = y\left[-y^2 + 3y - (2 + qE)\right] \equiv f_E(y)$$

assuming harvest rate is proportional to both effort level and population size ($h = qEy$) with $E = 0$ corresponding to the natural growth rate $f_0(y)$. The three critical points of this ODE are

$$y^{(1)} = 0 \quad \text{and} \quad \begin{pmatrix} y^{(2)} \\ y^{(3)} \end{pmatrix} = \frac{1}{2}\left[3 \pm \sqrt{1 - 4qE}\right]. \tag{8.1.4}$$

The three critical points can also be located by the graphical method.

Exercise 23 *Plot natural depensation growth rate curve $f_0(y)$ and the harvest curve $h(y) = qEy$ (for a fixed E) and locate the three critical points as well as determine their stability from the intersection of these two graphs.*

At $E_c = 1/4q$, two of the critical points coalesce. Further increase of effort level leaves us with a single equilibrium population at the origin. The result shows a bifurcation of the *saddle-node* (or a broken *pitchfork*) type at $E_c = 1/4q$. Here we are interested in the critical point $y^{(2)}$ which provides the fishery management a steady fish population for *sustainable yield* h_{SY} at the rate

$$h_{SY}(E) = qEy^{(2)} = \frac{1}{2}qE\left\{3 + \sqrt{1 - 4qE}\right\}. \tag{8.1.5}$$

The graph of the sustainable yield $h_{SY}(E)$ as a function of E is no longer an upside down parabola. In particular, there is a finite nontrivial sustainable yield at bifurcation $E = E_c$ but there is no nontrivial sustainable yield for $E > E_c$. Unlike the logistic growth case with $f_0(y) = y(1-y)$, the disappearing of a sustainable yield (8.1.5) is abrupt, not gradual.

Exercise 24 *Plot $y_{SY}(E)$ and $h_{SY}(E)$ as functions of E.*

Exercise 25 *For the depensation natural growth rate, calculate the maximum sustainable yield h_{MSY} and the corresponding y_{MSY} and E_{MSY} ($\neq E_c$).*

8.1.2.2   General natural growth rate

Why would the fishery owner/management be concerned with the maximum sustainable yield h_{MSY}? An obvious answer is to be able to catch more fish for more consumption whether it is for a higher level of profit (in the case of a privately owned fishery) or a higher level of satisfaction (utility) of the consumers (in the case of a state owned one), or both. Since we cannot fish at too high a level of harvest rate (to avoid driving the fish population to extinction), we need to discuss what looms as a major factor that determines the prudent limit of the fishing effort of a commercial fishery, namely the economics of commercial fishing.

8.2   Economic Overfishing vs. Biological Overfishing

8.2.1   Revenue for a price taker

Fish stocks are harvested for different reasons, one is for commercial purpose, selling the catches for revenue. For a given fish population, most small fisheries are *price takers*. Their catches are not big enough to have an influence on the market price for the fish they harvest. For a price taker, the price per unit biomass of the fish harvested is a given constant price p. In the development of this chapter, we focus on a fish stock whose natural growth is adequately characterized by the logistic growth model to bring out the issues in this area for a price taker. At equilibrium, let $y^{(2)}$ denote the fish population corresponding to the largest asymptotically stable critical point point given by (8.1.2), i.e., $y^{(2)} = 1 - qE$. For a fixed fishing effort E, the sustainable yield h_{SY} has also been found in (8.1.3) to be $qE(1 - qE)$. The corresponding sustainable revenue is then simply

$$R(E) = ph_{SY} = pqE(1 - qE). \qquad (8.2.1)$$

Its graph is just a scaled up version of the graph for $h_{SY}(E)$ since p is a constant.

Fortuitously, the graph of the yield-effort relation or the revenue-effort relation (8.2.1) is also an upside down parabola. But it is only indirectly related to the parabola of the scaled logistic growth rate $f_0(y) = y(1 - y)$ (or its un-scaled version $ay(1 - y/Y_c)$. The latter parabola is about the biological growth of the fish population while the former is about the financial return for the effort invested by the fishery.

While more will be said about this later, we make the following observations:

- With $R(0) = 0$, there would be no harvest and hence no revenue if there is no effort.
- The $R(E)$ curve has a positive slope at the origin so that even a low level of effort produces some revenue.
- However, revenue does not increase indefinitely with more effort. Beyond a certain effort level E_{MSY}, more effort leads to less revenue.
- In fact, the gross revenue reduces to zero at the *bifurcation point* $E_c = 1/q$ (with the bifurcation being of the transcritical type as we found in Chapter 2).

At $E = E_c = 1/q$, the two critical points coalesce to one $y^{(2)} = y^{(1)} = 0$; and the nontrivial steady state population dwindles to zero so that there is no fish available eventually for harvesting at that (or a higher) level of fishing effort. Evidently, the sustainable yield first increases and then decreases as effort increases heading toward extinction as E tends toward E_c from below.

8.2.2 Effort cost and net revenue

By committing itself to a certain level of fishing effort E, the fishery has incurred a certain cost, C, for its fishing activities. This cost certainly varies with the level of effort committed as it requires more money to purchase more fishing equipment and pay more fishermen to work on its fishing boat(s). While C may depend also on other factors, we consider here for the purpose of illustration the simple case of $C = cE$ where the known constant c corresponds to the cost of a unit of effort. The net revenue

$$P(E) = R(E) - cE = pqE \left\{ 1 - \frac{c}{pq} - qE \right\}$$

then constitutes the actual profit for the fishery.

Exercise 26 *Plot the straight line cE on the same plot as the revenue-effort curve, $R(E)$.*

Exercise 27 *Plot Revenue curve $R(E)$ and the linear effort cost line $C = cE$ and locate the effort E_0 for zero net revenue.*

Exercise 28 *Put the plot from Exercise 19 side by side with the plot from Exercise 27 and observe how $y^{(2)}$ and E_0 change with E.*

For a fixed effort level E, $P(E)$ is the vertical distance between the straight line cE and the revenue curve $R(E)$ for that particular effort level. For a given unit cost factor c, the special effort level E_0 for which the line cE intersects the revenue curve $R(E)$ is the *break even* effort level; there is no net revenue (no profit and hence the subscript 0) if the fishery fishes with effort E_0. For the logistic growth case, we have

$$E_0 = \frac{1}{q} \left(1 - \frac{c}{pq} \right). \tag{8.2.2}$$

What is the point of calling attention to this special level of effort, you may ask; the fishery owner or management would surely not fish with effort E_0 since there is no profit to be made. A simple answer is that investment in equipment and contracting fishermen are fixed commitment that cannot be changed instantaneously. So it is important to know in advance what is the break even effort level E_0 in order to stay below it.

However, there is a more serious and painful but realistic observation regarding the break even effort E_0 in the case of open access fish populations, not those in a fish farm but the fish in sea or ocean. Beyond some finite distance offshore fishing limit, be it 12 miles or 200 miles, there is no authority having sole control of the fish in open sea to regulate the fishing effort level. As long as there is profit to be made, some new fishery would enter and fish (or an existing fishery would increase its effort to fish) in that region and thereby increase the overall fishing effort level for the fish there. This increase in fishing effort would continue (at least) until we reach the zero profit effort level E_0 when further increase in fishing effort would no longer be profitable.

While there appears to be no incentive to increase fishing effort beyond the break even point, there is also no incentive to reduce it from E_0. If E drops below E_0, then there would be profit to be made and effort level would increase by some entity to take advantage of the opportunity to reap the available profit. Of course, there may be a better return for investment by fishing elsewhere or getting into another business; the fact remains that there is always some unemployment and/or under-utilized fishing effort (already paid for or contracted) which provides a source of new effort for fishing at our particular fish habitat.

8.2.3 *Economic overfishing vs. biological overfishing*

At $E_0(c)$, the net revenue is zero so that there is no profit to be made at this level of fishing effort. For still higher value of E, net revenue would turn negative. All the while, a profit can be made at a lower fishing effort than $E_0(c)$. For that reason, fishing at an effort level $E \geq E_0(c)$ is said to be overfishing economically. Evidently, it is not advisable to engage in *economic overfishing*. On the other hand, overfishing economically may still lead to a positive sustainable yield in contrast to fishing at $E \geq E_c$. For this latter higher level of fishing effort, not only would the net revenue be negative (and hence overfishing economically) but the fish population is being driven to extinction eventually. In this latter case, the fish population is said to be overfished biologically and we have what is known as *biological overfishing*.

In principle, a small degree of economic overfishing (say at the level of $E_0(c)$) may be tolerable since it maintains a positive sustainable yield. But with demand for fish increasing with time and unit fish price rising with the demand, the expression for $E_0(c)$ given by (8.2.2) tells us that a higher unit fish price increases the break even point fishing effort $E_0(c)$ and drives the sustainable yield lower and eventual to extinction. No government preaching of public good or conservation could overcome the forces of economics to prevent overfishing. Central planning and regulations seem necessary for more sensible management of *assets in the common*.

8.2.4 *Regulatory controls*

The fish population of the east coast of North America has been severely depleted by overfishing in the last half of the 20th century. This is particularly true for the more valuable fish stocks such as cod, haddock, and red fish. Various estimates suggested that it would take a 20-year program of intensive experimental management, including a fishing moratorium, to rebuild these fisheries to the carrying capacity of the region [12, 13]. The west coasts of the Americas are also not free from the problem of overfishing. There were the collapses of the sardine fishery near California, the anchovy fishery in Peru and the halibut population in the Pacific Northwest. Many similar situations around the world have given considerable impetus to the insistence of a 200-mile offshore territorial rights for fishing by various countries with a significant fishing industry such as U.S. and Canada.

The simple fishery model at the start of this chapter when applied to open access fish habitats has given us some insight to the reason for overfishing of the fish populations in such habitat. When left unregulated there, open access inevitably leads to overfishing both economically and biologically with the latter leading to extinction or a significantly lower maximum sustainable yield. The push for an extended offshore limit distance (from 12 miles to 200 miles) is intended to allow the host country along the shore to impose some regulatory policies to prevent overfishing. Options of such regulations include i) limiting the number of fishing days in a year and ii) imposing a progressive catch tax, proportionately or disproportionately higher for larger catches. The former has the effect of reducing the fishing effort E; the latter would increase the unit cost of effort. Either way (and there are others) would result in a larger sustainable yield, at least according to our simple model above. However, with equipment in place, fishing staff already under contract (and hence the unit cost of effort decreasing with larger total catch), the net effect of a reduction of effort level through limitation of a fishing season is less clear. Because of a larger sustainable yield with the reduced fishing effort possibly concurrent with some reduction of the harvesting cost $C = cE$ (depending on the product of a smaller E and (possibly) a larger c), it is expected to be generally beneficial. Depending on the existing committed effort, a short fishing season could lead to a collapse of the fishery (or the fishing industry). Under the same conditions, the imposition of a (not so aggressive) catch tax with no restriction on fishing effort is likely to reduce the sustainable yield and hence economic overfishing (and possibly biological overfishing as well).

Still, our model suggests that if unit fishing effort cost c is not affected by the actual fishing effort E, regulations such as the two mentioned above may have the desirable effect of increasing the sustainable yield (at least in principle) and that is good. But experience with such regulations suggests that the expected outcomes are often not realized; sooner or later, public pressure may force changes or deviation from these regulatory policies. In the next section, we discuss one main source of the forces that work against regulations (such as two above) based on sustainable yield (steady state) considerations. When the important effect of discounting is appreciated, we then show in the next few chapters through some more sophisticated mathematical modeling, how we address this important effect through planning over a period. The optimal resource management strategy over a planning period, with or without discounting, will then lead to a better understanding and hopefully more reasonable policy for the utilization of fish and other resources that can be subject to regulations or some form of control.

8.3   Discounting the Future

8.3.1   *Current loss and future payoff*

Why do fishermen and fisheries object to regulations that are beneficial to them with some degree of public support (when there is no readily available substitute for the benefit from a regulated fish population)? One reason that readily comes to mind is the immediate loss of current revenue. Suppose fishing is allowed only for a limited number of days in a year, known as a fishing season (and usually significantly fewer than the days in a calendar year). We can see from our model that this reduction of effort below E_0, say to $E_1 < E_0$, would lead to a positive net revenue through both a higher sustainable yield and a smaller fishing effort cost. However, it would take time for the fish population to adjust to the lower effort level before reaching a higher asymptotically stable equilibrium population size $y_1^{(2)} > y_0^{(2)}$. Here $y_0^{(2)}$ is the nontrivial equilibrium state $y^{(2)}$ associated with break even effort level E_0. It is important to remember that E_0 depends on the cost (for unit effort) to price (for unit fish harvested) ratio c/p as found in (8.2.2) for our model. While there would be an immediate increase in the fish population, it would take a significantly long time (infinitely long in principle) to get from $y_0^{(2)}$ to $y_1^{(2)}$. The central planning board that imposes the effort reduction must find ways to assure the fisheries that when the time comes, the payoff would more than offset the revenue lost. Even then, can the fisheries wait and suffer the economic loss before then?

The drop of yield due to the reduced effort from E_0 to E_1 is

$$\Delta h_L = q y_0^{(2)} (E_0 - E_1) = q(1 - qE_0)(E_0 - E_1)$$

while the gain when we are at (or near) the new steady state $y_1^{(2)}$ is

$$\Delta h_G = qE_1(y_1^{(2)} - y_0^{(2)}) = qE_1 q(E_0 - E_1).$$

At the very least, we should have the future revenue gained from the effort reduction be no less than the loss of current revenue. For a price taker, this requires $p\Delta h_G \geq p\Delta h_L$ assuming that the price for unit fish does not change over the period of time involved. This imposes a lower bound on the new effort level E_1, namely

$$qE_1 \geq 1 - qE_0(c). \tag{8.3.1}$$

To be fair, there is also the savings from the reduction of cost effort since it has been reduced from cE_0 to cE_1. A more appropriate requirement would be

$$p\Delta h_G + c(E_0 - E_1) \geq p\Delta h_L. \tag{8.3.2}$$

8.3.2 *Interest and discount rate*

The current loss of revenue Δh_L only accounts for the one time loss at the moment of effort reduction. To the extent that it takes time to reach the new steady state $y_1^{(2)}$, there would be repeated losses over time as long as $y(t)$ is not yet $y_1^{(2)}$. Assuming fishing continues indefinitely at the effort level E_1, the net revenue loss would decrease with time as $y(t)$ increases toward $y_1^{(2)}$. Before we try to calculate these intervening losses, there is another issue that needs to be addressed.

If you win a lottery, you would be given a choice of receiving the lottery payment P_0 up front or in a number of installments over a period of time. Typically the total of all the installment payments would be somewhat more than the lump sum payment P_0. If you select the lump sum payment, you can put all of it in the bank and earn interest at some annual interest rate of δ percents. At the end of the installment payment period, say t years, the money in the bank would grow to

$$P(t) = P_0(1 + \delta)^t$$

if compounded annually.

You may get a better deal at a different bank offering you the same interest rate per annum but compounded monthly (or even daily). With this new deal, your account at the end of t years would be

$$P(t) = P_0 \left[\left(1 + \frac{\delta}{N} \right)^N \right]^t = P_0 \left(1 + \frac{\delta}{N} \right)^{Nt}$$

where $N = 12$ for monthly compounding and $N = 365$ for daily compounding. A mathematician would drive a hard bargain by insisting on instantaneous compounding to get

$$P(t) = \lim_{N \to \infty} \left[P_0 \left(1 + \frac{\delta}{N} \right)^{Nt} \right] = P_0 e^{\delta t} \tag{8.3.3}$$

so that a dollar today is worth $e^{\delta t}$ dollars at the end of t years (assuming an annual interest rate of $\delta\%$ compounded instantaneously).

Exercise 29 *Prove (8.3.3).*

Turning it around, the result (8.3.3) can be taken to say that a dollar payment/income received t years from now is equivalent to only the fraction $e^{-\delta t}$ of a dollar given to me now. As such, the effort reduction requirement (8.3.2) should be modified to read

$$e^{-\delta t} [p\Delta h_G + c(E_0 - E_1)] \geq p\Delta h_L \tag{8.3.4}$$

keeping in mind that only the loss incurred initially (when the effort level is reduced from E_0 to E_1) is considered in arriving at this requirement. To account for the ensuing losses prior to reaching $y_1^{(2)}$ (or being close enough to it for all practical purposes) will be addressed in the next chapter.

Chapter 9

Optimization over a Planning Period

9.1 Calculus of Variations

9.1.1 *Present value of total profit*

Whether you are the sole owner of a fish farm or the central planning board of a regulated fishing industry fishing over an open access fish habitat, it seems more sensible to develop a strategy for the optimal exploitation of the fish population ab initio (from scratch), unencumbered by existing practices (except for the size of the fish population at hand). The goal of this section is to examine the best policy for that purpose instead of trying to tinker with current practices piecemeal while trying to anticipate possible adverse effects downstream. When we have such a policy for optimal exploitation of the resource, we can then adopt appropriate strategies and regulations to implement that policy.

For such an approach we start with the (by now) familiar notion of the discounted net revenue (known as the *present value* of future return for fishing effort) from the harvest at some future time t:

$$L = e^{-rt}\,(p-c)\,h. \tag{9.1.1}$$

As before, $h(t)$ is the harvest rate (also called catch, yield, etc.) of our fishing effort at time t, p is the price for a unit biomass of fish harvested and r is the known discount rate. We continue to consider the fishery involved to be a price taker so that p remains a constant throughout our discussion of fish harvesting policy. However, c here is the cost of harvesting a unit of fish biomass and may vary with t, directly or indirectly through the harvest rate $h(t)$ and/or the population size $y(t)$. Until further notice, we take

$$c = c_0 + c_1 h \equiv c(h) \tag{9.1.2}$$

where c_0 and c_1 are constants.

The present value from the stream of profits over a period of T years (starting from $t = 0$) is

$$PV = \int_0^T L(t)dt = \int_0^T e^{-rt}\,[p - c(h)]\,hdt. \tag{9.1.3}$$

171

As before, the harvest rate h in this expression is related to the fish population by the growth rate relation

$$y' = f(y) - h \qquad (9.1.4)$$

where $f(y)$ is the natural growth rate of the unharvested fish population. At the initial time $t = 0$ of the planning period $[0, T]$, the initial fish population is known to be

$$y(0) = Y_0. \qquad (9.1.5)$$

Several scenarios are possible at the end of the planning period. Until further notice, we limit our discussion to the case of leaving a prescribed stock of fish for the next planning period (beyond year T) so that

$$y(T) = Y_T. \qquad (9.1.6)$$

The goal is to maximize the present value of the total profit from the revenue stream over the planning period $[0, T]$ subject to the growth dynamics (9.1.4), and the end conditions (9.1.5) and (9.1.6). Such a problem can be transformed into one in an area of mathematics known as the *calculus of variations*.

9.1.2 *A problem in the calculus of variations*

In the integrand of the integral (9.1.3), the harvest rate $h(t)$ is related to the fish population by the growth rate relation (9.1.4). By writing (9.1.4) as

$$h = f(y) - y', \qquad (9.1.7)$$

we may use this relation to eliminate h from the integrand of PV given by (9.1.3), wherever it may appear, to get

$$PV = \int_0^T e^{-rt} \left[p - c(f(y) - y') \right] [f(y) - y'] dt. \qquad (9.1.8)$$

The alternative form (9.1.8) of the present value of future profit stream shows that the integrand to depend only on the independent variable t, the unknown $y(t)$ and its first derivative $y'(t)$ (and various system parameters such as the unit price p, the unit cost c and the discount rate r). The maximization problem is now free of any side relation except for the one prescribed condition in terms of fish population at the two ends of the time interval. This is typical of what is known as the *basic problem* of the calculus of variations.

   In general, the maximum (or minimum) value of J depends on the set S of eligible comparison functions specified. For example, we may limit the eligible comparison functions $y(x)$ to n times continuously differentiable functions, denoted by C^n, or continuous and piecewise smooth functions, denoted by PWS. Given that y' appears in the integrand $F(x, y, y')$ of $J[y]$, we begin the theoretical development rather naturally with C^1 comparison functions, also known as *smooth comparison functions*. The choice of comparison functions may be modified and enlarged in different ways to be discussed below. A comparison function is called an *admissible comparison function* if it also satisfies all the prescribed end conditions (9.1.5) and (9.1.6).

9.1.3 The basic problem

In many areas of science and engineering, we often need to find a function $Y(x)$ defined in the interval $[a, b]$ so that the value of an integral of the form

$$J[y] \equiv \int_a^b F(x, y(x), y'(x)) \, dx, \quad (\;)' \equiv \frac{d(\;)}{dx} \qquad (9.1.9)$$

is extremized (maximized or minimized) when $y(x) = Y(x)$. In the integrand, we have used x as the independent variable (to convey the message that the independent variable needs not be time only) and designated the integral involved as $J[y]$ to indicate the fact that the value of the integral varies as we change the function $y(x)$. The quantity $J[y]$ is called the *performance index* and the integrand $F(x, y, v)$ is called the *Lagrangian* of the problem. The problems of maximizing PV in (9.1.8) is of this form; many others can be recast into the same form. The mathematical theory for this type of optimization problems is called the *calculus of variations*. In most cases, the admissible comparison functions $y(x)$ are specified at one end so that we have the initial condition:

$$y(a) = A. \qquad (9.1.10)$$

The situation at the other end, $x = b$, is less standardized. For this chapter, we limit our discussion to a prescribed value for $y(x)$ at the other end of the solution domain $[a, b]$, i.e.,

$$y(b) = B. \qquad (9.1.11)$$

Whenever the problem of extremizing $J[y]$ is subject to the two end constraints (9.1.10) and (9.1.11), we have a *basic problem* in the calculus of variations.

For definiteness, we (go against the popular practice and) talk specifically about minimization problems unless the context of a specific problem requires that we do otherwise. (Note that a maximization problem for $J[y]$ may be restated as a minimization problem for $-J[y]$.) In that case, the basic problem is stated more succinctly as

$$\min_{y \in S} \{ J[y] \mid y(a) = A, \; y(b) = B \} \qquad (9.1.12)$$

where S is the set of eligible *comparison functions* to be specified by the problem. Given that $y'(x)$ appears in the integrand, the set of eligible comparison functions must be at least differentiable (either continuously differentiable denoted by C^1 or piecewise smooth denoted by PWS). A comparison function is called an *admissible comparison function* if it also satisfies all the prescribed constraints on $y(x)$ such as the end conditions.

As stated in (9.1.12), the minimization problem poses a humanly impossible task. Taking it literally, the problem requires that we compare the values of $J[y]$ for all admissible comparison functions to find a minimizer $Y(x)$. In general, there are infinitely many admissible comparison functions as it is for the case of smooth (C^1) functions that satisfy the prescribed end conditions. What makes the calculus of variations a vibrant approach for science and engineering is that it manages to reduce this impossible search problem to another task that is humanly possible.

9.2　An Illustration of the Solution Process

9.2.1　*The action of a linear oscillator*

The reduction process mentioned above is rather lengthy to implement for some problems including the fishery problem. For the purpose of illustrating the solution process, we consider first the linear mass-spring oscillating system with an object of mass m at one end of a (linear) spring fixed at the other end. This is an ODE problem that is studied in all elementary differential equations courses whether or not you are a student of physics or chemistry.

Suppose $y(t)$ is the distance of the mass stretched from the at rest position at time t and the spring resists the stretching with a restoring spring force proportional to the length of the stretch y, i.e., a restoring force $= -ky$ for some spring constant k. There are generally two approaches to (a mathematical model for) determining the motion of the mass as it oscillates about its resting position $y = 0$. One is to apply Newton's laws of motion to formulate the mathematical model in the form of the differential equation (already encountered in any course on differential equations)

$$my'' + ky = 0, \qquad (\ )' = \frac{d(\ )}{dt}. \tag{9.2.1}$$

Together with some initial conditions (such as position and velocity at $t = 0$) on y, the motion of the mass is completely determined by the solution of the corresponding IVP.

An alternative approach to the same problem (advocated by mathematicians-physicists Hamilton and Lagrange) is to minimize an energy expression for this spring-mass system. The *kinetic energy* of the oscillating mass is *mass* $\times$ $(velocity)^2 /2 = m(y')^2/2$, while the *potential energy* $(= -$ work done by the resisting spring force) is $ky^2/2$. Hamilton's principle asserts that among all possible modes of oscillations that start at $y(0) = Y_0$ and end in $y(T) = Y_T$, the one that actually takes place *minimizes* the *action* (*integral*) of the system:

$$\min_y \left\{ J[y] = \frac{1}{2} \int_0^T \{m[y'(t)]^2 - k[y(t)]^2\}dt \right\} \tag{9.2.2}$$

with

$$y(0) = A, \qquad y(T) = B \tag{9.2.3}$$

where the *Lagrangian function* $F(t, y, y') = \left\{ m (y')^2 - ky^2 \right\}/2$ is the difference between the kinetic energy and potential energy.

The problem of minimizing the integral $J[y]$ in (9.2.2) subject to the end conditions $y(0) = A$ and $y(T) = B$ is similar to that for fish harvesting problem. This new problem may appear simpler since there is no added constraint of a differential equation (such as (9.1.4) for PV) on the process of extremization of $J[y]$. They are however of the same type of problem in the calculus of variations after we use (9.1.7) to express h in terms of y and y' in PV. Also, the mathematical modeling part of the oscillating mass-spring systems appears to be much more straightforward than the fishery problem. In the first instance, we have Newton's laws of motion known to be applicable to such a system. The second law of motion, immediately gives the governing differential equation (9.2.1) after the usual assumption on the restoring force. The alternative (action minimization) approach is more subtle. It requires the introduction of the concept of energy and characterization of kinetic and potential energy and what combination of these is to be minimized. However, the comparison left out the then new and equally subtle concept of momentum, reaction to action, etc., which took years of education for Newton's laws of motion to gain general acceptance. In any case, kinetics and potential energy are now basic concepts in physics and engineering familiar to most undergraduate in the physical sciences so that the action integral of a dynamical systems is no more difficult to write down than the relevant second law of motion for the problem.

9.2.2   *The condition of stationarity*

To motivate and illustrate the general (feasible) method of solution for the min-imization problem, we consider the basic problem for the linear mass-spring sys-tem with the performance index (9.2.2) and the two end conditions (9.1.5) and (9.1.6), working with x instead of t as the independent variable. Suppose we have found a C^1 function $Y(x)$ which satisfies the two end conditions and renders $J[y]$ a minimum value so that $J[y] \geq J[Y]$ for all admissible comparison functions. Let $y(x) = Y(x) + \varepsilon z(x)$, where ε is a (small) parameter and $z(x)$ is a smooth function in $(0, T)$ with

$$z(0) = z(T) = 0. \tag{9.2.4}$$

The stipulation that z vanishes at the end points is necessary for any comparison function $y(x)$ to satisfy the two end conditions (9.1.10) and (9.1.11). For any par-ticular $z(x)$ (and prescribed m and k), the value of $J[y]$ now depends on the value of the parameter ε, in fact only on ε after we have executed the integration with respect to x. We indicated this dependence by writing $J[y] \equiv J(\varepsilon)$:

$$J[y] = \frac{1}{2} \int_0^T \left\{ m \left[Y' + \varepsilon z' \right]^2 - k \left[Y + \varepsilon z \right]^2 \right\} dx \equiv J(\varepsilon). \tag{9.2.5}$$

Since $Y(x)$ minimizes $J[y]$, we have $J(\varepsilon) \geq J(0)$. For $\varepsilon = 0$ to be a local minimum point of the ordinary function $J(\varepsilon)$, we know from elementary calculus that $\varepsilon = 0$ must be a stationary point of $J(\varepsilon)$, i.e., $J'(\varepsilon) \equiv dJ/d\varepsilon$ must vanish at $\varepsilon = 0$. For $J[y]$ given by (9.2.5), we have

$$J'(0) = \left. \frac{dJ}{d\varepsilon} \right|_{\varepsilon=0} = \int_0^T \left\{ mY'z' - kYz \right\} dx \tag{9.2.6}$$

$$= \left[mY'z \right]_0^T - \int_0^T \left\{ (mY')' + kY \right\} z \, dx$$

$$= - \int_0^T \left\{ (mY')' + kY \right\} z \, dx = 0$$

where we have used the end conditions $z(0) = z(T) = 0$ to eliminate the terms outside the integral sign. To proceed to our principal result for determining $Y(x)$, we need a mathematical tool, called *the fundamental lemma of calculus of variations* that is probably new to most of the readers. It deserves a separate section of development. Here, we merely apply the lemma to (9.2.6) to conclude that the minimizing function $Y(x)$ must satisfy the second order ODE

$$(mY')' + kY = 0 \tag{9.2.7}$$

and the two end conditions in (9.2.3) if (9.2.6) is to hold for all admissible $z(t)$. As such, we have, with the help of a general mathematical result (Lemma 1 below), accomplished rather spectacularly a seemingly impossible task posed by (9.2.2). More specifically, we have narrowed the search for $Y(x)$ among an uncountably many candidates to among the solutions of the BVP defined by the *Euler differential equation* (9.2.7) for the given performance index and the two (Dirichlet) end conditions (9.2.3).

It is important to note that a problem in the calculus of variations may or may not have a solution as shown in some examples and exercises (see also Chapter 11 of this volume). If a solution exists, the BVP defined by (9.2.7), (9.1.5) and (9.1.6) can be solved, numerically if necessary.

9.2.3 The Euler differential equation

A solution of the simple *BVP* for $Y(x)$ associated with the Hamilton's principle for the linear mass-spring system defined by (9.2.7) and (9.2.3) is easily found to be

$$\hat{y}(x) = \frac{A\sin(\omega(T-t)) + B\sin(\omega t)}{\sin(\omega T)}, \qquad \omega = \sqrt{\frac{k}{m}} \tag{9.2.8}$$

provided $\omega T \neq n\pi$ for any integer n. The problem has no solution otherwise. In addition, the solution is unique for $T < \pi$. It would be tempting to conclude for this range of T that the $\hat{y}(x)$ given by (9.2.8) minimizes $J[y]$ since it is the only candidate that meets the requirements of $J'(0) = 0$ and the prescribed end conditions. However, $J'(0)$ is only a stationarity condition and $J(0)$ may be a maximum, a minimum or an inflection point. To see whether a solution $\hat{y}(x)$ of (9.2.7), called an *extremal* of the $J[y]$, renders J at least a local minimum, we examine the sign of $J''(0)$. With

$$J''(0) = \frac{d}{d\varepsilon} \left\{ \int_0^T \{m[\hat{y}' + \varepsilon h']h' - k[\hat{y} + \varepsilon h]h\} \, dx \right\} \Bigg|_{\varepsilon=0}$$

$$= \int_0^T \{m(h')^2 - kh^2\} \, dx,$$

it appears that $J''(0)$ can be of either sign. However, an application of Jacobi's rather ingenious technique shows that $J''(0)$ is in fact positive (see Chapter 5 of [34]) so that $J[\hat{y}]$ is a local minimum at least for $T < \pi$. Given that the solution $Y(t)$

of the BVP (9.2.7) and (9.2.3) is unique so that there is no other extremal for the problem, $J[Y]$ is actually a global minimum relative to the set of admissible comparison functions for the mass-spring problem.

The more subtle aspect of the solution obtained above for the linear mass-spring system pertains to the relevant set S of comparison functions. We agreed earlier to take S to be the set of C^1 functions. These are functions that has a continuous first derivative, also called smooth functions for brevity. However, in applying Lemma 1 below, we require that $g = -[(mY')' + kY]$ to be continuous in the interval $(0, T)$. For a constant mass m, this mean $Y''(t)$ should be continuous, a more stringent requirement for the comparison functions than just being smooth. For this specific spring-mass problem, it turns out that $Y(t)$ being a C^2 function is a consequence of the requirement of a continuous first derivative and the process of minimization; it does not require g to be continuous (as we did in Lemma 1). More specifically, the property of $Y(t)$ being twice continuously differentiable is not an additional restriction on the admissible comparison functions or the minimizer. The situation is analogous to the fact that analyticity of a function of a complex variable follows from the requirement of the existence of a first derivative [6].

9.3 The General Basic Problem

9.3.1 *The fundamental lemma*

We now undertake the task of proving the fundamental lemma of calculus of variations, the main mathematical tool that enables us to deduce the Euler DE (9.2.7) from the requirement that $J'(0)$ vanishes for all admissible comparison functions. For the illustrative example with an extremal $Y(x)$, this requirement can be rewritten as (9.2.6) which is of the form

$$\int_a^b g(x)h(x)\,dx = 0 \qquad (9.3.1)$$

with $g(x) = -[(mY'(x))' + kY(x)]$ and any *admissible comparison function* $h(x)$ that is C^1 (short for continuously differentiable) in $(a, b) = (0, T)$ with $h(a) = h(b) = 0$. We want to conclude that $g(x) = 0$ for all x in (a, b). Since this is a very fundamental and useful result, we want to state it clearly and precisely as the following lemma:

Lemma 1 *If $g(x)$ is continuous in (a, b) and (9.3.1) is satisfied for every choice of admissible comparison function, then $g(x) \equiv 0$ in (a, b).*

Proof Suppose $g(x)$ is not identically zero but is positive for some point $\bar{x}$ in (a, b). Since $g(x)$ is continuous, we have $g(x) > 0$ for all x in some interval (c, d) with $a \leq c < \bar{x} < d \leq b$. Now, (9.3.1) must hold for any C^1 function $h(x)$ with $h(a) = h(b) = 0$. In particular, it holds for the function

$$h(x) = \begin{cases} (x - c)^2 (d - x)^2, & c \leq x \leq d \\ 0, & \text{otherwise} \end{cases} \qquad (9.3.2)$$

which is C^1 with $h(a) = h(b) = 0$ and therefore an admissible comparison function. For this particular $h(x)$, the integral on the left-hand side of (9.3.1) is positive which contradicts the hypothesis (9.3.1). Hence, $g(x)$ cannot be positive in (a, b). By a similar argument, $g(x)$ also cannot be negative in (a, b). The *fundamental lemma* of the calculus of variations is proved.   □

It is not difficult to see that the proof applies also to a general continuous function $g(x)$. It remains valid if h is C^n for $n \geq 2$; it is only necessary to choose a similar but smoother comparison function $h(x)$ in (9.3.2).

Unfortunately, the minimizer $\hat{y}(x)$ is not always a C^2 function for other problems, not even C^1 function in some cases. As we will encounter shortly, the minimizer for some problems may be what we call a *PWS* (*piecewise smooth*) function:

Definition 15 *A function $y(x)$ is PWS in the interval (a, b) if it is continuous in [a,b] and its first derivative is C^1 in (a, b) except for a finite number of points in the interval where it has a finite jump discontinuity.*

Figure 9.1 shows the graph of a typical PWS function $y(x)$. Its derivative $y'(x)$ is shown in Figure 9.2.

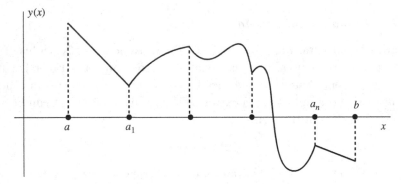

Fig. 9.1   Typical graph of a PWS function.

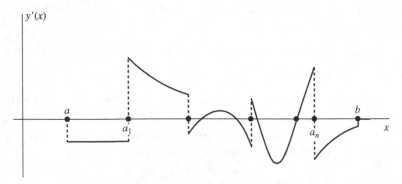

Fig. 9.2   Graph of the derivative of a PWS function.

We note that there are other definitions of PWS functions in the literature. One alternative definition would allow $y(x)$ to have a finite jump discontinuity at a finite number of points in the interval (a, b). For now, we limit our discussion to comparison functions that allow applications of the version of the fundamental lemma proved in this subsection and refer readers to the literature (e.g., [34]) for the more general case of PWS comparison functions.

9.3.2 Euler differential equation for the basic problem

The reduction of the basic problem to solving a Euler differential equation carried out for the spring-mass problem is also possible for a basic problem with a general Lagrangian $F(x, y, y')$. For this purpose, we narrow our set S of comparison functions to C^2 functions and again set $y(x) = Y(x) + \varepsilon z(x)$ in (9.1.9) so that

$$J[y] = \int_a^b F(x, Y + \varepsilon z, Y' + \varepsilon z')\, dx = J(\varepsilon).$$

We assume in the following calculations (as we will throughout this volume) that all second partial derivatives of F are continuous. With $F_{,\xi}$ indicating a partial derivative of F with respect to an argument ξ, we have

$$J'(0) \equiv \left.\frac{dJ}{d\varepsilon}\right|_{\varepsilon=0} = \int_a^b \left[F_{,y}\left(x, Y, Y'\right) z + F_{,y'}\left(x, Y, Y'\right) z' \right] dx$$

$$= \int_a^b \left\{ \hat{F}_{,y}\left(x\right) - \frac{d}{dx}\left[\hat{F}_{,y'}\left(x\right) \right] \right\} z\, dx \equiv \int_a^b G(x, Y, Y') z\, dx = 0$$

where

$$\hat{F}_{,y}\left(x\right) \equiv F_{,y}\left(x, Y, Y'(x)\right)$$

and

$$\hat{F}_{,y'}\left(x\right) \equiv F_{,y'}\left(x, Y, Y'\right).$$

Note that we have made use of the end conditions $z(a) = z(b) = 0$ (that must hold for all comparison functions) to eliminate the integrated terms from the expression for $J'(0)$ above. With S restricted to C^2 functions and F also C^2 in all arguments, $\hat{F}_{,y} - \left(\hat{F}_{,y'} \right)' \equiv \hat{G}(x)$ is continuous in (a, b). Hence, Lemma 1 applies and we have proved the following fundamental result in the calculus of variations:

Theorem 11 *Under the hypotheses that $S = C^2$ and $F(x, y, v)$ is C^2 in the infinite layer $L_\infty = \{a \leq x \leq b,\ y^2 + v^2 < \infty\}$, a minimizer $Y(x)$ for the basic problem (9.1.9)–(9.1.12) with the Lagrangian $F(x, y, y')$ must satisfy the Euler differential equation*

$$\left[(F_{,y'})' - F_{,y} \right]_{y=Y(x)} = 0 \tag{9.3.3}$$

and the two prescribed end conditions $Y(a) = A$ and $Y(b) = B$ for the basic problem.

Corollary 4 *If* $\hat{F}_{,y'y'}(x) = [F_{,vv}(x, Y(x), v)]_{v=Y'(x)}$ *exists and is continuous, then (9.3.3) can be written in its ultra-differentiated form:*

$$\left(\hat{F}_{,y'y'}\right) Y'' + \left(\hat{F}_{,y'y}\right) Y' + \hat{F}_{,y'x} = \hat{F}_{,y} \ .$$

At any point x where $\hat{F}_{,y'y'} \neq 0$, we can write the above ODE for $Y(x)$ as

$$Y'' = \frac{1}{\hat{F}_{,y'y'}} \left[\hat{F}_{,y} - \hat{F}_{,y'x} - \left(\hat{F}_{,y'y}\right) Y'\right] = f(x, Y, Y').$$

It should be noted that $Y(x)$ and $Y'(x)$ are not known until we have solved the BVP. Hence, $\hat{F}_{,y'y'}(x)$ is known to be nonvanishing a priori only for some rather special Lagrangians, e.g., $F(x, y, v) = e^{-(x+y+v)}$.

Though we have informally introduced the term *extremal* previously in this chapter, it seems appropriate at this point to define it formally:

Definition 16 *A solution $Y(x)$ of the Euler DE (9.3.3) is called an extremal of the given Lagrangian.*

Just as in the special case of the mass-spring system, an extremal is the counterpart of a stationary point in ordinary optimization problems in calculus. It may render the performance index a minimum, a maximum or a stationary point. In many applications, what kind of an extremum is the extremal may be clear from context or can be checked by direct verification. In others where the situation may be more delicate, a more general theory will need to be developed to give results corresponding to the second order conditions for the optimization of functions in calculus. Different aspects of such a theory may be found in [34] and elsewhere.

9.4   Integration of Special Cases

The Euler DE (9.3.3) is generally a nonlinear non-autonomous second order ODE and not amenable to an explicit solution in terms of known functions. To see what progress can be made toward such a solution, we examine below different special types of Lagrangian $F(x, y, y')$.

9.4.1 $F(x, y, y') = F(x)$

A Lagrangian that depends only on the independent variable x is of no interest since there is not a function $y(x)$ available to optimize the performance index. It is not a calculus of variations problem.

9.4.2 $F(x, y, y') = F(y)$

For this case, the Euler DE becomes $F_{,y}(y) = 0$ which is just an equation for the roots of a function of y. If one or more roots exist, they are constants, independent of x, i.e.,

$$Y(x) = \bar{Y}^{(k)} \tag{9.4.1}$$

for one or more known constant $\bar{Y}^{(k)}$. Since the solution must satisfy the two end conditions, a solution of the original problem exists only if $A = B$ and one of the roots equal to A; that is $\bar{Y}^{(k)} = A = B$ for some k. It is the only solution since we have restricted admissible comparison functions to C^1 functions.

There are problems in application areas where the solution is not restricted at the end points [34]. In other words, no end conditions are prescribed. In that case, any solution of the form (9.4.1) of the Euler DE is a candidate for the solution of the original problem. If there are more than one solutions of the form (9.4.1), e.g., $y^{(1)} = C$ and $y^{(2)} = D$, it is possible to consider a solution of the original problem that is a combination of the multiple solutions. For example, we may want to consider

$$Y(x) = \begin{cases} y^{(1)} = C & (a \leq x < c) \\ y^{(2)} = D & (c < x \leq b) \end{cases}$$

as a solution for the problem. But such combinations are usually inappropriate for the class of admissible comparison functions. The combination above, for example, would require us to enlarge the class of admissible comparison function and allow them to be discontinuous functions with a finite jump discontinuity.

More significantly, there are other necessary (and sufficient) conditions for optimality for any solution candidate; the Euler DE is only one of them. It should also be pointed out that optimization of a performance index without some or all prescribed end conditions is, technically speaking, no longer a *basic problem* of calculus of variations. As we shall see in later chapters, stationarity of non-basic problems typically leads to imposition of certain end conditions involving the Lagrangian of the problem. These (necessary) end conditions required for optimality are known as *Euler boundary conditions*.

9.4.3 $F(x, y, y') = F(y')$

For this case, the Euler DE becomes $[F_{y'}(y')]' = 0$ or $F_{y'}(y') = C$ where C is an arbitrary constant of integration. This last relation is just an equation for determining y' as a root of $F_{y'}(y') - C \equiv f(y') = 0$, so that $y' = c_1$. We have then the following general result to be used to avoid unnecessary calculations:

Theorem 12 *For a Lagrangian that depends only on y', the extremals must be of the form $Y(x) = c_0 + c_1 x$ with the two constants of integration to be determined by appropriate auxiliary conditions.*

Example 25 *Find the curve $y(x)$ in the (x, y)-plane with the shortest distance between the points $(0, A)$ and $(1, B)$.*

From elementary calculus, we know that the arclength between the two points in question is given by

$$L = \int_0^1 \sqrt{1 + (y')^2} dx, \quad \text{with} \quad y(0) = A, \quad y(1) = B.$$

Given the theorem, we can immediately write down $Y(x) = c_0 + c_1 x$. The first end condition requires $y(0) = c_0 = A$. The second end condition then becomes

$$Y(1) = A + c_1 \cdot 1 = B \quad \text{or} \quad c_1 = (B - A)$$

so that

$$Y(x) = A + (B - A)x.$$

The solution is a straight line between the two end points, confirming what we already knew without calculus of variations.

It is important to emphasize that we should make use of the theorem above the moment we see a Lagrangian of the form $F(y')$ and immediately write down the linear extremal

$$Y(x) = c_0 + c_1 x$$

without going through the Euler DE. In the case of a plane curve with the shortest arclength connecting two points, carrying out the differentiation in $(F_{,y'})'$ would lead to

$$\frac{y'y''}{\sqrt{1 + (y')^2}} = 0.$$

The analysis of this more complicated condition is not much more involved and would lead to the same result. However, working with the Euler DE for other more complicated Lagrangian of the form $F(y')$ may lead to a difficult (and possibly intractable) second order ODE and should be avoided.

9.4.4 $F(x, y, y') = F(x, y')$

This case is similar to the previous case except that the first integral $F_{y'}(x, y') = C$ is now a first order non-autonomous ODE with the roots of $F_{y'}(x, y') - C \equiv f(x, y') = 0$ now depends on the independent variable x:

$$y'(x) = v^{(k)}(x; C)$$

where $v^{(k)}(x; C)$ is some known function of x and the constant C. We have then the following extension of Theorem 12:

Theorem 13 *For a Lagrangian that depends only on x and y', the extremals for the basic problem are given by*

$$Y(x; C) = A + \int_a^x v^{(k)}(t; C)\,dt$$

where $v^{(k)}(x; C)$ is a root of the equation $F_{y'}(x, y') - C = 0$ with the constant C determined by $Y(b) = B$ if the solution for the same $v^{(k)}(x; C)$ applies throughout the interval (a, b).

It should be noted that a solution of $Y(b; C) = B$ for C may not exist. If so, the original basic problem of calculus of variations does not have a solution. Interesting applications of this class of basic problems when a solution does exist can be found in [34].

9.4.5 $F(x, y, y') = F(y, y')$

For this class of Lagrangians, the Euler DE can also be integrated once to get a first order ODE. To derive this first integral, we start by multiplying both sides of the Euler DE by y' and re-arranging the result to read

$$
\begin{aligned}
0 &= y' \left\{ F_{,y} \left(y, y' \right) - \left[F_{,y'} \left(y, y' \right) \right]' \right\} \\
&= y' F_{,y} \left(y, y' \right) - \left[y' F_{,y'} \left(y, y' \right) \right]' + y'' F_{,y'} \left(y, y' \right) \\
&= \left[F(y, y') - y' F_{,y'} \left(y, y' \right) \right]'.
\end{aligned}
$$

Upon integrating both sides with respect to the independent variable, we get the following useful result:

Theorem 14 *For a Lagrangian that does not depend explicitly on x, the Euler DE has a first integral*

$$
F(y, y') - y' F_{,y'} \left(y, y' \right) = C. \tag{9.4.2}
$$

The first integral (9.4.2) is essentially a first order autonomous ODE for $y(x)$. Its many applications will be illustrated in the next section. Unlike the special case $F(x, y, y') = F(y')$, the use of the first integral should not be automatic. For some simple Lagrangian such as $F(y, y') = (y')^2 - y^2$, the relation (9.4.2) becomes

$$
y^2 + (y')^2 = C.
$$

Finding an extremal $Y(x)$ involves solving a nonlinear first order ODE

$$
y' = \pm \sqrt{C - y^2}
$$

which is possible but nontrivial. This should be compared with the corresponding Euler DE:

$$
y'' + y = 0.
$$

The solution of the latter can be written down immediately.

9.4.6 $F(x, y, y') = F(x, y)$

For this last special case of the general Lagrangian for the basic problem of calculus of variations, there is no general method for simplification or reduction toward its solution. Not surprisingly, the same can be said about the general case where the Lagrangian depends on all three arguments. No recommendation can be made about a method of solution without knowing more about the Lagrangian itself.

9.5   The Fishery Problem with Catch-Dependent Cost of Harvest

9.5.1   The Euler DE and BVP

We now apply the general results of the previous section to find the extremal(s) for the optimal harvesting policy for the fishery problem for the case where the unit harvest cost only depends linearly on the harvest rate h. With

$$c = c_0 + c_1 h$$

where c_0 and c_1 are two known non-negative constants, the Lagrangian for the fishery problem (9.1.9) becomes

$$F(t, y, y') = e^{-rt} (p - c_0 - c_1 h) h, \quad \text{with} \quad h = f(y) - y'.$$

For brevity we have kept h in the Lagrangian F instead of expressing it in terms of y and y'. To set up the Euler DE, we need the following two expressions:

$$F_{,y'} = -e^{-rt} (p - c_0 - 2c_1 h),$$
$$F_{,y} = e^{-rt} f^{\cdot}(y) (p - c_0 - 2c_1 h)$$

obtained with the help of the chain rule. Similarly, we can also obtain

$$e^{rt} (F_{,y'})' = r(p - c_0) + 2c_1 [f^{\cdot}(y)y' - y'' - rh].$$

The Euler DE is then formed from $(F_{,y'})' = F_{,y}$ to give for the extremal $Y(t)$

$$r(p - c_0) = f^{\cdot}(Y)(p - c_0) + 2c_1 [Y'' - rY' + f(Y)\{r - f^{\cdot}(Y)\}]. \qquad (9.5.1)$$

The relation (9.5.1) is evidently a second order ODE generally nonlinear only in the unknown $Y(t)$. A nontrivial general solution for this equation is possible but likely to be complicated and requires some careful analysis. We contend ourselves here with examining a few special cases of the general problem to show possible complications that require more work for an appropriate solution for the optimal harvest policy.

9.5.2   Linear growth rate

Consider first the special case $f(y) = ay$ where $a > 0$ is the growth rate constant, a model appropriate for a fish population that is small compared to the carrying capacity of the environment (such as the high sea or a major river). In that case, we have $f^{\cdot}(y) = a$ and, assuming that $c_1 > 0$, the Euler DE becomes

$$Y'' - rY' - a(a - r)Y = \frac{1}{2c_1}(r - a)(p - c_0) \qquad (9.5.2)$$

which is a second order linear ODE with constant coefficients and with a constant forcing. The known method of solution gives the following general solution:

$$Y(t) = D_1 e^{at} + D_2 e^{(r-a)t} + p^* \quad \text{with} \quad p^* = \frac{p - c_0}{2ac_1}$$

assuming (the biologically realistic case of) $r \neq 2a$. (Otherwise, the solution would take a different form but still possible.) The two constants of integration D_1 and D_2 are then chosen to satisfy the two boundary conditions (9.1.5) and (9.1.6).

Though it is possible to do so, we will not write down the complete solution for the problem after determining the two constants of integration. Instead, we examine presently two special cases of the prescribed terminal fish stock.

9.5.2.1 $Y_T = Y_0$

If the current generation of consumers wants to be fair to the next generation, it would leave behind a fish stock at least equal to their own endowment from the previous generation. For the special case $Y_T = Y_0$, we have from the boundary conditions

$$Y(0) = D_1 + D_2 + p^* = Y_0, \tag{9.5.3}$$

$$Y(T) = D_1 e^{aT} + D_2 e^{(r-a)T} + p^* = Y_T = Y_0. \tag{9.5.4}$$

Together they determine D_1 and D_2 to be

$$D_1 = (Y_0 - p^*) \frac{1 - e^{(r-a)T}}{e^{aT} - e^{(r-a)T}}, \qquad D_2 = (Y_0 - p^*) \frac{e^{aT} - 1}{e^{aT} - e^{(r-a)T}}.$$

The corresponding extremal $Y(t)$ is well defined. However, (as an exercise) we must check to be sure that both $Y(t)$ and $h(t) = aY - Y'$ should be nonnegative (as the fish population cannot be negative nor can there be a negative harvest rate).

9.5.2.2 $Y_T = 0$

At the other extreme, the current generation may be so selfish that the entire fish stock is so heavily harvested and consumed so that there will be nothing left for the next generation beyond the current planning period. In that case, we have from (9.5.3)

$$D_2 = (Y_0 - p^*) - D_1$$

and upon using this result in (9.5.4) with $Y_T = 0$

$$D_1 = \frac{(Y_0 - p^*) e^{(r-a)T} + p^*}{e^{(r-a)T} - e^{aT}}.$$

With the result for D_1, the expression for D_2 becomes

$$D_2 = -\frac{(Y_0 - p^*) e^{-aT} + p^*}{e^{(a+r)T} - e^{-aT}}.$$

Again the extremal is well defined and infinitely continuously differentiable. However, we need to verify that both $Y(t)$ and $h(t)$ are nonnegative.

9.5.3 *Constant unit harvest cost ($c_1 = 0$)*

Consider a different special case with $c_1 = 0$ (so that $c = c_0$) but now the growth rate is not restricted (to be ay) or specified (except for the purpose of illustrations). The Euler DE for a constant unit harvest cost degenerates to

$$f'(Y) = r. \tag{9.5.5}$$

With all the derivative terms disappeared, the Euler DE reduces to an equation determining one or more levels of optimal population size Y_s that do not change with time, with the solution given by (9.4.1).

For example, if the growth rate is (normalized) logistic with $f(y) = y(1 - y)$, we have $f'(y) = 1 - 2y$ with the "Euler DE" (9.5.5) having the unique solution

$$Y_s = \frac{1}{2}(1 - r)$$

and the corresponding optimal harvest rate

$$h_s = f(Y_s) - Y_s' = f(Y_s) = \frac{1}{4}(1 - r^2).$$

If the discount rate is zero ($r = 0$), the expression for h_s is just the *maximum sustainable yield* for the logistic growth we encountered a while back.

While all that is satisfying, this extremal is not acceptable as an optimal solution unless the initial and terminal fish stocks are both exactly Y_s. For $Y_0 \neq Y_s$, it is not difficult to conclude intuitively what the correct solution for this rather pathological problem should be. For the case $Y_T < Y_s < Y_0$, we should

- fish at an infinite harvesting rate to drive the fish population down to the optimal population Y_s instantaneously;
- fish at the optimal harvesting rate h_s for planning period $0 < t < T$, and then
- fish at an infinite harvesting rate to drive the fish population down to the prescribed terminal population Y_T instantaneously to complete the optimal harvesting program for maximum profit.

However, such an optimal policy is not realistic since harvesting capacity is finite so that we cannot fish at infinite harvesting rate to reduce Y_0 to Y_s and then Y_s to Y_T.

To the extent that the harvest rate is constrained by the inequality,

$$0 \leq h \leq h_{\max}, \tag{9.5.6}$$

how do we deduce mathematically the optimal policy given the constraints? For the present relatively simple problem, we can again figure out the optimal policy intuitively. But for more complicated problems when our intuition fails to lead to the correct solution, we would need a deductive process for obtaining the solution and showing that it is optimal. For the fishery problem, we show in the next section what a rigorous mathematical solution for the problem entails. In the last few chapters of this volume, we will introduce and illustrate the more generally applicable theory of *optimal control* for a broad class of constrained optimization problem. But before doing that, we need to develop the next two chapters some additional mathematical tools needed for that purpose.

9.5.4   The method of most rapid approach

For the fishery problem, the relevant performance index is the present value of the profit over a planning period,

$$J[u] = \int_0^T e^{-rt}[p - c]u \, dt$$

(where we now use $u(t)$ for the harvest rate) with the growth dynamics of the fish population governed by

$$y' = f(y) - u, \qquad y(0) = Y_0, \qquad y(T) = Y_T. \tag{9.5.7}$$

Before applying a general solution technique to such problems a more limited method known as the Most Rapid Approach that is uniquely suited for this problem.

For simplicity, we consider here only p and c being given functions of time t. In order to turn a profit over the planning period, we stipulate that $p(t) - c(t)$ should be positive for some segments of $[0, T]$. In that case the method of calculus of variation would use the growth equation to eliminate the harvest rate to get

$$J = \int_0^T e^{-rt}[(p - c)] [f(y) - y'] \, dt.$$

The corresponding Euler DE is

$$\left[-e^{-rt}(p - c)\right]' = e^{-rt}[(p - c)]f\,'(y)$$

or

$$(p - c) [f\,'(y) - r] + (p - c)' = 0. \tag{9.5.8}$$

Since p and c are known functions of t (and do not involve y or y'), the Euler DE is an equation (and not a differential equation) that determines y as a function of t, denoted by $Y_i(t)$.

Generally, an extremal with two constants of integration determined by the second order Euler DE is known as an *interior solution* of the problem and denoted by $Y_i(t)$. When the Euler DE degenerates to an equation without any derivative of the unknown, it is known as a *singular solution* of the problem and denoted by $Y_s(t)$. A singular solution generally cannot be made to satisfy the prescribed conditions at the two ends of the solution domain since it does not contain any constant of integration, not being a solution of an ODE. By itself, it is usually not the solution of the original optimization problem. This can be seen from the extreme case where $p = p_0$ and $c = c_0$ are both constants with $p_0 - c_0 > 0$. For the logistic growth case where $f(y) = y(1 - y)$, we have $Y_s = (1 - r)/2$ which does not satisfy either end condition unless $Y_0 = Y_T = (1 - r)/2 \; (= Y_s)$.

To be comprehensive, we need to consider the general case $Y_0 \neq Y_s(0) \neq Y_T$ whether or not the singular solution is a constant. For the optimal fishing strategy, we may assume that the maximum harvest rate available is such that Y_T can be reached in finite time not longer than T. Otherwise, the problem has no solution (or a solution that harvests at the maximum rate and get as close to Y_T as possible by time T). The other requirement from the Euler DE is that the optimal policy should have $Y(t) = Y_s(t)$ whenever possible.

For $Y_0 > Y_s(0)$, the optimal harvesting strategy would be to drop the fish stock from Y_0 to $Y_i(0)$ to get on the singular extremal $Y_s(t)$. But this cannot be done instantaneously given an upper bound on the maximum harvest rate so that $U(t) \leq u_{max}$. If the natural fish growth rate of the fish population less than u_{max} (the finite harvesting capacity of the fishery), a realistic solution would be to fish at the maximum allowable harvest rate u_{max} from the start to drive Y_0 down to $Y_s(t_0)$ after a finite interval of elapsed time $[0, t_0]$. (It is possible that $t_0 > T$ in which case the optimal harvest strategy would fish at u_{max} for the planning period.) For $t_0 < T$, the optimal policy calls for $Y(t) = Y_s(t)$ whenever possible and we should stay with that policy as long as possible once we reach the optimal stock level $Y_s(t)$ at $t = t_0$. We should get off the optimal stock level at the last possible moment and fish at the maximum harvest rate u_{max} starting at (the latest possible moment) t_T that would drive the optimal fish stock at that instant to the prescribed lower terminal stock Y_T (or not harvesting so that $Y(t)$ grows to the required terminal population if $Y_T > Y_s(t_T)$). Such a policy, known as *the most rapid approach* strategy, is obviously *feasible* and intuitively *optimal* given the practical constraints. For $Y_T < Y_s(t_T)$, this most rapid approach strategy is

$$U(t) = \begin{cases} u_{max} & (0 \leq t < t_0) \\ U_s(t) & (t_0 < t < t_T) \\ u_{max} & (t_T < t \leq T) \end{cases} \qquad (9.5.9)$$

where $U_s(t) = f(Y_s(t)) - Y_s'(t)$ with $Y_s(t)$ being a solution of (9.5.8). The first switch time t_0 is determined by $x(t_0) = Y_s(t_0)$ where $x(t)$ is the solution of the IVP

$$y' = f(y) - u_{max}, \qquad y(0) = Y_0.$$

The second switch time t_T is determined by the condition $z(t_T) = Ys(t_T)$ where $z(t)$ is the solution of the terminal value problem

$$z' = f(z) - u_{max}, \qquad z(T) = Y_T.$$

That (9.5.9) is the optimal control for maximizing $J[u]$ is proved by an appropriate application of Green's theorem in integral calculus (see [33] and Proposition 34 of Chapter 13).

It is not difficult to see how this same approach may also apply to problems with other combinations of the prescribed initial and terminal value of $y(t)$. It is important to note also that a solution of the optimization problem may not exist with a singular solution phase or may not exist at all. We have already encountered one example when the method of most rapid approach fails to include the singular solution, namely when $t_0 > t_T$ for the problem with $Y_0 > Y_s(0)$ and $Y_s(T) > Y_T$. Another problem with no solution is for $Y_T > Y_0 > Y_s(0)$ and a natural growth rate too slow to grow the fish population from Y_0 to Y_T by the terminal time T.

To illustrate other possible complications, we note that the situation is problematic in a different way if the initial fish stock is below the singular level so that we have $Y_0 < Y_s(0)$, or when the terminal fish stock is above the optimal level so that $Y_s(T) < Y_T$. To apply the method of *most rapid approach* and also meet the stock constraints at the two ends in this case, we need to harvest at a negative rate (by adding fish to the fishing ground for example). An infinite negative harvest rate (for instantaneous growth to the optimal fish stock level) is clearly not an option; and a finite negative harvest rate is also not an option for fishing in the open sea. The best we can do is not to fish until the fish population increases to the level of $Y_s(t_0)$ for some $t_0 > 0$ near the start and to stop fishing at some $t_T < T$ so that fish population would increase to Y_T at the terminal time. The most rapid approach solution would not include a singular solution phase if $t_0 > t_T$.

If fishing is regulated, a realistic strategy for the case $Y_0 < Y_s(t) < Y_T$ would be to impose a fishing moratorium so that the fish stock can grow from the initial stock Y_0 up to the optimal level $Y_s(t_0)$ over a finite time interval $[0, t_0]$, hopefully $t_0 < T$. Once Y_s is reached, we should again stay on the optimal trajectory $Y(t) = Y_s(t)$ as long as possible until the very last moment t_T when we must get off by not fishing in order for the fish stock to grow to the prescribed terminal level at $t = T$. If the initial fish stock cannot be brought up to the stationary fish stock level $Y_s(t)$ or the prescribed terminal stock level Y_T by a fishing moratorium, there would be no solution that meets all the requirements of our optimal harvest strategy, re-affirming a previous discussion of possible non-existence of a solution for BVP.

At the other extreme, we have a different kind of problem for the case $Y_T < Y_s(T)$ and/or $Y_s(0) < Y_0$. Instantaneous reduction from Y_0 to $Y_s(0)$ or $Y_s(T)$ to Y_T is not possible; we can only drive the fish stock down to the desired level by fishing at maximum harvest rate u_{max}. If this cannot be done with the prescribed u_{max}, then the problem does not have a solution that meets all the conditions of optimality. While we certainly can (and often have to) settle for the best we can do under the circumstances, a higher value of the performance index (the discounted stream of profits over the planning period in the fishery case) could have been attained should the constraints be relaxed.

Evidently, for cases when the prescribed initial or terminal fish stock is not the optimal stock level required by the extremal, there may or may not be a feasible harvesting policy under the inequality constraint:

$$0 \le u(t) \le u_{max}. \tag{9.5.10}$$

As mentioned above, the intuitive strategy has been proved rigorously to be optimal for the fishery problem by the Green's Theorem [3, 34] (see also Proposition 34 of Chapter 13). The method is not always applicable to other related problems such as those involving higher order systems of growth dynamics. For this reason, other solution processes will be discussed in the last few chapters to address optimization problems with inequality constraints on the controls. We begin this effort by developing in the next chapter additional mathematical tools for problems in calculus of variations to extend its range of applications and meet the needs of optimal control theory for problems in optimization with inequality constraints.

Chapter 10

Modifications of the Basic Problem

10.1 Smooth and PWS Extremals

In writing down a Lagrangian that involves the first derivative of the optimizer $Y(x)$, we naturally seek a solution among the class of *smooth* (continuously differentiable) functions denoted by (the class of) C^1 functions. In the derivation that led to the Euler differential equations as a necessary condition to qualify as an optimizer, i.e., an extremal, we actually imposed a more restrictive constraint on the class of admissible functions. They are not just C^1 functions, but twice continuously differentiable (or C^2) functions since the Euler DE involves two differentiations and, after carrying out the differentiation of $F_{,y'}$ with respect to x would contain two differentiations. Unfortunately, in many applications, the solution of the problem turns out not to meet these restrictions, not even a C^1 function. We illustrate this situation with a simple example below.

Suppose we would like to find the minimizer $Y(x)$ for the performance index

$$J[y] = \int_a^b y^2(1-y')^2 dx, \quad \text{with} \quad y(a) = A, \quad y(b) = B. \tag{10.1.1}$$

The Lagrangian is of the type $F(y, y')$ which admits a first integral for an extremal $Y(x)$:

$$F - y'F_{,y'} = y^2\left[1 - (y')^2\right] = C_0. \tag{10.1.2}$$

This first order ODE can be re-arranged to read

$$\frac{yy'}{\sqrt{y^2 - C_0}} = \pm 1 \quad \text{or} \quad y^2 = C_0 + (C_1 + x)^2 \tag{10.1.3}$$

where C_0 and C_1 are two constants of integration to be determined by the prescribed end conditions.

Example 26 $y(2) = 1, \quad y(3) = \sqrt{3}$.

For this set of end conditions, we find $C_0 = 3/4$ and $C_1 = -3/2$ so that

$$Y^2 - \left(x - \frac{3}{2}\right)^2 = \frac{3}{4}. \tag{10.1.4}$$

Of the two branches $Y = \pm\sqrt{(3/4) + (x - 3/2)^2}$ of the smooth hyperbola between $x = 2$ and $x = 3$, the end conditions are met only by the positive branch (see Figure 10.1).

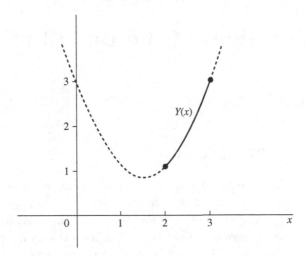

Fig. 10.1   A piece of the hyperbola.

Example 27 $y(0) = 0$, $y(2) = 1$.

For this set of end conditions, we find from (10.1.3) $C_0 = -9/16$ and $C_1 = -3/4$ so that a formal solution is given by

$$\left(x - \frac{3}{4}\right)^2 - Y_f^2 = \frac{9}{16}. \tag{10.1.5}$$

In principle, the positive branch of the two square roots $Y_f = \pm\sqrt{3/4 + (x - 3/2)^2}$ of the smooth hyperbola between $x = 0$ and $x = 1$ should be selected in order to satisfy the two end conditions. However, we see from Figure 10.2 that there is no part of either branch of hyperbola (10.1.5) lies in the interval $(0, 3/2)$! There is a smooth piece of the hyperbola in the interval $(3/2, 2)$ available should we wish to use it. The end condition $y(0) = 0$ is satisfied by the apex of the other branch of the hyperbola. But there is no parts of the formal solution lies in the interval $(0, 3/2)$, and hence no solution for the problem.

To salvage a solution for the problem, we may consider taking the extremal to be

$$\bar{y}(x) = \begin{cases} 0 & (0 \le x < 3/2) \\ \sqrt{\left(x - \frac{3}{4}\right)^2 - \frac{9}{16}} & (3/2 < x \le 2) \end{cases} \tag{10.1.6}$$

keeping in mind that $\bar{y}(x) = 0$ is a solution of the first integral of the Euler DE with $C_0 = 0$. Even if it is appropriate, the adoption of (10.1.6) also requires that

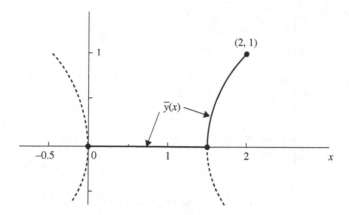

Fig. 10.2 Incorrect solution.

we relax the restriction on the class of admissible comparison functions, allowing them to be continuous but only piecewise smooth as the function $\bar{y}(x)$ defined by (10.1.6) has a kink at $x = 3/2$ and is therefore not differentiable there.

Less we think that $\bar{y}(x)$ as given by (10.1.6) is our minimizer, it is not! It is not difficult to verify that $J[\bar{y}(x)] > 0$. On the other hand, it is clear from the performance index that min $[J] = 0$ since the integrand is nonnegative. From the Lagrangian, we see that there are two ways to render $J = 0$. One is setting $y(x) = 0 \equiv y_0(x)$; but that does not satisfy the terminal condition of $y(2) = 1$. The other is to set $y'(x) = 1$ or $y(x) = x - 1 \equiv y_2(x)$ where we have specified the constant of integration to satisfy the boundary condition $y(2) = 1$. But with $y_2(0) = -1$, this choice of $y(x)$ does not satisfy the boundary condition $y(0) = 0$ at the other end of the solution interval.

In order to find a suitable $Y(x)$, we take a page from our choice of $\bar{y}(x)$ in the previous paragraph and set

$$Y(x) = \begin{cases} y_0(x) = 0 & (0 \leq x \leq c) \\ y_2(x) = x - 1 & (c \leq x \leq 2). \end{cases} \qquad (10.1.7)$$

This choice of the minimizer satisfies both end conditions; it also satisfies the first integral (10.1.2) of the Euler DE with $C_0 = 0$. To ensure that $Y(x)$ is at least continuous at the switch point c, we pick c so that $x - 1 = 0$ at $x = c$ or $c = 1$. With this choice of c, $Y(x)$ is continuous, piecewise smooth (abbreviated as the class of PWS functions) in the interval $(0, 2)$, satisfies the Euler DE where it is differentiable and satisfies the two end conditions. It clearly minimizes $J[y]$. To the extent that no C^1 function can be found to do the same, we have to settle for this PWS function (10.1.7) as the solution for the problem and thereby requiring us to enlarge the class of admissible comparison functions to be the class of PWS functions. In doing so, the proof of the fundamental lemma of calculus of variations no longer applies. Readers are referred to [34] for a generalization of Lemma 1 to allow for PWS comparison functions.

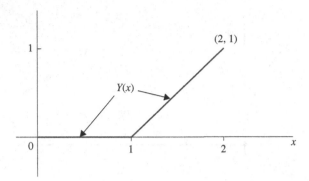

Fig. 10.3   The true minimizer.

10.2   Unspecified Terminal Unknown Value

10.2.1   *Euler boundary conditions*

When we use the fishery problem as an example to introduce the basic problem of the calculus of variations, we stipulated a known initial fish population at the start of the fishing program and also a terminal fish population to reflect a desire to leave an appropriate amount of fish for the next planning period or the next generation of consumers. The last of these stipulations is somewhat idealistic. Even an Act of the Congress of the United States does not bind future Congress from having a different agenda and repealing it. A more realistic assumption may be to leave the terminal fish population unspecified and let the optimal policy dictate what to leave behind for the next planning period. Not specifying the terminal fish population enlarges the collection of admissible comparison functions and should generate a present value of the total profit not less than that for the case of a prescribed terminal population and generally larger.

Mathematically, one issue arises immediately in connection with the solution of the Euler DE for this modified problem with only one prescribed (initial) end condition. With the expectation that the same Euler DE continues to apply (and we shall demonstrate this presently) as a necessary condition for optimality, the second order ODE would still need two auxiliary conditions for a complete specification of the solution. Where and how do we get this second boundary condition to accompany the known initial condition $y(a) = A$?

For an answer to that question and other effects of not having a terminal condition, we re-work the development leading to the stationary condition $J'(0) = 0$ and obtain, in addition to the usual Euler DE, another necessary condition for stationarity in the form of a Euler boundary condition. Suppose again $Y(x)$ is the optimizer and $y(x) = Y(x) + \varepsilon z(x)$ a nearby admissible comparison function with $z(a) = 0$ (which follows from $y(0) = Y(0) = A$). The stationary condition $J'(0) = 0$

requires

$$J'(0) \equiv \left.\frac{dJ}{d\varepsilon}\right|_{\varepsilon=0} = \int_a^b \left[F_{,y}\left(x,Y,Y'\right)z + F_{,y'}\left(x,Y,Y'\right)z' \right] dx$$

$$= \left[F_{,y'}\left(x,Y,Y'\right)z(x) \right]_{x=a}^b + \int_a^b \left\{ \hat{F}_{,y}\left(x\right) - \frac{d}{dx}\left[\hat{F}_{,y'}\left(x\right) \right] \right\} z\,dx$$

$$\equiv \left[\hat{F}_{,y'}\left(b\right) \right] z(b) + \int_a^b \hat{G}(x)z(x)\,dx = 0 \qquad (10.2.1)$$

where we have made use of the initial condition $z(a) = 0$ and the simplifying notation

$$\hat{G}(x) = \hat{F}_{,y}\left(x\right) - \frac{d}{dx}\left[\hat{F}_{,y'}\left(x\right) \right]$$

with $\hat{P}(x) = P(x, Y(x), Y'(x))$ for any $P(x, y, v)$.

The difference between the expression $J'(0)$ here and that for the basic problem is the addition of the first term involving $z(b)$ in the expression. Since there is no prescribed condition at $x = b$, the class of admissible comparison function $z(x)$ is now larger; they include those with $z(b) = 0$ as well as those with $z(b) \neq 0$. If we restrict comparison at first to those $z(t)$ that vanishes at the end point $x = b$ so that $z(b) = 0$, the first term in (10.2.1) disappears and the situation is the same as that for the basic problem encountered in Section 3 of Chapter 4. Application of the fundamental lemma of calculus of variations again requires the Euler DE to be satisfied as a necessary condition for stationarity.

Returning now to the more general case that includes $z(x)$ not vanishing at $x = b$, the Euler DE as a necessary condition for extremum leaves us with

$$J'(0) = \left[\hat{F}_{,y'}\left(b\right) \right] z(b) = 0.$$

Since $J'(0) = 0$ must hold also for comparison function with $z(b) \neq 0$ (as well as those with $z(b) = 0$), we must have the following requirement for the extremals of the problem:

Theorem 15 *In the extremization of the performance index $J[y]$ given by (9.1.9) subject to (9.1.10), the extremal $Y(x)$ must satisfy also the Euler boundary condition (in addition to the Euler DE)*

$$\hat{F}_{,y'}\left(b\right) = F_{,y'}\left(b, Y(b), Y'(b)\right) = 0 \qquad (10.2.2)$$

at the end point $x = b$ where there is not a prescribed boundary condition.

The Euler boundary condition (10.2.2) provides us with a second BC needed for the second order Euler DE. In general, there should be an Euler BC whenever an end condition is *not* prescribed (whether it is at $x = b$ or at $x = a$).

Example 28 *Find the optimal curve $Y(x)$ and the optimal y coordinate B of the terminal point $(1, B)$ with the shortest distance between the fixed point $(0, A)$ and the terminal point. (In other words, the value B is not specified and is a part of the solution.)*

Similar to the previous example with $(1, B)$ specified, we know that the arclength of the curve starting at $(0, A)$ is given by

$$L = \int_0^1 \sqrt{1 + (y')^2}dx, \qquad \text{with} \qquad y(0) = A.$$

With $F(x, y, y') = F(y')$, we can immediately write down $Y(x) = c_0 + c_1 x$. The first end condition requires $y(0) = c_0 = A$. The second end condition comes from the Euler BC $F_{y'}(Y'(1)) = 0$ or

$$\frac{Y'(1)}{\sqrt{1 + (Y'(1))^2}} = \frac{c_1}{\sqrt{1 + c_1^2}} = 0$$

so that $c_1 = 0$ and

$$Y(x) = A$$

(as expected from other more elementary methods of solution).

10.2.2   Terminal payoff

A firm is contracted to produce flu vaccine units to be delivered at the start of the flu season next year $(t = T)$ with a payoff B per unit of vaccine delivered. In general, more vaccines would be produced (with a diminishing return) if more resources dedicated to making them. The rate of vaccine production is taken to be

$$y'(t) = \alpha\sqrt{u(t)}, \qquad y(0) = 0 \tag{10.2.3}$$

where $u(t)$ is the resource investment rate dedicated to vaccine making and α is a proportional constant. No vaccine is produced until the firm begins its investment. For a more general vaccine production rate, we have $y' = f(u)$, where the function $f(u)$ typically has the properties i) $f(0) = 0$, ii) $f'(u) > 0$ and iii) $f''(u) < 0$, the last signifying diminishing return of a higher rate of investment.

To summarize, the investment is over a fixed period T with whatever completed units to be delivered at $t = T$. At the time of delivery, there will be a payoff B for each unit completed and not before. The optimization problem is to find an optimal spending/investment rate $U(t)$ that maximizes the net return, i.e., profit, for the total investment:

$$J[u] = Be^{-rT}y(T) - \int_0^T e^{-rt}u(t)\, dt. \tag{10.2.4}$$

As stated, the problem can be solved by the method of calculus of variations. For this purpose, we use the ODE (10.2.3) to eliminate u from J to get

$$J[y] = Be^{-rT}y(T) - \int_0^T e^{-rt}\left[y'/\alpha\right]^2\, dt. \tag{10.2.5}$$

With no terminal end condition, the problem is not a basic problem in the calculus of variations. Because of the terminal payoff term $Be^{-rT}y(T)$ in the expression for $J[y]$, it is also not quite the simple problem of no terminal condition. To find the solution of the problem, we again convert the problem to an optimization problem for an ordinary function $J(\varepsilon)$. Suppose $Y(t)$ is the maximizer and $y(t) = Y(t) + \varepsilon z(t)$ is an admissible comparison function. The integral $J[y] = J(\varepsilon)$ becomes a function of ε. The stationary condition may be rearranged to read

$$J'(0) = Be^{-rT}z(T) - \left[\frac{2}{\alpha^2}e^{-rt}Y'(t)z(t)\right]_0^T + \int_0^T \left[\frac{2}{\alpha^2}e^{-rt}Y'\right]' z(t)dt$$

$$= e^{-rT}\left[B - \frac{2}{\alpha^2}Y'(T)\right]z(T) + \int_0^T \left[\frac{2}{\alpha^2}e^{-rt}Y'\right]' z(t)dt = 0$$

after integration by parts and application of the end condition $z(0) = 0$.

The condition $J'(0) = 0$ holds for all admissible comparison functions now widened to allow for those $z(t)$ with $z(T) \neq 0$ (as well as those comparison functions with $z(T) = 0$). Suppose we apply the stationary condition to those with $z(T) = 0$ (and there are plenty of those). In that case, we apply the fundamental lemma of calculus of variations to get the expected Euler DE

$$\left[\frac{2}{\alpha^2}e^{-rt}Y'\right]' = 0. \tag{10.2.6}$$

It can be integrated twice to get

$$Y(t) = c_1 + c_2 e^{rt}$$

where c_1 and c_2 are two constants of integration. The initial condition $y(0) = 0$ then allows us to eliminate one of the two constants leaving us with

$$Y(t) = c_2 \left(e^{rt} - 1\right).$$

To determine the second constant, we need another end condition. For that we return to $J'(0) = 0$ now simplified by the Euler DE to

$$J'(0) = e^{-rT}\left[B - \frac{2}{\alpha^2}Y'(T)\right]z(T) = 0,$$

keeping in mind that the condition must also hold for all comparison functions with $z(T) \neq 0$. This requires

$$Y'(T) = \frac{1}{2}B\alpha^2 \tag{10.2.7}$$

since $e^{-rT} > 0$. The condition (10.2.7) is known as an inhomogeneous Euler boundary condition. It provides the auxiliary condition needed to determine the remaining constant of integration to be

$$Y'(T) = c_2 re^{rT} = \frac{1}{2}B\alpha^2 \tag{10.2.8}$$

or

$$c_2 = \frac{1}{2r}B\alpha^2 e^{-rT}$$

so that

$$Y(t) = \frac{1}{2r} B\alpha^2 e^{-rT} \left(e^{rt} - 1 \right),$$ (10.2.9)

$$Y'(t) = \frac{1}{2} B\alpha^2 e^{-r(T-t)}.$$ (10.2.10)

The corresponding optimal investment rate is then obtained from (10.2.3):

$$U(t) = \frac{1}{4} B^2 \alpha^2 e^{-2r(T-t)}.$$ (10.2.11)

It is always a good idea to check to see if the optimal results make sense. In this problem, we can calculate the total investment by evaluating the integral

$$\int_0^T e^{-rt} U(t)\, dt = \frac{1}{4} B^2 \alpha^2 e^{-2rT} \int_0^T e^{rt}\, dt$$ (10.2.12)

$$= \frac{1}{4r} B^2 \alpha^2 e^{-2rT} \left[e^{rT} - 1 \right].$$

The expression (10.2.12) is only half of the terminal payoff for the vaccine units already produced given by

$$B e^{-rT} Y(T) = \frac{1}{2r} B^2 \alpha^2 e^{-2rT} \left(e^{rt} - 1 \right).$$ (10.2.13)

As such the optimal policy is definitely worth pursuing. While it remains to be proved, $J[Y(t)]$ is expected to be the highest profit from the firm's investment.

In order for the payoff to double our investment, we must invest at the optimal rate given by (10.2.11). What if there is a limit on your bank account or credit that is below the required investment rate. Since the highest investment rate occurs at the terminal time T, what should the optimal investment rate be if

$$\frac{1}{4} B^2 \alpha^2 > u_{\max}$$

where $u_{\max}$ is the maximum investment rate allowed?

At the other end of the spectrum, it is possible that with a different production function $f(u)$, even the best investment strategy cannot earn a net profit. Since we cannot have negative investment (at least no one would send you cash on demand), the best we can do under the circumstances would be not to invest. In practice, there are typically constraints on the controlling mechanism of the problem. For the investment for vaccine production problem, what we can control is the investment rate. In the simplest situation, we would have the inequality constraint

$$u_{\min} \leq u(t) \leq u_{\max}$$ (10.2.14)

with the possibility that the lower bound be some positive number greater than zero. For example, the U.S. tax law requires any non-profit organization to spend at least 5% of its asset each year to qualify for tax-exempt status. In a later chapter, we develop a more general theory that takes into account inequality constraints such as (10.2.14) (which also applies to the fishery problem).

10.2.3 *Euler boundary conditions with terminal payoff*

When the problem involves a terminal payoff (known as a salvage value) such as the one in the optimal investment problem above, the performance index typically contains a non-integrated term evaluated at an end point:

$$J[y] \equiv \int_a^b F(x, y(x), y'(x)) \, dx + Q(b, y(b))$$

where $Q(b, y(b))$ is a known function of b and $y(b)$. The Euler BC can be derived in the same way as we did for the particular example of vaccine production. In this way, we obtain an inhomogeneous Euler *BC*:

Theorem 16 *In the extremization of the performance index $J[y]$ with the Lagrangian $F(x, y, y')$ as given by (9.1.9) subject to (9.1.10), the extremal $Y(x)$ must satisfy the inhomogeneous Euler boundary condition*

$$F_{,y'} (b, Y(b), Y'(b)) + Q_{,y_b} (b, Y(b)) = 0 \tag{10.2.15}$$

(with $y_b \equiv y(b)$) at the end point $x = b$ where there is no prescribed boundary condition.

For the optimal investment rate problem, we have $F(t, y, y') = - e^{-rt}(y'/\alpha)^2$ while $Q(T, y(T)) = Be^{-rT}y(T)$. The inhomogeneous Euler BC (10.2.15) becomes

$$-\frac{2}{\alpha^2}e^{-rT}Y'(T) + Be^{-rT} = 0, \quad \text{or} \quad Y'(T) = \frac{1}{2}\alpha^2 B$$

identical to what we found previously in (10.2.7).

10.3 Higher Derivatives and More Unknowns

10.3.1 *Higher derivatives equivalent to more unknowns*

Up to now, we have restricted our discussion to Lagrangians that involve a single unknown optimizer $y(x)$ and its first derivative as well as the independent variable x. Many problems in applications involved higher derivatives of the unknown and/or more than one unknown optimizers. We need to extend our theory to deal with these more complicated problems.

An important problem in biomechanics is the buckling of a column (which may be an actin filament or a microtubule). In a variational formulation, the problem is to find an optimizer $Y(x)$ that extremizes the performance index

$$J[y] = \frac{1}{2} \int_0^L \left[EI \left(y''\right)^2 - P \left(y'\right)^2 \right] dx. \tag{10.3.1}$$

In this expression, L, E, I and P are prescribed constants.

Generally speaking, problems with derivatives higher than first order can be transformed into one with only first derivatives. For the buckling problem above, we can set $z = y'$ the performance index may be rewritten as

$$J[y, z] = \frac{1}{2} \int_0^L \left[EI \left(z'\right)^2 - P \left(z\right)^2 + \lambda(z - y') \right] dx$$

where the relation between y and z is incorporated into the performance index by adding something to J that does not change its value since $z - y' = 0$. The multiplicative parameter λ corresponds to the Lagrange multiplier encountered in the optimization of ordinary functions in elementary calculus. The transformed problem now involves more than one unknown but with only first derivative of the unknowns in the Lagrangian. As such, it can be handled in a way similar to that for the basic problem of one unknown especially if we treat the multiple unknowns as components of a single vector unknown.

While we can always transform a problem involving higher order derivative into a problem with a vector unknown and its first derivative only, it is sometimes advantageous to stay with the higher order derivative formulation, as it relates more easily to the science of the problem. For scientists and engineers who work in the biomechanics of deformable bodies, EIy'' is the bending moment of the slender body (called "beam" for brevity) proportional to (an approximate expression of) the curvature of the bent beam. This interpretation may be lost when we replace y'' by z'. But for the problems discussed in these notes, we will not pursue further discussion of problems with higher derivatives in the Lagrangian and convert any such problems to problems for multiple unknowns and their first derivatives.

Exercise 30 *By forming $J(\varepsilon) = J[Y(x) + \varepsilon z(x)]$, determine the Euler DE of the performance index $J[y]$ in (10.3.1) from the stationary condition $J^{\cdot}(0) = 0$.*

10.3.2   *Euler differential equations for several unknowns*

For the case of two unknowns $x(t)$ and $y(t)$, now with t as the independent variable, we have as a performance index

$$J[x,y] \equiv \int_a^b F(t, x(t), y(t), x'(t), y'(t))\, dt \qquad (10.3.2)$$

for a basic problem with

$$x(a) = X_a, \qquad x(b) = X_b, \qquad y(a) = Y_a \quad y(b) = Y_b. \qquad (10.3.3)$$

To reduce the problem to solving Euler DE, we again suppose that $X(t)$ and $Y(t)$ are the relevant optimizers and set $x(t) = X(t) + z_x(t)$ and $y(t) = Y(t) + z_y(t)$. Upon carrying out the integration involved, we are left with $J(\varepsilon)$ with $\varepsilon = 0$ being the stationary point of $J(\varepsilon)$.

The usual calculations lead to

$$J^{\cdot}(0) = \left[\hat{F}_{,x'}(t)\, z_x(t) + \hat{F}_{,y'}(t) z_y(t) \right]_{t=a}^b$$
$$+ \int_a^b \left\{ \left[\hat{F}_{,x}(t) - \hat{F}_{,x'}(t) \right] z_x(t) + \left[\hat{F}_{,y}(t) - \hat{F}_{,y'}(t) \right] z_y(t) \right\} dt, \quad (10.3.4)$$

where $\hat{G}(t) = G(t, X(t), X'(t), Y(t), Y'(t))$. With $J^{\cdot}(0) = 0$ to hold for all admissible comparison functions $z_x(t)$ and $z_y(t)$, adroit applications of (an extended version of) the fundamental lemma of calculus of variations result in the following necessary conditions for $(X(t), Y(t))$ to render $J[x,y]$ an extremum:

Theorem 17 *For all admissible C^2 vector comparison functions $\mathbf{y}(t) = (y_1(t), \cdots, y_n(t))$ and a C^1 Lagrangian $F(t, \mathbf{y}, \mathbf{v})$, any vector optimizer $\mathbf{Y}(t)$ for the basic problem (10.3.2)–(10.3.3) must satisfy the following n Euler differential equations*

$$\left(\hat{F}_{,y_i'}(t) \right)' - \hat{F}_{,y_i}(t) = 0 \qquad (i = 1, \ldots, n). \tag{10.3.5}$$

A vector function $\mathbf{Y}(t)$ that satisfies the Euler DE (10.3.5) is a vector extremal of the given Lagrangian.

10.3.3 *Some examples*

Example 29 *Suppose a firm is contracted to produce two different types of vaccines with different unit payoff and different level of diminishing returns. Determine the optimal investment policy for maximum profit from the investment.*

Let the production rates x' and y' for the two vaccines, respectively, as functions of u and v, the two respective investment rates for their production, be given by

$$y' = \alpha u^{1/\beta}, \qquad x' = \gamma v^{1/\sigma}, \qquad y(0) = x(0) = 0. \tag{10.3.6}$$

Suppose we want to maximize

$$J[y, x] = [B_y y(T) + B_x x(T)] e^{-rT} - \int_0^T (u + v) e^{-rt} dt.$$

The problem is a simple generalization of the single vaccine case; it is straightforward to deduce the following two Euler DE for the problem:

$$\left[\frac{\beta}{\alpha} \left(\frac{Y'}{\alpha} \right)^{\beta-1} e^{-rt} \right]' = 0, \qquad \left[\frac{\sigma}{\gamma} \left(\frac{X'}{\gamma} \right)^{\sigma-1} e^{-rt} \right]' = 0$$

and the inhomogeneous Euler BC

$$\left(\frac{Y'(T)}{\alpha} \right)^{\beta-1} = \frac{\alpha}{\beta} B_y, \qquad \left(\frac{X'(T)}{\gamma} \right)^{\sigma-1} = \frac{\gamma}{\sigma} B_x.$$

It follows that the extremals for the problem are

$$Y(t) = \frac{\alpha(\beta-1)}{r} \left(\frac{\alpha B_y}{\beta} \right)^{1/(\beta-1)} \left[e^{-r(T-t)/(\beta-1)} - e^{-rT/(\beta-1)} \right],$$

$$X(t) = \frac{\gamma(\sigma-1)}{r} \left(\frac{\gamma B_x}{\sigma} \right)^{1/(\sigma-1)} \left[e^{-r(T-t)/(\sigma-1)} - e^{-rT/(\sigma-1)} \right].$$

For the special case of $\beta = 2$ and $\sigma = 3$, these expressions simplify to

$$Y(t) = \frac{\alpha^2 B_y}{2r} \left[e^{-r(T-t)} - e^{-rT} \right],$$

$$Z(t) = \frac{2\gamma}{r} \left(\frac{\gamma B_x}{3} \right)^{1/2} \left[e^{-r(T-t)/2} - e^{-rT/2} \right]$$

with

$$Y(T) = \frac{\alpha^2 B_y}{2r} \left[1 - e^{-rT}\right],$$

$$Z(T) = \left(\frac{4\gamma^3 B_x}{3r^2}\right)^{1/2} \left[1 - e^{-rT/2}\right].$$

For $\alpha = 2\sqrt{\gamma}$ and $3B_y^2 > \gamma B_x$ it follows from

$$\left[1 - e^{-rT}\right] > \left[1 - e^{-rT/2}\right]/\sqrt{3}$$

that the optimal strategy is to produce more vaccine type y as investment for production diminishes less than type x.

Example 30 *When the vaccine delivery date is the start of the next flu season, the firm will have to store the vaccine produced for a period until delivery. It may choose to delay production somewhat to save some storage cost. But it cannot wait until the last minute for then the production cost may either be much too high or the production rate too limited to meet the deadline. To investigate the effect of storage cost, we start with the simplest case of the two-vaccine problem with production rates (10.3.6). Storage cost is included in the performance index which in this case is the profit from the vaccine production:*

$$J[y, z] = [B_y y(T) + B_x x(T)]e^{-rT} - \int_0^T \left[(u + v) + (c_y y + c_x x)^m\right] e^{-rt} dt.$$

Here c_y and c_x are the storage costs per unit time for a unit of vaccine type y and type x, respectively. We are interested in the change of investment strategy (from the one obtained in the previous example) due to the presence of storage costs to get the maximum profit for the investment.

For $m = 1$, the Euler DE for this problem are

$$\left[\frac{\beta}{\alpha}\left(\frac{Y'}{\alpha}\right)^{\beta-1} e^{-rt}\right]' = c_y e^{-rt}, \qquad \left[\frac{\sigma}{\gamma}\left(\frac{X'}{\gamma}\right)^{\sigma-1} e^{-rt}\right]' = c_x e^{-rt}.$$

Except for the inhomogeneous terms $c_y e^{-rt}$ and $c_x e^{-rt}$ on the right-hand side, the problem is not very different from the previous one without storage cost. Mathematically, it remains a calculus problem and will be left as an exercise. However, the presence of the storage cost would change the production strategy somewhat.

For $m \geq 2$, the two Euler DE become coupled and must be solved simultaneously

$$\left[\frac{\beta}{\alpha}\left(\frac{Y'}{\alpha}\right)^{\beta-1} e^{-rt}\right]' = mc_y(c_y Y + c_x X)^{m-1} e^{-rt},$$

$$\left[\frac{\sigma}{\gamma}\left(\frac{X'}{\gamma}\right)^{\sigma-1} e^{-rt}\right]' = mc_x(c_y Y + c_x X)^{m-1} e^{-rt}.$$

A vector version of the first integral for Lagrangians of the form $F(y, y')$ will be useful to reduce the fourth order system to a second order system. The needed results will be deduced through a parametric representation of the relevant extremals to be discussed in the next section.

10.4   Parametric Form and Free End Problems

10.4.1   Basic problem in parametric form

Suppose the solution $y(x)$ of the basic problem for one unknown is represented parametrically as $x = \phi(t)$ and $y = \psi(t)$, $\alpha \leq t \leq \beta$, with $\phi(\alpha) = a$, $\phi(\beta) = b$, $\psi(\alpha) = A$, and $\psi(\beta) = B$. In that case, the performance index may be rewritten as

$$J = \int_\alpha^\beta F(x, y, \frac{\dot{y}}{\dot{x}}) \dot{x}\, dt = \int_\alpha^\beta f(x, y, \dot{x}, \dot{y})dt = J[x, y]$$

with

$$f(x, y, \dot{x}, \dot{y}) = F(x, y, \frac{\dot{y}}{\dot{x}})\dot{x}, \qquad (\ )^{\cdot} = \frac{d(\ )}{dt}.$$

A direct application of Theorem 17 leads to the two Euler DE

$$\left(\hat{f}_{,\dot{x}}\right)^{\cdot} = \hat{f}_{,x}, \qquad \left(\hat{f}_{,\dot{y}}\right)^{\cdot} = \hat{f}_{,y}. \tag{10.4.1}$$

In terms of $F(x, y.\dot{y}/\dot{x})$, we have the following two Euler DE for the basic problem in parametric form:

Theorem 18 *The extremals $\{X(t), Y(t)\}$ of the parametric form of the general performance index $J[y(x)]$ of the basic problem must satisfy the following two Euler DE:*

$$\left[\hat{F} - \frac{\dot{Y}}{\dot{X}}\hat{F}_{,y'}\right]^{\cdot} = \dot{X}\cdot\hat{F}_{,x}, \qquad \left[\hat{F}_{,y'}\right]^{\cdot} = \dot{X}\cdot\hat{F}_{,y}. \tag{10.4.2}$$

Proof By the chain rule of partial differentiation, we have

$$\hat{f}_{,\dot{y}} = \left[\frac{\partial}{\partial\dot{y}}\left(F(x, y, \frac{\dot{y}}{\dot{x}})\right)\dot{x}\right]_{x=X(t),\ y=Y(t)} = \hat{F}_{,y'}(t)$$

and

$$\hat{f}_{,y} = \left[\frac{\partial}{\partial y}\left(F(x, y, \frac{\dot{y}}{\dot{x}})\right)\dot{x}\right]_{x=X,\ y=Y} = \hat{F}_{,y}(t)\dot{X},$$

where

$$\tilde{G}(t) = [G(x, y, \dot{y}/\dot{x})]_{x=X(t), y=Y(t)}.$$

These relations enable us to write the second Euler DE in (10.4.1) as the second Euler DE in (10.4.2).

Similarly, the first Euler DE in (10.4.2) is established from the first relation in (10.4.1) with the help of

$$
\hat{f}_{,x^{\cdot}} = \left[\frac{\partial}{\partial x^{\cdot}} \left(F(x, y, \frac{y^{\cdot}}{x^{\cdot}}) \right) x^{\cdot} + F\left(x, y, \frac{y^{\cdot}}{x^{\cdot}} \right) \right]_{x=X,\ y=Y}
$$

$$
= \left[\frac{\partial}{\partial y'} \left(F(x, y, \frac{y^{\cdot}}{x^{\cdot}}) \right) \left(-\frac{y^{\cdot}}{x^{\cdot}} \right) + F\left(x, y, \frac{y^{\cdot}}{x^{\cdot}} \right) \right]_{x=X,\ y=Y}
$$

$$
= \hat{F}(t) - \frac{Y^{\cdot}}{X^{\cdot}} \hat{F}_{,y'}(t)
$$

and

$$
\hat{f}_{,x} = \left[\frac{\partial}{\partial x} \left(F(x, y, \frac{y^{\cdot}}{x^{\cdot}}) \right) x^{\cdot} \right]_{x=X,\ y=Y} = X^{\cdot} \hat{F}_{,x}(t). \qquad \square
$$

10.4.2 A first integral and Erdmann's conditions

With the help of the chain rule of differentiation, the two Euler DE (10.4.2) may be written in terms of x and $Y(x)$ alone,

$$
\frac{d}{dx} \left[\hat{F} - Y'\hat{F}_{,y'} \right] = \hat{F}_{,x}, \qquad \frac{d}{dx} \left[\hat{F}_{,y'} \right] = \hat{F}_{,y} . \qquad (10.4.3)
$$

The second of these ODE is the original Euler DE for the basic problem. The first is a new relation when we allow x to vary as well. For this new condition, we deduce a previous assertion (that was verified but not deduced):

Corollary 5 *If the Lagrangian of the original basic problem $F(x, y, y')$ does not depend explicitly on the independent variable x, then $\hat{F}(x) - Y'\hat{F}_{,y'}(x)$ is invariant along an extremal $Y(x)$ of the Lagrangian, i.e., $\hat{F}(x) - Y'\hat{F}_{,y'}(x) = C_0$.*

A different type of useful information can also be extracted from the parametric form of the Euler DE (10.4.3). $F_{,y}$ is generally a function x, y and y' and $y(x)$ is taken to be a continuous and piecewise smooth function of x with only a finite number of simple jump discontinuities in (a, b) (to accommodate solutions for problems such as the second example in the first section of this chapter). Hence, $\hat{F}_{,y}(x)$ as a function of x has at worst some simple jump discontinuities at one or more location in the solution interval. It follows from the integrated form of the second relation of (10.4.3),

$$
\hat{F}_{,y'}(x) = C_1 + \int_a^x \hat{F}_{,y}(s) ds \qquad (10.4.4)
$$

that $F_{,y'}$ must be a continuous function of x (since the right-hand side of (10.4.4) is continuous). A similar argument leads to the conclusion that $\hat{F} - Y'\hat{F}_{,y'}$ must also be a continuous function of x. Together, we have the two well-known Erdmann corner conditions:

Corollary 6 *If $y(x)$ is restricted to the class of PWS functions, then $\hat{F}_{,y'}$ and $\hat{F} - Y'\hat{F}_{,y'}$ are both continuous along an extremal (even if the extremals themselves are only PWS).*

These two Erdmann conditions have very important applications in determining the smoothness of extremals. We illustrate with the following simple example:

Example 31 $F(x, y, y') = \left[(y')^2 - y^2 \right] / 2.$

The first Erdmann condition requires $\hat{F}_{,y'}(x) = Y'(x)$ be continuous along an extremal. In other words, even if we allow the comparison functions to be PWS, the admissible extremals can only be those that are smooth (continuously differentiable). Note that for this problem the second Erdmann condition does not yield any new information since both Y and Y' are already known to be continuous.

What is more interesting is the consequence of the first Euler DE in (10.4.3) for the same example:

$$-\frac{1}{2}\frac{d}{dx}\left[(Y')^2 + Y^2 \right] = -Y'(Y'' + Y) = 0.$$

We know from the first Erdmann that any extremal is smooth. If $Y'(x_0) \neq 0$ at a point x_0, then $Y'' + Y = 0$ in some interval centered at x_0. It follows from $Y'' = -Y$ that the extremal is twice continuously differentiable (C^2 for short) in the interval (since Y is continuous there). The same relation then implies that y'' itself is C^2. Repeating the same argument shows that any extremal for our Lagrangian must be C^∞ in the interval centered at x_0 where $Y'(x_0) \neq 0$.

10.4.3 Free end problems

For some problems, the location of an end point of the solution domain is not specified but chosen for optimality. We will see a substantive example of this type in two later chapters of these notes. Here we mention only the simple example of finding the shortest path $y(x)$ from a prescribed end point (a, A) to a point on a line away from the point. For this shortest path, the location of the far end b is clearly not prescribed but depends on the location and orientation of the line. To illustrate, consider the following well-known example from plane geometry:

Example 32 *Find the shortest path $y(x)$ from the prescribed point $(a, y(a)) = (a, A)$ to the line $y = B > A$.*

From simple geometrical consideration, the shortest path from (a, A) to the line $y = B > A$ is obviously the vertical line from (a, A) to (a, B), or the line defined by $x(y) = a$ for $A \leq y \leq B$. In the calculus of variations setting, we want to minimize

$$J = \int_a^b \sqrt{1 + (y')^2}\, dx$$

subject to the end conditions $y(a) = A$ and $y(b) = B$ where a, A and B are fixed but b can vary.

If b is prescribed and B not, the missing second end point (b, B) was found previously to be replaced by the Euler BC $[F_{,y'}(x, y, y')]_{x=b} = 0$. For the present problem, B is known and we need an auxiliary condition to take the place of the missing (or incomplete) second end condition. To derive this replacement condition, we return to the expression (10.3.4) (now using variational notations with (a,b) there replaced by (α, β) here, etc.) for δJ but now with $\delta b = \delta x(\beta)$ arbitrary (since b is not prescribed). By first allowing the variation to have vanishing end values, $\delta J = 0$ again leads to the two Euler DE (10.4.2) or (10.4.3). But for the broader set of comparison functions with $\delta x(\beta) \neq 0$ (and hence possibly a lower minimum J), we are left with

$$J^{\cdot}(0) = \left[F_{,y'}(X,Y,Y^{\cdot}/X^{\cdot})z_y(t) + \left\{ F\left(X,Y,\frac{Y^{\cdot}}{X^{\cdot}}\right) - \frac{Y^{\cdot}}{X^{\cdot}}F_{,y'}\left(X,Y,\frac{Y^{\cdot}}{X^{\cdot}}\right) \right\} z_x(t) \right]_\alpha^\beta$$

$$= \left[F\left(X,Y,\frac{Y^{\cdot}}{X^{\cdot}}\right) - \frac{Y^{\cdot}}{X^{\cdot}}F_{,y'}\left(X,Y,\frac{Y^{\cdot}}{X^{\cdot}}\right) \right]_{t=\beta} z_x(\beta) = 0$$

since $z_x(\alpha) = 0$, $z_y(\alpha) = 0$, and $z_y(\beta) = 0$. The following *transversality condition* in lieu of the missing end condition follows from the necessary vanishing of the coefficient of $z_x(\beta)$:

Theorem 19 *For a modified basic problem with an unspecified end location (but a specified terminal value for the extremal there), the following transversality condition must be met at the unknown terminal end location b:*

$$\left[\hat{F}(x) - Y'(x)\hat{F}_{,y'}(x) \right]_{x=b} = 0. \tag{10.4.5}$$

To apply this result to the shortest path Example 32 above, we know that any extremal of the problem must be the linear function

$$Y(x) = A + c(x - a)$$

with

$$B = A + c(b - a) \quad \text{or} \quad b = a + \frac{1}{c}(B - A). \tag{10.4.6}$$

The constant c is chosen to meet the necessary (transversality) condition (10.4.5):

$$\left[\hat{F}(x) - Y'(x)\hat{F}_{,y'}(x) \right]_{x=b} = \frac{1}{\sqrt{1 + c^2}} = 0$$

which requires $c = \infty$. It follows from (10.4.6)

$$b = a$$

consistent with the conclusion reached by geometrical consideration (that starting from the point (a, A), the curve to the point (b, B) with the shortest arclength is $X(y) = a$ for $A \leq y \leq B$.

Example 33 *For the more interesting case of finding the shortest path from (a, A) to a point on the line $g(x) = mx + n$. Neither coordinate of the terminal point (b, B) is prescribed.*

While geometrical consideration would have the shortest path being perpendicular to the line $g(x)$, so that the extremal $Y(x)$ should be $A - (x - a)/m$, we are interested in a calculus of variations solution for the problem. For such a solution, we need a replacement for the missing condition similar to (10.4.5) for the case when B is known. Since neither b nor B is known (hence $z_x(\beta)$ and $z_y(\beta)$ generally do not vanish) but related by $y(b) = g(b) = g(x(\beta))$ along the prescribed line. In that case, the expression for $J^{\cdot}(0) = 0$ after simplification by the Euler DE now takes the form

$$J^{\cdot}(0) = \left[F_{,y'}(X, Y, Y^{\cdot}/X^{\cdot}) z_y(t) + \left\{ F\left(X, Y, \frac{Y^{\cdot}}{X^{\cdot}} \right) - \frac{Y^{\cdot}}{X^{\cdot}} F_{,y'}\left(X, Y, \frac{Y^{\cdot}}{X^{\cdot}} \right) \right\} z_x(t) \right]_{t=\beta}.$$

Let $\hat{b}$ be the optimal value for b with $b = \hat{b} + \varepsilon z_x(\beta)$. Then $y(b) = g(b) = g(\hat{b} + \varepsilon z_x(\beta)) = g(\hat{b}) + g'(\hat{b})\varepsilon z_x(\beta) + O(\varepsilon^2)$. But we also have $y(b) = Y(\hat{b}) + \varepsilon z_y(\beta) + O(\varepsilon^2)$; it follows that

$$z_y(\beta) = g'(\hat{b}) z_x(\beta)$$

and the nonvanishing boundary terms in $J^{\cdot}(0)$ may be rewritten as

$$J^{\cdot}(0) = \left[\hat{F}_{,y'}(\hat{b}) g'(\hat{b}) + \left\{ \hat{F}(\hat{b}) - \frac{Y^{\cdot}}{X^{\cdot}} \hat{F}_{,y'}(\hat{b}) \right\} \right] z_x(\beta).$$

For the stationary condition $J^{\cdot}(0) = 0$ to hold for any nonvanishing $z_x(\beta)$, we must have

$$\hat{F}_{,y'}(\hat{b}) g'(\hat{b}) + \left\{ \hat{F}(\hat{b}) - Y'(\hat{b})(\hat{F}_{,y'}(\hat{b}) \right\} = 0. \qquad (10.4.7)$$

For $g(x) = mx + n$ and $y(x) = A + c(x - a)$, we obtain from (10.4.7)

$$\frac{mc}{\sqrt{1 + c^2}} + \frac{1}{\sqrt{1 + c^2}} = 0$$

so that

$$c = -\frac{1}{m}$$

just as we expect from geometrical consideration.

Chapter 11

Boundary Value Problems are More Complex

11.1 Boundary Value Problems and Their Complications

11.1.1 *Second order equations*

For the ODE systems studied in Parts 1 and 2, all supplementing auxiliary conditions are prescribed at the same value (t_0 or x_0) of the independent variable. As such, they are called initial conditions (IC) and the resulting problem for the ODE system an initial value problem (IVP). In many other problems in applications, some of the auxiliary conditions are prescribed at one end of a finite interval of the independent variable and the others are prescribed at the other end of the same interval. This is the case for the auxiliary conditions at the end points of the solution interval for the Euler DE for the basic problem of calculus of variations. They are called boundary conditions (BC) for the ODE; together, the ODE and the BC define a boundary value problem (BVP).

While you can have as few as a single initial condition in an IVP (for a first order ODE), we need at least a second order ODE to have a BVP since there must be at least two auxiliary conditions, one at each of the two ends of an interval. As an example, we may wish to find a function $T(x)$ that satisfies the simple ODE

$$\frac{d^2T}{dx^2} = f(x), \quad (0 < x < L), \tag{11.1.1}$$

where $f(x)$ is a known function, and the auxiliary conditions

$$T(x = 0) = T_0, \quad T(x = L) = T_l. \tag{11.1.2}$$

The two auxiliary conditions (11.1.2) for the second order ODE (11.1.1) are the *boundary conditions* for the differential equation. Together, the ODE (11.1.1) and the boundary conditions (11.1.2) define a BVP.

Typically (but not exclusively), BVP are for differential equations whose independent variable characterizes location in physical space. For this reason, we use x instead of t as the independent variable. While physical space is three-dimensional in nature, we limit discussion in this part of the notes mainly to spatially one-dimensional problems leading to BVP for ODE and not PDE (an abbreviation for partial differential equations).

As usual, we can work with a first order system as in IVP with the order of the system $n \geq 2$. However, many problems appear naturally in the form of a system of second order ODE; these include equations of motion of the classical many-body problem in planetary science, astrophysics and molecular dynamics. For these problems and diffusive phenomena in the biological sciences occurring in a finite spatial domain, the relevant ODE systems typically appear naturally in the form of second order equations that can be written as a vector ODE in the form:

$$\mathbf{y}'' = \mathbf{f}(x, \mathbf{y}, \mathbf{y}'), \qquad (\ )' = \frac{d(\ )}{dx}, \tag{11.1.3}$$

for $\mathbf{y}(x) = (y_1(x), \ldots, y_n(x))^T$ with $n > 1$. (Obviously, an odd order system of ODE cannot be written in this form and we would have to work with a form appropriate for the problem.) Our focus in these notes is mainly on the scalar second order ODE

$$y'' = f(x, y, y'), \qquad (x_0 < x < x_T), \tag{11.1.4}$$

to convey some basic information about BVP. For the special case of a single linear second order equation, it is advantageous to take it in the so-called *self-adjoint* form

$$L[y] = f(x), \qquad (x_0 < x < x_T), \tag{11.1.5}$$

with the linear differential operator $L[\ ]$ defined by

$$L[y] \equiv (p(x)y')' - q(x)y \tag{11.1.6}$$

for some known functions $p(x)$ and $q(x)$. (Note that $L[\ ]$ here is unrelated to the interval length L in (11.1.1) for the independent variable.) More will be said about self-adjoint ODE in subsequent developments.

11.1.2   *No solution or too many solutions*

Compared to boundary value problems, initial value problems in ODE are relatively uncomplicated. Under some mild conditions on the vector growth rate function $\mathbf{f}(x, \mathbf{y})$ (e.g., differentiability in all arguments of the components of $\mathbf{f}$), the IVP $\mathbf{y}' = \mathbf{f}(x, \mathbf{y})$, $\mathbf{y}(x_0) = \mathbf{y}^o$ is known to have a unique solution for some finite interval that includes x_0 in the interior (see Appendix B). This is not always the case for boundary value problems. We illustrate the nature of the difficulties with the simple ODE (11.1.1) made even simpler by setting

$$f(x) = 0, \qquad x_0 = 0 \quad \text{and} \quad x_T = 1.$$

The method of solution for this BVP is straightforward. Integrate the ODE (11.1.1) twice to get

$$T(x) = C_0 + C_1 x.$$

The two boundary conditions (11.1.2) require

$$T(0) = C_0 = T_0, \quad T(1) = C_0 + C_1 = T_l$$

which completely determine C_0 and C_1 so that

$$C_0 = T_0, \quad C_1 = T_l - T_0, \qquad T(x) = T_0 + (T_l - T_0)x.$$

If the boundary condition $T(1) = T_l$ is replaced by $T'(1) = Q_l$, we can repeat the solution process to get

$$C_0 = T_0, \quad C_1 = Q_l, \qquad T(x) = T_0 + Q_l x.$$

Now suppose the two original boundary conditions (11.1.2) are replaced by $T'(0) = Q_0$ and $T'(1) = Q_l$. The same solution process leads to the following two conditions on the solution of the ODE:

$$T'(0) = C_1 = Q_0, \quad T'(1) = C_1 = Q_l.$$

These two conditions cannot be met by any choice of C_1 unless $Q_1 = Q_0$. Hence, the problem has no solution if $Q_1 \neq Q_0$. If $Q_1 = Q_0$, then we get $C_1 = Q_l (= Q_0)$ and $T(x) = C_0 + Q_0 x$ is a solution of the BVP for any C_0 since the problem imposes no other requirement on the solution. As such the solution of the new *BVP* is not unique in this case as it has a one-parameter family of solutions, each corresponding to a different real value for C_0.

The example above shows that BVP for ODE are much more complex than IVP. The existence and uniqueness of the solution is not assured even for the simplest BVP which would have caused no concern if the boundary conditions should be replaced by initial conditions. While the higher level of complexity inherent in BVP will continue to pose challenges to their researchers and users, the physical systems modeled by the BVP often provide an explanation for the relevant complexity (such as nonexistence or non-uniqueness of its solution). It is therefore critical that we have a complete grasp of the phenomenon being characterized by the model. Whereas a unique solution is dictated for some problems, the occurrence of non-unique solutions is sometimes beneficial or even necessary for progress as we shall see later.

11.2 One-Dimensional Heat Conduction

11.2.1 *Rate of change of heat*

In order to understand and explain the complications encountered in solving BVP associated with the ODE (11.1.1), it would be helpful to digress from the development of mathematical theories and solution techniques and discuss the problem of heat conduction in a straight wire. In addition to providing insight on possible reasons for the non-existence and non-uniqueness of solutions for the BVP in ODE, a mathematical formulation of the heat conduction problem will enable us to embark on a discussion of eigenvalue problems in ODE systems in a later chapter and how they enable us to solve PDE problems. Also, heat is characteristically diffusive and diffusion, heat and temperature all play important roles in the development of biological organisms. Investing a little time to acquire some understanding of phenomena pertaining to them should contribute to our repertoire of research tools in mathematical and computational biology. More discussion of the role of heat, temperature and more generally diffusion in biology can be found in [4,5] and elsewhere (including the second volume of these notes).

Here we consider a straight metallic electrically conducting wire of length L, mass density (per unit volume) ρ, and uniform cross-sectional area A. The cylindrical surface of the wire is assumed to be insulated from heat flow. The wire is sufficiently slender so that the temperature distribution is uniform over a cross section at any point along the central axis of the wire as a function of position and time. Let the central axis be positioned along the x-axis extending from $x = 0$ to $x = L$. To study the temperature distribution $T(x,t)$ in the wire, we look at a segment of the wire between $x = a\,(> 0)$ and $x = b\,(< L)$ at time t (see Figure 11.1).

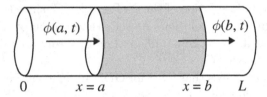

Fig. 11.1   Diagram of heat flux.

Let c_0 be the *specific heat* of the wire material defined to be the amount of heat energy needed to raise the temperature of a unit mass of material by one degree Kelvin $(1^\circ K)$. The specific heat is a material property and can be measured experimentally. The rate of change of the *heat (thermal) energy* $H(x,t)$ in this segment is given by

$$\frac{dH}{dt} = \frac{d}{dt}\int_a^b c_0 T(x,t)\rho A\, dx. \tag{11.2.1}$$

There is another way of calculating the same rate of change of heat energy. Let $\phi(x,t)$ be the amount of heat per unit time flowing from left to right across a unit area of the cross section at location x and at time t, called *heat flux* for short. Let $F(x,t)$ be the rate of heat generated by a unit mass of the wire within the insulated cylindrical surface by chemical, electrical, or other processes. In that case, we have

$$\frac{dH}{dt} = A\left[\phi(a,t) - \phi(b,t)\right] + \int_a^b F(x,t)\rho A\, dx. \tag{11.2.2}$$

The bracketed expression is the difference between the heat going into one end of the wire segment and the heat leaving the other end. The integral corresponds to the amount of heat generated internally within the segment.

The two expressions (11.2.1) and (11.2.2) for dH/dt must be the same. The rate of increase in heat in the segment (calculated from the temperature change in the segment) must be equal to the rate of heat gained (calculated by the net heat gain between the two cross section of the segment). We have therefore

$$\frac{d}{dt} \int_a^b c_0 \rho T(x,t)\, dx = \phi(a,t) - \phi(b,t) + \int_a^b \rho F(x,t)\, dx \qquad (11.2.3)$$

where c_0 and ρ may vary with x and t. Suppose both T and ϕ are smooth functions and both c_0 and ρ are independent of t. In that case, we can write (11.2.3) as

$$\int_a^b \left[c_0 \rho \frac{\partial T}{\partial t} - \rho F(x,t) + \frac{\partial \phi}{\partial x} \right] dx = 0. \qquad (11.2.4)$$

If the integrand is continuous in x and the segment (a,b) is arbitrary, we conclude that

$$c_0 \rho T_{,t} = -\phi_{,x} + \rho F\,, \qquad (\ )_{,z} = \frac{\partial (\ )}{\partial z} \qquad (11.2.5)$$

where we have used a less cumbersome notation $(\ )_{,z}$ to denote partial differentiation with respect to z (similar to using a prime to denote differentiation with respect to the only independent variable in ODE). Equation (11.2.5) actually holds for all x in $(0, L)$ and all $t > 0$. It may be considered a local conservation law for the heat energy in the wire.

11.2.2 *Fourier's law and the heat equation*

If the internal heat generation rate per unit mass $F(x,t)$ is known, then (11.2.5) is one relation for the two unknown functions $T(x,t)$ and $\phi(x,t)$ and their derivatives. The functions involved are of two variables x and t and the derivatives involved are partial derivatives. Such a differential equation is said to be a *partial differential equation* (PDE) to be distinguished from an ODE which involves only functions of one independent variable. Now (11.2.5) is one equation for two unknown functions. In order to determine both unknowns, we need a second relation for the same two quantities. Intuitively, we expect that there is a relation between the temperature gradient at a point in a body and the heat flow rate (per unit area) across that position: a steeper gradient should generate more flow in the direction of lower temperature. A mathematical characterization of this relation is the *Fourier law*:

$$\phi = -K_0 T_{,x} \qquad (11.2.6)$$

where the *coefficient of thermal conductivity* K_0 is a measure of the ability of a material to promote flow, i.e., to diffuse or conduct heat. K_0 has to be determined experimentally and it varies from material to material. For some materials, it may even vary with the temperature of the material. Upon substitution of (11.2.6) into (11.2.5), we get a single second-order PDE for $T(x,t)$ with the known heat generating term $F(x,t)$ acting as a forcing term for the PDE:

$$c_0 \rho T_{,t} = (K_0 T_{,x})_{,x} + \rho F. \qquad (11.2.7)$$

We limit ourselves in this chapter to the case where K_0 is a known quantity. If, in addition, the wire is homogeneous in its ability to conduct heat, then K_0 is a constant and (11.2.7) simplifies to read

$$c_0 \rho T_{,t} = K_0 T_{,xx} + \rho F. \tag{11.2.8}$$

If we let $\alpha = K_0/\rho c_0$ be the thermal diffusivity of the conducting wire (see Table 11.1 of this section for a list of diffusivities in common materials), then (11.2.8) becomes

$$T_{,t} = \alpha T_{,xx} + f(x,t), \qquad f(x,t) = \frac{1}{c_0} F(x,t). \tag{11.2.9}$$

Table 11.1   Values of the thermal diffusivity
for some common materials.

Material	α^2 (cm^2/sec)
Silver	1.71
Copper	1.14
Aluminum	0.86
Cast iron	0.12
Granite	0.011
Brick	0.0038
Water	0.00144

In either form, (11.2.7) or (11.2.9), the PDE is supplemented by an initial condition (in the time variable t) and two boundary conditions (in the space variable x). We normally know the temperature distribution of the wire at some reference time $t = 0$ so that

$$T(x,0) = T_0(x), \quad (0 \le x \le L). \tag{11.2.10}$$

Since the PDE is first order in t, this is all we can prescribe (and in fact all we know) at the initial time. The end $x = 0$ of the wire may be kept at some ambient temperature of its surroundings (which may change with time) so that

$$T(0,t) = T_0(t), \quad (t > 0). \tag{11.2.11}$$

Other physically meaningful end conditions are also possible. For example, the end may be insulated from heat loss (or gain) so that, by Fourier's law, we have

$$T_{,x}(0,t) = 0, \quad (t > 0). \tag{11.2.12}$$

The situation for the other end $x = L$ is similar.

11.2.3   *Steady state temperature distributions*

As in the case of an IVP in ODE, one way to delineate the solution of the problem is to locate critical points. For PDE, this corresponds to the time independent solutions of the problem, which is possible only if $F(x,t)$ is time invariant, i.e., only if the rate of (spatially distributed) internal heat generation does not change with t. Whereas setting all time derivatives to zero for the ODE problem to find critical points results in equations that do not involve derivatives of the unknowns, setting the time derivative of the (spatially) one-dimensional heat (*aka* diffusion) equation to zero leaves us with a second order ODE supplemented by two boundary conditions. (The initial condition needs not be a part of this steady state problem since the phenomenon in question has evolved from its initial state for quite some time before approaching its steady state.) This BVP is of the form (11.1.1)–(11.1.2) of the previous section, possibly with the end conditions replaced by others such as a prescribed value for the first derivative of $T(x)$ corresponding to a prescribed heat flux in the heat conduction problem.

In the context of heat conduction, the reason why there should be no solution for

$$T'' = 0, \quad \text{with} \quad T'(0) = Q_0 \quad \text{and} \quad T'(1) = Q_l \qquad (11.2.13)$$

when $Q_0 \neq Q_l$ is now understandable. If there is a net heat flux into the wire, positive or negative, then we have

$$Q_l - Q_0 = \int_0^L \frac{d\phi}{dx} dx = - \int_0^L \frac{d}{dx} \left(K_0 \frac{dT}{dx} \right) dx \neq 0.$$

But by the heat energy change calculation of Section 1, $(K_0 T,_x),_x \neq 0$ requires $c_0 \rho T,_t \neq 0$. It follows that the temperature in the rod cannot be time-invariant and hence not possible to have a (time independent) steady state solution.

It turns out that a prescribed temperature condition on either end of the wire suffices to ensure a unique solution for the steady state temperature $T(x)$. To show this, we consider the general steady state problem defined by

$$0 = (K_0 T,_x),_x + \rho F(x) \qquad (11.2.14)$$

with one of the following sets of end conditions:

$$(i) \ T(0) = T_0, \quad T(L) = T_l, \qquad (11.2.15)$$
$$(ii) \ T(0) = T_0, \quad T'(L) = Q_l, \qquad (11.2.16)$$
$$(iii) \ T'(0) = Q_0, \quad T(L) = T_l. \qquad (11.2.17)$$

Theorem 20 *If the steady state heat conduction problem defined by (11.2.14) and one of the three sets of end conditions (11.2.15)–(11.2.17) has a solution, it is unique.*

Proof Suppose it has two different steady state solutions $T_1(x)$ and $T_2(x)$. Let $T = T_2 - T_1$; then T satisfies (11.2.14) with $F \equiv 0$ and with no flux or zero temperature at each end corresponding to the prescribed set of end conditions. Multiply (11.2.14) by T and integrate over $(0, L)$ to get

$$0 = \int_0^L [K_0 T,_x],_x T \, dx = [K_0 T,_x T]_0^L - \int_0^L K_0 (T,_x)^2 \, dx. \qquad (11.2.18)$$

The first term on the right vanishes because of the vanishing boundary conditions for $T(x)$, either $T = 0$ or $T_{,x} = 0$ at an end point. Since K_0 is positive, the integrand of the second term on the left is nonnegative. In order for (11.2.18) to hold, we must have $T_{,x} = 0$ so that $T(x) = C_0$ for all x in $(0, L)$. Given $T = 0$ at one of the two ends, we must have $C_0 = 0$ resulting in $T(x) = 0$ or $T_1(x) = T_2(x)$ for all x in $(0, L)$. Hence the solution of any of the three steady state problems is unique.   □

The result above can be extended to more general boundary conditions such as the mixed condition

$$\alpha T + \beta T_{,x} = \gamma$$

for some constants α, β, and γ (see exercises), not all zero. This condition characterizes a leaky end with complete insulation ($\alpha = \gamma = 0$, $\beta \neq 0$) and a fixed end temperature ($\beta = 0$) as special cases.

For the excluded case of a prescribed flux at both ends as in (11.2.13), we saw already that the problem has no solution unless $Q_0 = Q_l$. If the flux at the ends are not the same so that there is a net heat flux at each instance of time, the thermal energy in the interior of the wire would continue to change. Correspondingly, the temperature in the wire would continue to change with time and the wire would not reach a (time-independent) steady state. For complex problems without an explicit analytical solution, an indirect proof of existence or nonexistence, if possible, would be helpful before embarking on any kind of approximate solution as numerical solutions for PDE are much more costly (and sometimes impractical even with the current level of computing power).

If the flux at the two ends are the same for all time, there would be no net heat flux throughout the wire. Whatever temperature we started out with would not change with time. However, nothing has been said in the statement of the steady state problem about the initial temperature of the wire. Hence, we would have a steady state solution with an undetermined temperature level (reflected in the undetermined constant of integration in the actual solution for $T(x)$). The proof of uniqueness fails in this case as neither $T(0)$ nor $T(L)$ is specified by the problem (where $T(x)$ is now the difference between two different solutions). In that case the constant C_0 in the proof above remains unspecified and the two solutions can differ by a constant, as found in the actual solution process. For more complex problems without an exact solution, a theoretical result such as Theorem 20 which does not rely on an explicit solution of the problem is important and valuable as it provides useful information that adds to our understanding of the mathematical model for the phenomenon being investigated.

11.2.4   *The time dependent problem*

We now know that when the steady state problem fails to have a solution (in the case of different prescribed heat fluxes at the two ends), it is because the physical problem does not reach a steady state. In that context, we expect the corresponding time dependent problem may still have a solution. In general, the question of existence of a solution for either the transient problem or steady state problem is much more difficult to settle. On the other hand, results on uniqueness are more accessible. We illustrate this with a uniqueness theorem for the linear initial-boundary value problem defined by the PDE (11.2.7), the initial condition (11.2.10) and one boundary condition (such as (11.2.11) or (11.2.12)) each at the end $x = 0$ and $x = L$ (including a prescribed heat flux at both ends). We limit consideration to problems with wire properties not varying in time which is mostly the case in applications.

Theorem 21 *If the initial-boundary value problem for one dimensional heat conduction with boundary conditions*

$$T(0,t) = T_0, \qquad T(L,t) = T_l \tag{11.2.19}$$

has a solution, it is unique.

Proof Suppose it has two different solutions $T_1(x,t)$ and $T_2(x,t)$. Let $T = T_2 - T_1$. Then T satisfies the PDE (11.2.7) with $F \equiv 0$,

$$c_0 \rho T_{,t} = (K_0 T_{,x})_{,x} \quad , \tag{11.2.20}$$

the homogeneous initial condition and boundary conditions:

$$T(x,0) = 0 \qquad (0 < x < L), \tag{11.2.21}$$

$$T(0,t) = T(L,t) = 0 \qquad (t > 0). \tag{11.2.22}$$

Multiply the homogeneous PDE (11.2.20) by T and integrate over $(0, L)$ to get

$$\int_0^L c_0 \rho T T_{,t} \, dx = \int_0^L [K_0 T_{,x}]_x T \, dx$$

$$= [K_0 T_{,x} T]_0^L - \int_0^L K_0 (T_{,x})^2 \, dx. \tag{11.2.23}$$

The first term on the right vanishes because of (11.2.22). Since c_0 and ρ are independent of t and $T_{,t}$ is assumed not to vanish identically, the integrand on the left may be written as $c_0 \rho (T^2)_{,t} / 2$ so that (11.2.23) becomes

$$\frac{1}{2} \frac{d}{dt} \int_0^L c_0 \rho T^2 \, dx = -\int_0^L K_0 (T_x)^2 \, dx \equiv -P(t) \tag{11.2.24}$$

where $P(t)$ is nonnegative. Upon integrating both sides over $(0, t_f)$, we get for any $t_f > 0$

$$\left[\frac{1}{2} \int_0^L c_0 \rho T^2 \, dx \right]_{t=t_f} = -\int_0^{t_f} P(t) \, dt \tag{11.2.25}$$

with the integral inside the brackets on the left vanishing at $t = 0$ because of (11.2.21). The left side of (11.2.25) is nonnegative as $T^2 \geq 0$ while the right side is non-positive because $P(t) \geq 0$. The only way (11.2.25) can be satisfied is

$$\int_0^L c_0\rho T^2(x, t_f)\, dx \equiv 0. \tag{11.2.26}$$

With $c_0\rho > 0$ and $T^2 \geq 0$, (11.2.26) holds only if $T(x, t_f) \equiv 0$. As t_f is arbitrary, we must have $T(x, t) = 0$ or $T_1(x, t) = T_2(x, t)$ for all x in $0 \leq x \leq L$ and $t \geq 0$. In other words, the two supposedly different solutions must in fact be the same. Given the initial temperature distribution throughout the wire and the prescribed end temperature at the two ends, *the temperature of any point of the rod is uniquely determined for all future time.*   □

Corollary 7 *The time dependent solution of the initial-boundary value problem (for wires with time invariant properties) remains unique if the prescribed temperature end condition at either end (or both) is replaced by a prescribed flux condition.*

Proof The integrated term $[K_0 T_{,x} T]_o^L$ also vanishes at the prescribed flux end because $T_{,x} = 0$ at that end in this case.   □

In the statement and proof of (the uniqueness) Theorem 21, we carefully stipulated the existence of a non-steady state solution. The question of existence of a solution to a genuine PDE problems is complicated (and will be addressed briefly in the second volume of these notes). We note in passing that the proof of Theorem 21 above does not apply for solution independent of t. All manipulations beyond (11.2.24) are vacuous since $(\ )_{,t} = 0$.

11.3 Analytical Solutions for BVP

11.3.1 *Reduction of order for autonomous equations*

For autonomous ODE, we learned from a previous chapter that we can always reduce the order of the ODE system by one. For a single second order ODE $y'' = f(y, y')$, this reduction results in a first order ODE of the form

$$\frac{dv}{dy} = \frac{f(y, v)}{v}$$

where $v = y' = dy/dx$. Depending on the form of $f(y, v)$, we may be able to obtain an exact solution in terms of known functions in the form $v = F(y; C_1)$ where C_1 is a constant of integration. The relation between v and y is a first order separable ODE $dy/dx = F(y; C_1)$ to be solved to get $y = G(x; C_1, C_2)$. If we should be successful in obtaining an explicit solution $G(x; C_1, C_2)$, there would still be the task of determining the two constants of integration C_1, C_2 so that the prescribed boundary conditions are satisfied. The solution process described above is illustrated by the example below.

Consider the autonomous second order ODE

$$y'' - (y')^2 - y^2 = 0, \qquad (\ )' = \frac{d(\ )}{dx} \tag{11.3.1}$$

which may be rewritten as two autonomous first order ODEs

$$y' = v, \qquad v' = v^2 + y^2.$$

The method of reduction of order use the chain rule to write the system as a single first order ODE

$$\frac{dv}{dy} = \frac{v^2 + y^2}{v}$$

which is the Bernoulli equation with $\alpha = -1$:

$$\frac{dv}{dy} - v = y^2 v^{-1}.$$

The method developed in Appendix A for this ODE gives the exact solution

$$v^2 = \left(\frac{dy}{dx}\right)^2 = c_0 e^{2y} - \left(y^2 + y + \frac{1}{2}\right).$$

The first order ODE for $y(x)$ is again separable and can be integrated to get

$$\pm \int_0^y \frac{du}{\sqrt{c_0 e^{2u} - \left(u^2 + u + \frac{1}{2}\right)}} = x + c_1 \tag{11.3.2}$$

with the constants c_0 and c_1 to be determined by the two auxiliary conditions augmenting the second order ODE to form a BVP.

To proceed further, we consider the two auxiliary conditions

$$y(0) = 0, \qquad y(1) = 1. \tag{11.3.3}$$

To use them to determine c_0 and c_1, we let $y'(0) = s$. In that case, the boundary condition at $x = 0$ gives

$$s^2 + \frac{1}{2} = c_0$$

and the integral relation (11.3.2) becomes

$$\int_0^y \frac{du}{\sqrt{c_0(s) e^{2u} - \left(u^2 + u + \frac{1}{2}\right)}} = x$$

where we have taken the positive square root since $y(x)$ is to increase from 0 to 1 as x increases from 0 to 1. The second boundary condition then requires

$$g(s) \equiv \int_0^1 \frac{du}{\sqrt{\left(s^2 + \frac{1}{2}\right) e^{2u} - \left(u^2 + u + \frac{1}{2}\right)}} = 1. \tag{11.3.4}$$

The last relation (11.3.4) requires that s be a zero of $G(s) \equiv g(s) - 1$ or the solution of

$$g(s) = 1.$$

A solution can be found by any of the available mathematical software such as Mathematica, MatLab or Maple, most of them use some form of Newton's method.

Exercise 31 *Obtain the initial slope s for the BVP defined by (11.3.1) and (11.3.3).*

11.3.2   A nonautonomous nonlinear ODE

The ODE

$$y'' = -\frac{y'}{x}(2y+1), \qquad (\ )' = \frac{d(\ )}{dx} \tag{11.3.5}$$

is of the type $y'' = f(x, y, y')$ with all three arguments present in f and hence nonautonomous. For this ODE, reduction of order is not applicable and there is no general recipe for reducing (11.3.5) to solving uncoupled scalar first order ODE. For this particular ODE however, rewriting it as

$$xy'' + (2y+1)y' = 0, \tag{11.3.6}$$

opens up some possible simplifications that help to reduce it to a calculus problem. This is accomplished by observing the two mathematical relations $xy'' + y' = (xy')'$ and $2yy' = (y^2)'$ so that the ODE can be written as

$$(xy' + y^2)' = 0 \qquad \text{or} \qquad xy' + y^2 = c_0^2.$$

The integrated equation containing the constant of integration c_0^2 is a first order separable ODE and can be solved exactly to get

$$y = \frac{c_2}{2}\frac{x^{c_2} - c_1}{x^{c_2} + c_1} \tag{11.3.7}$$

where we have set $c_0 = c_2/2$ and where c_1 is a second unknown constant.

Note that the constants of integration of a nonlinear ODE normally do not appear linearly in its general solution. This is in contrast to the solution being a linear combination of complementary solutions and a particular solution such as

$$y(x) = c_1\phi_1(x) + c_2\phi_2(x) + \phi_p(x)$$

for a second order linear inhomogeneous ODE by superposition principles (see Exercise 11). For nonlinear ODE, the linear system of algebraic equations associated with the boundary conditions of linear BVP for the determination of these constants are generally replaced by a system of nonlinear equations. Such a system will have to be solved by an iterative method such as the equivalence of Newton's method for several unknowns (called the Newton-Raphson method in most references).

To illustrate, suppose the boundary conditions for (11.3.5) are

$$y(1) = 0, \qquad y(2) = 5. \tag{11.3.8}$$

The first condition is met by $c_1 = 1$. (We discard the possibility of $c_2 = 0$ in order to have a nontrivial solution.) The second boundary condition at $x = 2$ then takes the form

$$g(c_2) \equiv \frac{c_2}{2}\frac{2^{c_2} - 1}{2^{c_2} + 1} - 5 = 0. \tag{11.3.9}$$

The nonlinear equation (11.3.9) for the single unknown c_2 may be solved by Newton's method. Having the nth iterate $c_2^{(n)}$, Newton's method determines the next iterate $c_2^{(n+1)}$ by

$$c_2^{(n+1)} = c_2^{(n)} - \frac{g(c_2^{(n)})}{g'(c_2^{(n)})}, \qquad g'(c_2) = \frac{dg}{dc_2} \tag{11.3.10}$$

for $n = 0, 1, 2, \ldots$. The method is effectively a two-term Taylor formula

$$g(z) = g(z^{(n)}) + g'(\xi)(z - z^{(n)})$$

with $\left| \xi - z^{(n)} \right| < \left| z - z^{(n)} \right|$. With $g(z) = 0$ for the exact solution, we get

$$z = z^{(n)} - \frac{g(z^{(n)})}{g'(\xi)}.$$

If $z^{(n)}$ is already close to the exact solution, we may approximate the (not completely known) quantity ξ by $z^{(n)}$ to get

$$z^{(n+1)} = z^{(n)} - \frac{g(z^{(n)})}{g'(z^{(n)})}, \qquad g'(z) = \frac{d[g(z)]}{dz} \qquad (11.3.11)$$

which is the same as (11.3.10) after a change of notation.

Exercise 32 *For (11.3.9), a reasonable initial guess is $c_2^{(0)} = 10$ since for large c_2, the relation $g(c_2) = 0$ is approximately*

$$0 = g(c_2) \sim \frac{c_2}{2} - 5.$$

i) Obtain by any numerical method the actual solution $c_2 = 10.01929042\ldots$. Hence, our initial guess provides a very accurate numerical solution.

ii) Show that $c_2^* = -10.01929042\ldots$ is also a solution of $g(c_2) = 0$.

iii) Show that the exact solution (11.3.7) with $c_1 = 1$ is an even function of c_2, i.e., $y(x; -c_2) = y(x; c_2)$, so that there is only one solution for the problem with $c_2 = 10.01929042\ldots$. As such, the second solution for c_2 does not lead to a new solution for the BVP.

Exercise 33 *Suppose the second boundary condition $y(2) = 5$ is replaced by $y(2) = -5$ (or any negative value), show by the graphical method that the new problem $g(c_2) = 0$ has no solution.*

The nontrivial example above shows again that in contrast to the situation for IVP, existence and uniqueness of solution are serious issues for BVP. It is important that we become familiar with some known results on how to determine whether or not a solution of a class of BVP exists and whether or not it is unique when it does exist.

11.4 The IVP Approach for Linear BVP

11.4.1 *Homogeneous linear ODE*

If both the ODE and the BC are linear, we can take advantage of the tools for IVP developed in Appendix B to obtain an accurate numerical solution for the BVP when an exact analytical solution is not available in terms of known functions. More specifically, we obtain an accurate solution for the problem by computing accurate numerical solutions for a number of related initial value problems and combining them for the actual solution of the BVP. We will illustrate this method of solution by a specific problem for the second order homogeneous (linear) ODE

$$y'' + p_0(x)y' + q_0(x)y = 0, \quad (a < x < b) \tag{11.4.1}$$

$$y(a) = \gamma_a, \qquad y(b) = \gamma_b \tag{11.4.2}$$

with $(\ )' \equiv d(\ )/dx$. For simplicity, we assume that $p_0(x)$ and $q_0(x)$ in (11.4.1) are both continuous for $a \le x \le b$. In that case, the homogeneous linear ODE (11.4.1) has a set of two complementary solutions. We take them to be the fundamental set $\{\phi_1(x; a), \phi_2(x; a)\}$ that satisfies the following initial conditions

$$\phi_1(a; a) = 1, \quad \phi_1'(a; a) = 0 \tag{11.4.3}$$

and

$$\phi_2(a; a) = 0, \quad \phi_2'(a; a) = 1. \tag{11.4.4}$$

Given the continuity of $p_0(x)$ and $q_0(x)$ in (a, b), $\phi_1(x; a)$ and $\phi_2(x; a)$ so defined may be generated numerically to a specified accuracy for $a < x \le b$ by any suitable method for IVP in ODE (see Appendix B).

By the superposition principle for complementary solutions (see Exercise 11), the solution of the linear ODE (11.4.1) may be expressed in terms of ϕ_1 and ϕ_2,

$$y(x) = y_0\phi_1(x; a) + v_0\phi_2(x; a) \tag{11.4.5}$$

where y_0 and v_0 are constants of integration to be determined by auxiliary conditions. The two boundary conditions (11.4.2) require

$$y(a) = y_0\phi_1(a; a) + v_0\phi_2(a; a) = y_0 \cdot 1 + v_0 \cdot 0 = \gamma_a, \tag{11.4.6}$$

$$y(b) = y_0\phi_1(b; a) + v_0\phi_2(b; a) = \gamma_b. \tag{11.4.7}$$

The condition (11.4.6) gives immediately

$$y_0 = \gamma_a \tag{11.4.8}$$

and, for $\phi_2(b) \ne 0$, the condition (11.4.7) then gives

$$v_0 = \frac{1}{\phi_2(b; a)} \left[\gamma_b - \gamma_a\phi_1(b; a)\right]. \tag{11.4.9}$$

With (11.4.8) and (11.4.9), the solution (11.4.5) of the BVP (11.4.1) and (11.4.2) is completely (and, by Theorem 26, uniquely) determined.

Example 34

$$y'' - y = 0, \qquad y(0) = 1, \qquad y(1) = 0.$$

With the linearly independent exact complementary solutions e^x and e^{-x}, we can construct $\phi_1(x) = \cosh(x)$ and $\phi_2(x) = \sinh(x)$ and write

$$y(x) = y^o \cosh(x) + v^o \sinh(x).$$

The boundary conditions then require

$$y(0) = y^o = 1, \qquad y(1) = y^o \cosh(1) + v^o \sinh(1) = 0$$

or

$$v^o = -\frac{\cosh(1)}{\sinh(1)} = -\coth(1).$$

Since $\sinh(1) \neq 0$, the exact solution of our simple BVP is

$$y(x) = \cosh(x) - \coth(1)\sinh(x).$$

For simple problems with solution in terms of known functions such as the one above, the solution process in terms of $\phi_1(x)$ and $\phi_2(x)$ is unnecessarily artificial and indirect. We could have simply made use of the superposition principles for linear ODE to write $y(x) = c_1 e^x + c_2 e^{-x}$ and apply the boundary conditions to determine the two constants c_1 and c_2. (The solution would then be in terms of e^x and e^{-x} and appears to be different from the one in terms of $\sinh(x)$ and $\cosh(x)$. However, the uniqueness theorem to be proved in Section 3 of this chapter assures us that they are in fact different form of the same solution.) But when there are not enough linearly independent solutions in terms of known functions for a fundamental set of solutions and we need to resort to numerical methods for an accurate approximate solution, the method of solution by way of the two *IVP* for the same ODE outlined above (namely, (11.4.1) with (11.4.3) and (11.4.1) with (11.4.4), respectively) provides a systematic way of generating a numerical solution for the BVP as illustrated by the next example.

Exercise 34

$$y'' + \sin(x)y' - (1 - e^{-x})y = 0, \quad (0 < x < 1), \tag{11.4.10}$$

with

$$y(0) = \gamma_a, \qquad y(1) = \gamma_b. \tag{11.4.11}$$

Explicit expressions for the fundamental solutions are not available and we need to rely on numerical methods to find them. But this can be done easily by available numerical software such as "NDSolve" in Mathematica (or "dsolve" in Maple or "ode23, ode45, etc." in MatLab) to get

$$\phi_1(1; 0) = 1.33221\ldots, \qquad \phi_2(1; 0) = 0.95059\ldots.$$

Upon inserting these into (11.4.9), we have

$$y_0 = \gamma_a, \qquad v_0 = \frac{1}{0.95059\ldots}[\gamma_b - 1.33221\ldots\gamma_a].$$

We can then use these values for y_0 and v_0 in (11.4.5) for the solution of the original linear BVP.

11.4.2   Inhomogeneous ODE

The situation for an inhomogeneous ODE

$$y'' + p_0(x)y' + q_0(x)y = f(x), \quad (a < x < b) \tag{11.4.12}$$

is very similar to the homogeneous case. In addition to the fundamental set $\{\phi_1(x;a), \phi_2(x;a)\}$ for the homogeneous equation defined by (11.4.3) and (11.4.4), we generate numerically a particular solution $\phi_p(x;a)$ of the inhomogeneous equation (11.4.12) with the initial conditions

$$\phi_p(a;a) = \phi_p'(a;a) = 0. \tag{11.4.13}$$

By the superposition principle of Exercise 11, the solution of the BVP defined by (11.4.12) and (11.4.2) can be expressed as the following linear combination of ϕ_1, ϕ_2, and ϕ_p:

$$y(x) = y_0\phi_1(x;a) + v_0\phi_2(x;a) + \phi_p(x;a). \tag{11.4.14}$$

The boundary conditions (11.4.2) require

$$y_0 = \gamma_a \tag{11.4.15}$$

$$y_0\phi_1(b;a) + v_0\phi_2(b;a) + \phi_p(b;a) = \gamma_b \tag{11.4.16}$$

with the second condition re-arranged to give

$$v_0\phi_2(b;a) = [\gamma_b - \phi_p(b;a) - \gamma_a\phi_1(b;a)]. \tag{11.4.17}$$

We may now summarize the results for this general case in the following proposition with the uniqueness claim assured by Theorem 26 in Appendix B or a stronger theorem in the next section of this chapter:

Proposition 16 *Let $\phi_1(x)$, $\phi_2(x)$ and $\phi_p(x)$ be as previously defined in (11.4.3)–(11.4.4).*

i) If $\phi_2(b) \neq 0$, the BVP defined by (11.4.12) and the end conditions (11.4.2) has a unique solution (11.4.14) with the constants y_0 and v_0 given by (11.4.15) and (11.4.17).

ii) If $\phi_2(b) = 0$, the same BVP either has no solution if $\gamma_b \neq \phi_p(b) + \gamma_a\phi_1(b)$ or has a one parameter family of solutions if $\gamma_b = \phi_p(b) + \gamma_a\phi_1(b)$.

The above solution process extends in a straightforward way to other boundary conditions. Suppose that instead of the (**Dirichlet**) boundary conditions (11.4.2), we have the (**Neumann**) boundary conditions

$$y'(a) = \gamma_a, \qquad y'(b) = \gamma_b. \tag{11.4.18}$$

Then we have instead of (11.4.15)–(11.4.16)

$$v_0 = \gamma_a, \qquad y_0\phi_1'(b;a) = \gamma_b - \gamma_a\phi_2'(b;a) - \phi_p'(b;a). \tag{11.4.19}$$

Proposition 17 *The Neumann BVP defined by (11.4.12) and (11.4.18) has a solution given by (11.4.14) and (11.4.19) if $\phi_1'(b; a) \neq 0$. If $\phi_1'(b; a) = 0$, then either there is no solution for the BVP if $\gamma_b \neq \gamma_a \phi_2'(b; a) + \phi_p'(b; a)$, or a one parameter family of solutions if $\gamma_b = \gamma_a \phi_2'(b; a) + \phi_p'(b; a)$.*

The result above generalizes what we encountered in the steady state temperature distribution in a straight wire with heat flux prescribed at both ends. Note that nothing was said about uniqueness even when $\phi_1'(b) \neq 0$. A constant can always be added to the solution $y(x)$ if $q_0(x) \equiv 0$.

The *method of IVP* outlined in this section can also be used for mixed end conditions

$$\alpha_a y(a) + \beta_a y'(a) = \gamma_a, \qquad \alpha_b y(b) + \beta_b y'(b) = \gamma_b \qquad (11.4.20)$$

as well as a higher order linear equation and linear systems. The details for these cases are left as exercises.

11.4.3 *Some possible complications*

Another advantage of the IVP method of solution of this section is that it clearly delineates the condition for which the problem has a solution. The following example does not satisfy the required condition for the existence of a solution:

$$y'' + \pi^2 y = 0, \qquad y(0) = 1, \qquad y(1) = 0.$$

For this problem, it is straightforward to construct $\phi_1^{(1)}(x) = \cos(\pi x)$ and $\phi_1^{(2)}(x) = \sin(\pi x)$ and write

$$y(x) = y^o \cos(\pi x) + v^o \sin(\pi x).$$

The boundary condition at $x = 0$ is satisfied (by taking $y^o = 1$). The second then requires

$$y(1) = \cos(\pi) + v^o \sin(\pi) = 0.$$

But $\phi_1^{(2)}(1) = \sin(\pi) = 0$ so that the second boundary condition requires $\cos(\pi) = 0$ which cannot be met by any finite constant v_0. The given BVP therefore has no solution.

11.5 The Shooting Method

For the general second order nonlinear ODE

$$y'' = f(x, y, y') \qquad (a < x < b) \qquad (11.5.1)$$

with boundary conditions

$$y(a) = \gamma_a, \quad y(b) = \gamma_b, \qquad (11.5.2)$$

an explicit solution such as those in Section 3 of this chapter is highly unlikely. In fact, we have learned throughout these notes that not many scalar first order ODEs admit an exact solution in terms of known functions. Even fewer second order ODE can be solved analytically. Prescribing the auxiliary conditions at different ends of the solution interval further complicates the deductive process for an exact analytical solution. With today's computing capacity, an accurate approximate solution is often possible by available mathematical software such as MatLab, Mathematica and Maple based on an appropriate numerical method. Among the many available methods for this purpose is the IVP approach so successful for linear problems.

An IVP approach to the solution of a nonlinear problem such as (11.5.1)–(11.5.2) does *not* have the use of superposition principles that enable us to reduce the solution process to solving a set of linear equations. Instead, it would consist of solving the ODE (11.5.1) with the initial conditions

$$y(a) = \gamma_a \quad \text{and} \quad y'(a) = s \tag{11.5.3}$$

for some initial slope s to get a (numerical) solution

$$y(x) = \phi(x; s) \tag{11.5.4}$$

and choosing s to satisfy the remaining boundary condition so that

$$\phi(b; s) = \gamma_b.$$

The appearance of s as a parameter in the solution $\phi(x; s)$ is to indicate that $y(x)$ would change with different values of s. We have from the theory of IVP the following existence and uniqueness theorem for $\phi(x; s)$:

Theorem 22 *Suppose $f(x, y, v)$ is continuous for (x, y, v) in the infinite layer $L_\infty = \{(x, y) \mid a \le x \le b, \; y^2 + v^2 < \infty\}$ in the (x, y, v)-space and is Lipschitz continuous for all (y, v) in L_∞ uniformly for all x in $[a, b]$ so that*

$$|f(x, y, v) - f(x, u, w)| \le K\,|y - u| + L\,|v - w|, \tag{11.5.5}$$

for some constant $K > 0$ and $L \ge 0$ and all (y, v) and (u, w) in the infinite layer L_∞. Then the IVP defined by (11.5.1) and the initial conditions (11.5.3) has a unique solution $\phi(x; s)$ defined on $[a, b]$ with

$$|\phi(x; s) - \phi(x; \sigma)| \le e^{M(x-a)}\,|s - \sigma|$$

for some $M = K + L$.

The theorem above is a stronger version (of the vector form) of Theorem 26 stated without proof in Appendix B. In that theorem, existence is guaranteed only for $a \le x \le \delta$ where δ may be less than b. With a more stringent requirement on f in Theorem 22, we are able to ensure existence (and uniqueness) for the entire interval $[a, b]$ of interest. A proof of the new theorem on (an even more general version of) the IVP can be found in [7].

With Theorem 22, the problem becomes one of choosing s so that

$$y(b) = \phi(b; s) = \gamma_b. \tag{11.5.6}$$

Unlike linear BVP of Chapter 7, the determination of the unknown slope s to satisfy the second boundary condition now involves the solution of a nonlinear equation of the form (11.5.6). As in the example (11.3.5)–(11.3.8), the nonlinear equation (11.5.6) may be solved by Newton's method (or some other numerical methods). Unlike the BVP (11.3.5)–(11.3.8), we generally do not have an analytical expression for $\phi(x; s)$; what Theorem 22 assures us is a numerical solution for any specific value of s. To implement the Newton algorithm (11.3.11), now in the form

$$s^{(n+1)} = s^{(n)} - \frac{g(b; s^{(n)})}{g'(b; s^{(n)})}, \tag{11.5.7}$$

we will have to recalculate $g(x; s) = \phi(b; s) - \gamma_b$ numerically for each iteration of Newton's algorithm as s changes from iteration to iteration. In addition, we do not have an explicit expression for the derivative $g'(b; s) \equiv d\,[\phi(b; s)]\,/ds$ needed in the algorithm (11.5.7).

We can obtain the missing information approximately by replacing the derivative of $g(s)$ by its finite difference analogue:

$$g'(b; s^{(n)}) = \frac{d\,[\phi(b; s)]}{ds} \simeq \frac{\phi(b; s^{(n)}) - \phi(b; s^{(n-1)})}{s^{(n)} - s^{(n-1)}}.$$

The resulting *secant* type method would reduce the order of accuracy slightly, (for those who knows) from quadratic convergence to super-linear convergence.

Alternatively, we can find $g'(b; s^{(n)})$ by setting

$$Z(x; s) \equiv \frac{\partial \phi}{\partial s}(x; s) \tag{11.5.8}$$

and solving the variational BVP

$$Z'' = f_{,y}\,(x, \phi(x; s), \phi'(x; s))Z + f_{,v}\,(x, \phi(x; s), \phi'(x; s))Z', \tag{11.5.9}$$

$$Z(a; s) = 0, \quad Z'(a; s) = 1 \tag{11.5.10}$$

for $Z(x; s)$ where $(\ )' = d(\ )/dx$. The linear ODE for $Z(x; s)$ is obtained by differentiate both sides of the ODE (11.5.1) for $y(x) = \phi(x; s)$ partially with respect to s. The two initial conditions for $Z(x; s)$ in (11.5.10) are obtained by differentiating the initial conditions

$$y(a) = \phi(a; s) = \gamma_a, \quad y'(a; s) = \phi'(a; s) = s$$

partially with respect to s keeping in mind (11.5.8). The problem for $Z(x; s)$ can be solved along with the BVP (11.5.1) and (11.5.2) in each iteration of (11.5.6). The solution obtained for $\phi(x; s)$ is then used in calculating the coefficients $f_{,y}\,(x, \phi(x; s), \phi'(x; s))$ and $f_{,v}\,(x, \phi(x; s), \phi'(x; s))$ in the new round of solution for $Z(x; s)$.

The boundary conditions (11.5.2) prescribe the values of the unknown of the BVP at the two ends of the solution interval. Each of these boundary conditions is known as a *Dirichlet* end condition. The solution process outlined above applied also if these boundary conditions are replaced by the more general mixed boundary conditions

$$B_a[y] \equiv \alpha_a y(a) - \beta_a y'(a) = \gamma_a, \quad B_b[y] \equiv \alpha_b y(b) + \beta_b y'(b) = \gamma_b \qquad (11.5.11)$$

with $\alpha_c \beta_c \neq 0$ for $c = a$ or $c = b$. For the special case of $\alpha_b = 0$ but $\beta_b \neq 0$, the second condition of (11.5.11) simplifies to $y'(b) = \gamma_b/\beta_b$. A boundary condition that prescribes an end value of the derivative of the unknown $y'(x)$ of the BVP is called a Neumann end condition. When $\alpha_a \neq 0$, we may take

$$\phi'(a; s) = s \text{ and } \phi(a; s) = \frac{1}{\alpha_a}[\gamma_a - \beta_a s] \qquad (11.5.12)$$

to generate $\phi(x; s)$, while the condition (11.5.6) for the determination of s is now replaced by

$$g(s) \equiv \alpha_b \phi(b; s) + \beta_b \phi'(b; s) - \gamma_b = 0. \qquad (11.5.13)$$

Whether it is for the special case of Dirichlet end conditions (11.5.2) or the more general mixed end conditions (11.5.11), the so-called *shooting method* developed in this section for two-point boundary value problems leads to a nonlinear equation $g(s) = \phi(b; s) - \gamma_b = 0$ for determining the unknown initial slope parameter s. The following proposition summarizes the developments above:

Proposition 18 *Under the same hypotheses on $f(x, y, v)$ as in Theorem 22, the BVP defined by (11.5.1) and (11.5.11) (which includes (11.5.2) as a special case) has a solution given by the solution (11.5.4) of the IVP defined by (11.5.1)–(11.5.3) for each zero s_k of $g(s) \equiv B_b[\phi] - \gamma_b = 0$.*

Though not discussed in these pages, the shooting method may also be extended to higher order ODE or a system of first order ODE. The more important questions for us here are: *i*) when are we assured of a solution for the nonlinear BVP, i.e., when does $g(s)$ have at least one zero, and *ii*) would there be more than one solutions? One answer to these questions is the next theorem:

Theorem 23 *Suppose in addition to the hypotheses of Theorem 22, $f(x, y, v)$ is continuously differentiable in the infinite layer $L_\infty = \{a \leq x \leq b, y^2 + v^2 < \infty\}$ in the (x, y, v)-space with*

$$f_{,y}(x, y, v) > 0, \qquad |f_{,v}(x, y, v)| \leq M$$

for some constant $M > 0$. Then $g(s) \equiv B_b[\phi] - \gamma_b = 0$ has exactly one zero and the nonlinear BVP has a unique solution.

The proof of this theorem can be found in [7] and elsewhere. We note here only that the shooting method approach to existence distinguishes problems like $y'' - y = 0$ (with $f_{,y}(x, y, v) > 0$) from those like $y'' + y = 0$ (with $f_{,y}(x, y, v) < 0$) and hence is not bound by the limitation of the latter. As such, it is an improvement over the contraction map method when both are applicable. On the other hand, Theorem 23 does not apply to a problem for $y'' + \sin(y) = 0$.

Part 4

Constraints and Control

Chapter 12

"Do Your Best" and the Maximum Principle

12.1 The Variational Notation

12.1.1 *Variations of a function*

Applications of the calculus of variations in science and engineering are conventionally done in the notations of "variations". There is one distinct advantage of the
variational notations; it allows us to operate like we do in differential calculus. In
this section, we introduce this popular notation and make use of it in subsequent
sections whenever there is an advantage to do so.

Suppose again that $Y(x)$ furnishes a local optimum for the basic problem. As
before, we compare $J[Y]$ with $J[y]$ for admissible comparison functions $y(x) =
Y(x) + \varepsilon z(x)$. The change $\varepsilon z(x)$ is called a *variation* from $Y(x)$ and is conventionally
denoted by δy:

$$\delta y \equiv \varepsilon z(x). \tag{12.1.1}$$

It is important to see the difference between δy and the more familiar quantity dy.
The variation δy is a small change in the shape of the function Y for all x in $[a, b]$,
whereas the differential dy is a small change in the value of Y at a point x due to
a small change in x.

Corresponding to the change δy from $Y(x)$ to $y(x) = Y(x) + \delta y$, the change in
$F(x, y, y')$ is an amount

$$
\begin{aligned}
F(x, y(x), y'(x)) &- F(x, Y(x), Y'(x)) \\
&= F(x, Y(x) + \varepsilon z(x), Y'(x) + \varepsilon z'(x)) - F(x, Y(x), Y'(x)) \\
&= \left[\hat{F}_y(x)\varepsilon z(x) + \hat{F}_{y'}(x)\varepsilon z'(x) \right] + O(\varepsilon^2)
\end{aligned}
\tag{12.1.2}
$$

where $O(\varepsilon^2)$ indicates a collection of terms, each proportional to ε^2 or higher power
of ε. Analogous to the definition of a differential, the two terms inside the brackets
on the right-hand side of (12.1.2) are defined to be the *first variation of F*, denoted
by

$$\delta F = \hat{F}_y(x)(\varepsilon z) + \hat{F}_{y'}(x)(\varepsilon z') = \hat{F}_y(x)\delta y + \hat{F}_{y'}(x)\delta(y').$$

In the special case when $F(x,y,y') = y'$, we have

$$\delta(y') = \varepsilon z' = (\varepsilon z)' = (\delta y)' \tag{12.1.3}$$

so that differentiation with respect to the independent variable and the δ operator are commutative and there is no ambiguity in writing $\delta y'$. In variational notations, the definition of δF can then be written as

$$\delta F = \hat{F}_y(x)\delta y + \hat{F}_{y'}(x)\delta y'. \tag{12.1.4}$$

Note that a more complete analogy with the definition of a differential would be $\delta F = \hat{F}_x(x)\delta x + \hat{F}_y(x)\delta y + \hat{F}_{y'}(x)\delta y'$. But the first term on the right drops out because the independent variable x is not varied in the basic problem so that $\delta x = 0$ (but is allowed to vary in the parametric form of the problem). The second variation is defined to be the group of terms proportional to ε^2 in (12.1.2). The term "variation" is often used for the first variation.

12.1.2 *Variations of a function of functions*

To extend the variational notation to terms proportional to ε^2 and higher powers of ε, the convention is to write

$$F(x,y,y') - F(x,Y(x),Y'(x)) \equiv \delta F + \delta^2 F + \delta^3 F + \cdots \tag{12.1.5}$$

with $\delta^k F$ consists of all terms proportional to ε^k. For example, we have

$$\delta^2 F = \frac{1}{2!}\left[\hat{F}_{,yy}(\delta y)^2 + \hat{F}_{,y'y'}(\delta y')^2 + 2\hat{F}_{,yy'}\delta y\delta y'\right], \tag{12.1.6}$$

etc. As before, we continue to use the notation $\hat{F}_{,yy} = F_{,yy}(x,Y(x),Y'(x))$, etc. It is important to note that $\delta^2 F$ does NOT mean doing the δ operation on F twice, i.e., $\delta^2 F \neq \delta(\delta F)$, given the two terms $\delta[\hat{F}_y(x)\delta y]$ and $\delta[\hat{F}_{y'}(x)\delta y']$ in $\delta(\delta F) = \delta\left[\hat{F}_y(x)\delta y + \hat{F}_{y'}(x)\delta y'\right]$ have no meaning in the context of the variational notations. (In fact, even $\delta(\delta y)$ and $\delta(\delta y')$ are not defined.)

From definition (12.1.4), it is a straightforward calculation to verify that the laws of variation of sums, products, quotient, powers, etc., are completely analogous to the corresponding laws of differentiation. We have, in particular,

$$\delta(c_1 F_1 \pm c_2 F_2) = c_1\delta F_1 \pm c_2\delta F_2,$$
$$\delta(F_1 F_2) = F_1\delta F_2 + F_2\delta F_1, \tag{12.1.7}$$
$$\delta\left(\frac{F_1}{F_2}\right) = \frac{F_2\delta F_1 - F_1\delta F_2}{F_2^2}.$$

If x and y are both functions of an independent variable t, then we have by the chain rule

$$\frac{dy}{dx} = \frac{dy/dt}{dx/dt} = \frac{\dot{y}}{\dot{x}}, \quad (\ )^{\cdot} \equiv \frac{d(\ )}{dt}.$$

It follows that

$$\delta\left(\frac{dy}{dx}\right) = \delta\left(\frac{y\dot{}}{x\dot{}}\right) = \frac{x\dot{}\,\delta y\dot{} - y\dot{}\,\delta x\dot{}}{(x\dot{})^2}.$$ (12.1.8)

But the operators $\delta(\ )$ and $d(\ )/dt$ commute; we can write the above relation as

$$\delta\left(\frac{dy}{dx}\right) = \frac{(\delta y)\dot{}}{x\dot{}} - \frac{dy}{dx}\frac{(\delta x)\dot{}}{x\dot{}} = \frac{d(\delta y)}{dx} - \frac{dy}{dx}\frac{d(\delta x)}{dx}.$$ (12.1.9)

We recover (12.1.3) only if x does not vary.

12.1.3   Variations of the performance index

When a and b are fixed constants, we have for the ordinary function $J(\varepsilon)$

$$J(\varepsilon) - J(0) = J\dot{}(0)\varepsilon + \frac{1}{2}J\ddot{}(0)\varepsilon^2 + \cdots .$$

For

$$J(\varepsilon) = J[Y(x) + \varepsilon z(x)] = \int_a^b F(x, Y(x) + \varepsilon z(x), Y'(x) + \varepsilon z'(x))\,dx,$$

we have

$$J(\varepsilon) - J(0) = \varepsilon \int_a^b \left[\hat{F}_y z(x) + \hat{F}_{y'} z'(x)\right]dx$$

$$+ \frac{1}{2}\varepsilon^2 \int_a^b \left[\hat{F}_{yy}(x)(z)^2 + 2\hat{F}_{yy'}(x)\,zz' + \hat{F}_{y'y'}(x)(z')^2\right]dx$$

$$+ \cdots .$$

With $\delta y \equiv \varepsilon z(x)$ and the definition of $\delta^k F$ (see (12.1.1), (12.1.4) and (12.1.6)), we take the (first) variation of the performance index $J[y]$ in (9.1.9) to be

$$\delta J = \int_a^b \delta F\,dx = \int_a^b \left[\hat{F}_y \delta y + \hat{F}_{y'}\delta y'\right]dx.$$ (12.1.10)

If $\hat{F}_y - \left(\hat{F}_{y'}\right)'$ is continuous, then integrating the second term of (12.1.10) by parts gives

$$\delta J = \left[\hat{F}_{y'}\delta y\right]_a^b + \int_a^b \left[\hat{F}_y - \left(\hat{F}_{y'}\right)'\right]\delta y\,dx.$$ (12.1.11)

Now, δy vanishes at the two boundary points for the basic problem because y cannot vary from its prescribed values at these points. By a similar argument to that used in the derivation of the Euler DE, now in variational notations, the basic lemma of calculus of variations may be applied with δy playing role of $z(x)$ (or more accurately $\varepsilon z(x)$) in the proof of that lemma: Any smooth optimizer $Y(x)$ of $J[y]$ (for which $\left(\hat{F}_{y'}\right)'$ is continuous in $[a, b]$) must be a solution of the Euler DE in $[a, b]$ for the given Lagrangian. With the solution of the Euler DE designated as extremals, we may re-state Theorem 11 in a form analogous to the stationary behavior of an ordinary function:

Proposition 19 *For any smooth extremal $Y(x)$ of $J[y]$, the first variation δJ vanishes for any admissible smooth variation δy.*

Proof This alternative statement of the necessity for an extremal to satisfy the Euler DE is the variational analogue of the vanishing of the differential df at an extreme point $\hat{x}$ of an ordinary function $f(x)$. With

$$J[Y + \delta y] - J[\delta y] = \delta J + \delta^2 J + \delta^3 J + \cdots \qquad (12.1.12)$$

where δJ is as previously given in (12.1.11) and

$$\delta^2 J = \frac{1}{2!} \int_a^b \left[\hat{F}_{yy}(x)(\delta y)^2 + 2\hat{F}_{yy'}(x)\,\delta y\,\delta y' + \hat{F}_{y'y'}(x)(\delta y')^2 \right] dx, \qquad (12.1.13)$$

etc., the second and higher-order variations, $\delta^k J$ ($k \geq 2$), involve higher powers of δy and $\delta y'$ and are smaller in magnitude by at least one power of ε (keeping in mind $\delta y = \varepsilon z$ and $\delta y' = \varepsilon z'$). The following proof of Proposition 19 is an analogue of the corresponding calculus result for functions. For sufficiently small $|\delta y|$ and $|\delta y'|$, the sign of the right-hand side of (12.1.12) is determined by δJ. If we have $\delta J > 0$, for some choice of δy, then for any $\alpha > 0$, however small, we have

$$\delta J[-\alpha\,\delta y] = \int_a^b \left[\hat{F}_y(x)(-\alpha\,\delta y) + \hat{F}_{y'}(x)(-\alpha\,\delta y') \right] dx$$

$$= -\alpha \int_a^b \left[\hat{F}_y(x)\,\delta y + \hat{F}_{y'}(x)\,\delta y' \right] dx < 0. \qquad (12.1.14)$$

Hence, $J[y] - J[\hat{y}]$ can be made to have either sign for a sufficiently small δy unless $\delta J = 0$. □

12.2   Adjoint Functions

12.2.1   *A minimum cost problem*

We are now ready to introduce the idea on how to handle the general class of problems (that includes the fishery problem) for which the stationary condition does not specify the relevant controlling mechanisms and/or the optimal solution violates inequality constraints on the controls. We develop presently a systematic approach that is generally applicable to this class of problems by working out a modified form of the single vaccine production problem discussed in Chapter 10. Here the firm is contracted to produce B units of the vaccine by $t = T$ and the production rate y' of these units is proportional to the investment rate u with

$$y' = u, \qquad y(0) = 0, \qquad y(T) = B.$$

The total cost of producing B units of vaccine to be delivered on time just before the flu season $t = T$ is

$$J[y] = \int_0^T (u^2 + 4y)dt$$

where the term $4y$ corresponds to a linear storage cost of y units of vaccines per unit time.

Similar to the harvest rate (or effort rate) for the fishery problem, the investment rate for the vaccine production problem is bounded by the resources available for investment. Here u is bounded above and below so that $0 \le u \le u_{max}$. For such a problem, the solution would be more readily revealed itself if we do not eliminate the control $u(t)$ from the performance index. Conceptually, it is $u(t)$ that we can choose and regulate directly, not $y(t)$. To see the actual benefit, we append the production rate relation to the performance index with the help of an *adjoint function* $\lambda(t)$ to get

$$I[u] = \int_0^T \left[u^2 + 4y + \lambda(u - y') \right] dt. \tag{12.2.1}$$

Two observations are appropriate at this point. For any extremal that satisfies the production rate relation, we only added zero to J so that I and J have the same numerical value. At this time, we have not specified $\lambda(t)$ except that it is to be a piecewise smooth (PWS) function (for reason we shall see later).

As in the calculus of variations, suppose $U(t)$ is the *optimal investment rate* and $Y(t)$ is the corresponding optimal vaccine units produced at time t. We consider admissible comparison functions nearby by setting $u = U + \delta u$ and $y = Y + \delta y$ with $\delta y(0) = \delta y(T) = 0$. The first variation δI is then given by

$$\delta I = \int_0^T \left[2u\delta u + 4\delta y + \lambda(\delta u - \delta y') \right] dt$$

$$= [\lambda(t)\delta y(t)]_{t=0}^T + \int_0^T \left[(2u + \lambda)\delta u + (4 + \lambda')\,\delta y \right] dt. \tag{12.2.2}$$

With $\lambda(t)$ still not yet specified, we can choose it to eliminate the term multiplying δy:

$$\lambda' + 4 = 0,$$

so that

$$\lambda(t) = -4t + \lambda_0 \tag{12.2.3}$$

for some unknown constant of integration λ_0. Since $y(t)$ is prescribed at the two ends, we have $\delta y(0) = \delta y(T) = 0$. Altogether, these actions leave us with

$$\delta I = \int_0^T \left[(2u + \lambda)\delta u \right] dt. \tag{12.2.4}$$

The stationary condition $\delta I = 0$ now leads to the Euler DE

$$u = -\frac{1}{2}\lambda(t) = \frac{1}{2}(4t - \lambda_0).$$

We can now use the expression above for the control $u(t)$ in the production rate relation $y' = u$ to obtain

$$y(t) = c_0 - \frac{1}{2}\lambda_0 t + t^2. \tag{12.2.5}$$

The new constant of integration c_0 must vanish since $y(0) = 0$. The remaining end condition $y(T) = B$ then determine λ_0 so that

$$y(t) = \frac{B - T^2}{T} t + t^2 \equiv Y_i(t), \qquad \lambda_0 = \frac{2(T^2 - B)}{T} \qquad (12.2.6)$$

with

$$u(t) = \frac{B - T^2}{T} + 2t \equiv U_i(t), \qquad \lambda(t) = 4\left(\frac{T^2 - B}{2T} - t\right) \equiv \Lambda_i(t).$$

The expressions for the *stationary solution* $Y_i(t)$ and $U_i(t) = Y_i'(t)$ are identical to the ones obtained by the method of calculus of variations (that uses the production rate relation to eliminate u from the performance index). Though it remains to be proved, they are expected to constitute the minimizer of the production cost given that there is no other admissible extremal. Note that $y(t) = 0$ and $u(t) = y'(t) = 0$ gives a smaller $J[y]$, but $y(t)$ does not satisfy the end condition $y(T) = B$.

12.2.2 Constraints on the investment rate

The unique stationary solution $Y_i(t)$ and $U_i(t)$ and the associated adjoint function $\Lambda_i(t)$ are also known as the *interior solution* in the optimal control literature with $U_i(t)$ known as the *interior control*. This interior solution applies as long as $B \geq T^2$. If $B \leq T^2$, then $U_i(t) < 0$ for $0 \leq t \leq t_0$ with

$$t_0 = \frac{T^2 - B}{2T}.$$

This is not acceptable since we cannot have negative investment in the context of this problem. For this reason, we have to restrict the investment rate to be nonnegative. Together with the limited resource constraint, we need to impose the inequality constraints $0 \leq u \leq u_{max}$ on the investment rate. In that case any solution must have $u(t) = 0$ for a range of t near the starting time $t = 0$.

It is natural and tempting to take the optimal investment rate to be

$$u(t) = \begin{cases} 0 & (0 \leq t \leq t_0) \\ U_i(t) & (t_0 \leq t \leq T) \end{cases}$$

with the corresponding $y(t)$ given by (12.2.5). However, λ_0 is already given in (12.2.6) to get $U_i(t)$, the only constant of integration c_0 in $y(t)$ is determined by $y(T) = B$ to give $c_0 = 0$ as before. In that case, we have

$$y(t_0) = \left(\frac{B - T^2}{T} + t_0\right)t_0 = \frac{B - T^2}{2T} t_0 < 0$$

which is not acceptable since vaccine units cannot be negative.

An alternative would be for the two constants of integration λ_0 and c_0 in (12.2.5) to be determined by $y(0) = 0$ and $y(T) = B$ to give

$$c_0 = \frac{T^2 - t_0^2 - B}{T - t_0} t_0 - t_0^2, \qquad \lambda_0 = \frac{T^2 - t_0^2 - B}{T - t_0}$$

so that

$$u(t) = \begin{cases} 0 & (0 \le t < t_0) \\ \frac{1}{2}(4t - \lambda_0) = u_0(t) & (t_0 < t \le T) \end{cases} \tag{12.2.7}$$

$$y(t) = \begin{cases} 0 & (0 \le t < t_0) \\ t^2 - t_0^2 - \frac{T^2 - t_0^2 - B}{T - t_0}(t - t_0) \equiv y_0(t) & (t_0 < t \le T) \end{cases}.$$

Whether this is the optimal solution remains to be investigated. The following observation at least makes it less than ideal. While $y(t)$ is now continuous at $t = t_0$, the corresponding investment rate,

$$u(t) = 2t - \frac{T^2 - t_0^2 - B}{2(T - t_0)} \qquad (t_0 < t \le T),$$

is generally not (since $u(t_0+) \ne 0$). The optimal control may well be among the discontinuous investment rates (leading to a PWS $y(t)$ and $\lambda(t)$); but it seems reasonable to investigate whether there are extremals that are continuously differentiable so that $u(t)$ is also continuous. Even if that should not be the case, it is still worth investigating whether there are other discontinuous investment rates that are superior to the one given by (12.2.7) selected rather arbitrarily without being subject to any kind of comparisons to test its merit.

12.2.3   The optimal control

To investigate more systematically the optimal control $U(t)$ for our problem, we return to the first variation of the performance index (12.2.4) with the adjoint function already chosen as given by (12.2.3). Our first action is to take the optimal control to vanish for some time interval adjacent to the starting time (since the interior control is negative for some stretch of time at and near the start of the program) so that

$$U(t) \begin{cases} = 0 & (0 \le t < t_s) \\ > 0 & (t_s < t \le T) \end{cases}. \tag{12.2.8}$$

In so doing, we do not assume that the switch time t_s is the same as t_0 (the instant when $U_i(t)$ vanishes as previously calculated) as there may be a better time to switch the investment rate. In that case, we need to rewrite (12.2.2) as

$$\delta I = \int_0^{t_s-} [(2u + \lambda)\delta u]\, dt + \int_{t_s+}^T [(2u + \lambda)\delta u]\, dt$$

$$= \int_0^{t_s-} [\lambda(t)\delta u]\, dt + \int_{t_s+}^T [(2u + \lambda)\delta u]\, dt$$

allowing for $U(t)$ to have a simple jump discontinuity at t_s (and therewith both $\lambda(t)$ and $y(t)$ continuous in the planning period). At this point, we may treat the unknown switch time t_s as a free boundary and use the free boundary condition to determine its value. However, we postpone a discussion along that line until later and point out instead that if the optimal control $U(t)$ does not vanish at $t = 0$ and thus by continuity for some interval $0 \le t \le t_s$, then the interior control must apply; but that is not acceptable because $U_i(t)$ is negative there and thereby violates the non-negative constraint on $u(t)$. Hence our assertion that the optimal control must vanish in some interval adjacent to the starting time holds.

Next, we note that $\lambda(t)$ in the first integral for δI must be positive, i.e., $\lambda(t) > 0$ for $0 \le t < t_s$. Otherwise, we can make $\delta I < 0$ (and thereby reduce the total cost $I = J$ currently attained with $u(t) = 0$) by taking $\delta u > 0$. In other words, we could do better by increasing $u(t)$ from 0 to $u > 0$. This contradicts the fact that the stationary solution $U_i(t)$ is negative and the nonnegative constraint keeps $U(t)$ from being $U_i(t)$ there). The opposite argument requires $\lambda(t) < 0$ for $t > t_s$. If not, then we would have $2u + \lambda > 0$ since we know from (12.2.8) $U(t) > 0$ for $t > t_s$. We could then reduce the cost by taking $\delta u < 0$ and $U(t)$ would not be optimal for $t > t_s$. With $\lambda(t)$ continuous in $(0, T)$, we have $\lambda(t_s) = 0$ and

$$\lambda(t_s) \begin{cases} > 0 & (t < t_s) \\ = 0 & (t = t_s) \\ < 0 & (t > t_s) \end{cases}.$$

The condition $\lambda(t_s) = 0$ ensures $\lambda(t)$ continuous at t_s and, together with (12.2.3), gives

$$\lambda(t) = 4(t_s - t), \qquad u(t) = 2(t - t_s).$$

The production rate relation can then be integrated to give

$$y(t) = (t - t_s)^2$$

where we have made use of the continuity condition $y(t_s) = 0$ to eliminate the constant of integration.

It remains to satisfy the terminal condition $y(T) = B$. The only freedom we have at our disposal is the unspecified switch time t_s which can be chosen so that

$$y(T) = (T - t_s)^2 = B$$

or

$$t_s = T - \sqrt{B}$$

which is positive because we want to address the range $B < T^2$ for which the interior solution violates the nonnegativity constraint on $u(t)$.

Altogether, we have for our optimal control

$$U(t) = \begin{cases} 0 & (0 \le t \le t_s) \\ 2(t - t_s) & (t_s \le t \le T) \end{cases},$$

which turns out to be continuous at the switch time, and

$$Y(t) = \begin{cases} 0 & (0 \le t \le t_s) \\ (t - t_s)^2 & (t_s \le t \le T) \end{cases}.$$

Strictly speaking, we still have to show that $J[U(t)]$ is a minimum among the admissible set of comparison functions. Readers are referred to [34] for the mathematics of establishing $J[U(t)]$ being a weak, strong, local or global minimum. We merely want to point out here that our systematic solution process provides a road map to one kind of solution that is more complicated than just a stationary solution for the performance index.

12.3   With Constraints, Do Your Best

12.3.1   *The augmented performance index*

The solution process for the problem of minimum production cost can be formulated as a general approach for optimization problems for which no interior control is appropriate for a part of, or the whole solution domain. We focus on the problem of maximizing

$$J[u] = \int_0^T F(t, y, u) \, dt + G(T, y(T)) \tag{12.3.1}$$

subject to the IVP

$$y' = f(t, y, u), \qquad y(0) = Y_0, \qquad (\ )' = \frac{d(\ )}{dt} \tag{12.3.2}$$

for a scalar unknown $y(t)$ and control $u(t)$ with the simple inequality constraints

$$u_{\min} \le u(t) \le u_{\max} \tag{12.3.3}$$

on the control $u(t)$ and possibly nonnegative constraints on the state variable $y(t)$. (There are other forms of constraints depending on the applications [2] but they will not be considered in these notes.) We stipulate that both F and f be C^1 functions in all of their arguments. (It is customary in the control literature to maximize with the most well known mathematical result known as the Maximum Principle in Optimal Control Theory.) A minimization problem may be converted to maximization by working with $-J$. The standard form of calculus variations problem may be converted to the form above by setting

$$y' = u$$

and using it to eliminate y' from the Lagrangian. Until further notice, we are concerned mainly with the case where T is prescribed but not $y(T)$. A typical salvage value function $G(T, y(T))$ is appended as a part of the performing index to reflex the value assigned to the state variable at the end of the planning period. In the terminal payoff problem of Chapter 10, we have the simple salvage value

$$G(T, y(T)) = Be^{-rT} y(T).$$

Our solution process begins by incorporating the dynamics into the performance index using an adjoint variable $\lambda(t)$ to get the following modified performance index

$$I[u] = G(T, y(T)) + \int_0^T \{F(t, y, u) + \lambda(t)[f(t, y, u) - y']\} \, dt. \tag{12.3.4}$$

After integrating the first variation δI by parts and choosing the adjoint variable to be the solution of the adjoint ODE

$$\lambda' = -\lambda(t)f_{,y}(t,y,u) - F_{,y}(t,y,u), \tag{12.3.5}$$

and the Euler BC

$$\lambda(T) = G_{,y(T)}(T,y(T)). \tag{12.3.6}$$

This choice of adjoint variable eliminates all terms involving δy in δI leaving us with

$$\delta I = \int_0^T \{F_{,u} + \lambda f_{,u}\}\delta u\ dt \equiv \int_0^T \sigma(t)\delta u\ dt \tag{12.3.7}$$

where $\sigma(t) = F_{,u} + \lambda f_{,u}$ is called the *sign function* of the problem.

12.3.2   The interior solution

Similar to determining the optimal value of an ordinary function, candidates for the optimizer are found from those satisfying the stationary condition. Stationarity of δI requires

$$\sigma(t) = F_{,u} + \lambda f_{,u} = 0 \tag{12.3.8}$$

(see (12.2.4) for the minimum cost problem for example). In principle, this condition can be used to determine an interior control $u_i(t)$ in terms of $y(t)$, $\lambda(t)$ and t. But this is sometimes not the case.

- For the minimum cost problem of the previous section, the interior control is not acceptable in a part of the solution domain.
- For the fishery problem when the unit price p and unit harvest cost c do not depend on the harvest rate u, we have

$$F(t,y,u) = e^{-rt}[p(t) - c(t.y)]\ u \quad \text{and} \quad f(t,y,u) = \bar{f}(y) - u$$

(where $\bar{f}(y)$ is the natural growth rate of the unharvested fish population) with $\sigma(t)$ independent of u.

- For the broad class of problem with both F and f linear in the control variable,

$$F(t,y,u) = F_0(t,y) + F_1(t,y)u \tag{12.3.9}$$

and

$$f(t,y,u) = f_0(t,y) + f_1(t,y)u, \tag{12.3.10}$$

the stationary condition (12.3.8) becomes

$$F_{,u} + \lambda f_{,u} = F_1(t,y) + \lambda f_1(t,y) = 0 \tag{12.3.11}$$

which does not involve $u(t)$ and hence does not determine $u_i(t)$ in terms of $y(t)$, $\lambda(t)$ and t. Instead, it gives a relation between y and λ with t as a parameter. Note that the fishery problem is a special case of this class of problem.

12.3.3 Do the best you can

While the classical method of calculus of variations fails to produce the correct solution for the problems described above, the correct solution for the minimum cost problem is accessible if we reformulate the steps taken toward the condition of stationarity for the optimal solution by introducing an adjoint function $\lambda(t)$ and reducing the change in the performance index to the first variation (see (12.3.7))

$$\delta I = \int_0^T \sigma(t)\delta u \ dt, \quad \sigma(t) = F_{,u} + \lambda f_{,u}.$$

For an optimizer, calculus of variations seeks an extremal that renders $\delta I = 0$ and, by the fundamental lemma, requires $\sigma(t) = 0$ for the determination of the extremal(s). When this requirement fails to determine an appropriate extremal, the minimum cost problem suggests that we proceed by examine the sign of $\sigma(t)$ in some range of the solution domain.

If $\sigma(x)$ is positive for our choice of $u(t)$, then δI as given by (12.3.7) can be made positive (and leading to a larger $J[u]$) by choosing a larger control, say $u(t) + \delta u$ for $\delta u > 0$ and we can continue to increase whatever u we have currently as long as $\sigma(x)$ is positive. The process stops either when we have $\sigma(t) = 0$ (since increasing u further does not improve on the existing $J[u]$) or when no further increase is allowed because of the constraint on the control.

Similarly, for $\sigma(t) < 0$, δI can again be made positive by choosing a smaller control, say $u(t) + \delta u$ for $\delta u < 0$, so that $I + \delta I > I$ and we should continue to reduce $u(t)$ as long as $\sigma(x)$ is negative. The process also stops either when we have $\sigma(t) = 0$ for the current control or when no further decrease in u is allowed because of constraint on the control. In other words, we would not improve on $J[u]$ by changing the current $u(t)$ either when $\sigma(t) = 0$ at stationarity or when no further change in $u(t)$ for additional improvement is allowed by the constraints on $u(t)$. In the first case, the optimal solution is an interior solution. In the latter case, the optimum is attained by an extreme value for the allowable range of the control, known as a *corner control* in the control literature.

The observation above on the relation among $\sigma(t)$, δu and δI in order to maximize $J[u]$ applies if $u_i(t)$ is not available or not applicable (typically because there are constraints on the control). For any (reference) $u(t)$, not necessary an interior control or a maximizer, and a comparison control $u + \delta u$, the relation (12.3.7) remains applicable. The following observations are critical toward finding a maximum $J[u]$:

1) $\sigma(t) > 0$: For this case, we can improve on $J[u]$ by taking $\delta u > 0$, i.e., by taking a new control $\tilde{u}(x) > u(x)$. The new control would change $\sigma(x)$ (usually making it smaller) since the new control begets new state and adjoint functions. We should keep increasing the control as long as $\sigma(t) > 0$. This process ends in either one of two ways: i) $\sigma(t) = 0$ when $u = u_i(t)$ in which case the interior control is optimal (so that $U(t) = u_i(t)$ with no further control changes), or ii) there is an *upper limit* $u_{\max}$ stipulated on a control so that we cannot increase $u(t)$ further and

we are left with $U(t) = u_{max}$. In this second case, with the control at its upper limit, the only allowable change on the control is in the wrong direction (as δu can only be negative) with

$$\delta I = \int_0^T \sigma(t)\delta u \, dt < 0$$

and therefore should not be implemented. Together, i) and ii) require that we continue to change $u(t)$ to improve the performance index until we either should not or could not improve further (because either we are at an interior control (stationary solution) or we have reached the upper bound on the control u_{max} and cannot use a larger control.

2) $\sigma(t) < 0$: For this case, we can improve on $J[u]$ by taking $\delta u < 0$, i.e., by taking a new control $\tilde{u}(t) < u(t)$. The new control would change $\sigma(t)$ (usually making it larger). We should keep decreasing the control as long as $\sigma(t) < 0$. This process ends in either one of two ways: i) $\sigma(t) = 0$ for some $u_i(t)$ and we have an interior control (and should not make further control changes) so that $U(t) = u_i(t)$, or ii) we have reached the *lower bound* u_{min} on the control so that we cannot reduce $u(t)$ further. We are then left with $U(t) = u_{min}$ so that any allowable change δu would be in the wrong direction (as it can only be positive) with

$$\delta I = \int_0^T \sigma(t)\delta u \, dt < 0$$

and therefore also should not be implemented. Together, i) and ii) mean that we cannot improve on the performance index further because either we are at an interior control (stationary solution) or we have reached the lower bound on the control $u(t) = u_{min}$ and cannot use a smaller control.

3) $\sigma(t) = 0$: This condition allows us to find the interior control $u_i(t)$ in terms of the state and control variables for a certain time interval. In that case, the Hamiltonian system for the state and adjoint variables determine the extremal(s) for the problem. It is possible that interior control is appropriate only for a part of the solution domain. In that case, we would have $\sigma(t) \neq 0$ for a segment of the solution domain and either Case 1 or Case 2 applies.

In addition to the three general cases, the following should be noted:

- Similar to optimization of an ordinary function, any stationary solution may be minimizing, maximizing or neither; it may be a local or global optimum. We need to investigate further even when there is only one stationary solution.
- Since $\sigma(t)$ may change sign more than once in the solution domain, the optimal control $U(t)$ is generally a combination of the interior control and the corner controls, a different type in a different subinterval of $[0, T]$.
- Even when the stationary condition $\sigma(t) = 0$ does not determine an extremal (e.g., when it does not involve the control), the consequence of this condition may still play a part in the solution and is known as a *singular solution* for the problem (having seen this in the basic fishery problem).

- When the appropriate solution is determined by the condition that we cannot improve upon the solution by any additional allowable change on the control (which may not be stationary), the solution process is effectively to do the best you can and will be designated as the "Do Your Best Principle".

We summarize the observations above in the following proposition:

Proposition 20 *The optimal control $U(t)$ for the problem of maximizing (12.3.1) subject to growth dynamics (12.3.2) and inequality constraints (12.3.3) is necessarily the interior control $U_i(t)$ whenever it exists and does not violate any constraints. When U_i is not available or not appropriate, $U(t)$ is necessarily the lower corner (or boundary) control $u_{\min}$ in any part of $[0,T]$ where $\sigma(t) < 0$ and the upper corner (or boundary) control $u_{\max}$ in any part of $[0,T]$ where $\sigma(t) > 0$.*

There is no simple recipe for determining the sign of $\sigma(t)$ in different part of the solution domain. In some cases (as in the Minimum Production Cost problem), it is seen directly from the interior solution of the problem that it is inappropriate for some portion of $[0,T]$ and what is the appropriate corner control for the problem. In many cases, the Euler BC $\lambda(T) = 0$ for problems without a prescribed terminal condition offers a suggestion for the optimal control adjacent to the terminal time. Such problems will be discussed later in this chapter as well as Chapters 13 and 14.

It is also important to keep in mind that similar to extremals in the calculus of variations, the one or more optimal controls determined by the Do Your Best Principle are only candidates for the extremizer since they are only consequences of the requirement $\delta I = \delta J = 0$. Similar to the consequences of stationarity of an ordinary function, they are only necessary conditions and are usually not sufficient for assurance of a maximum when the optimal control involves an interior control. As in the extremization of ordinary functions, second order or some other appropriate validation should be examined to confirm the extremity of the performance index for the relevant optimal control.

12.4 Fishery Problem without Prescribed Terminal Condition

12.4.1 *The "Do Your Best" approach*

We now apply the Do Your Best Principle to a more conventional version of the fishery problem that does not have a prescribed terminal condition on $y(T)$. Whether or not the fishery have a target total catch or a contracted fishing effort for the fishing period, it does not prescribe or know the fish population at terminal time. Without a prescribed terminal fish population, the new problem corresponds to the standard optimal control problem defined by (12.3.1), (12.3.2) and (12.3.3). In this section, we apply the Do Your Best Principle to this new problem. For the Do Your Best approach, we introduce an adjoint variable $\lambda(t)$ and the augmented performance index

$$I[u] = J[u] + \int_0^T e^{-rt}\lambda[f(y) - u - y'] \, dt$$

$$= \int_0^T e^{-rt}\{[p - c]u + \lambda[f(y) - u - y']\} \, dt.$$

(Here we artificially take the adjoint variable to be $e^{-rt}\lambda(t)$ instead of simply $\lambda(t)$ for the advantage that e^{-rt} becomes a common factor for all the terms in the adjoint DE and may cancel out.) In that case, the first variation of I is

$$\delta I = \int_0^T e^{-rt}[(p - c) - \lambda]\delta u \, dt \tag{12.4.1}$$

with $\lambda(t)$ having been chosen to satisfy the adjoint DE

$$\lambda' + (f'(y) - r)\lambda = 0. \tag{12.4.2}$$

For the problem with no prescribed terminal condition, the optimizer must also satisfy the Euler BC

$$\lambda(T) = 0. \tag{12.4.3}$$

12.4.2   *Interior control is singular*

If the control $u(t)$ should render I a stationary value, we must have $\delta I = 0$ and therewith

$$F_{,u} + \lambda f_{,u} = e^{-rt}[(p - c) - \lambda] = 0. \tag{12.4.4}$$

For problems with p and c being prescribed functions of t, the relation (12.4.4) does not involve the harvest rate $u(t)$ (or $h(t)$ in the earlier chapters) and therefore does not determine an interior control. If we insist on making use of (12.4.4), then we would have the singular solution $\lambda = p - c$ as a known function of time (because we have taken both p and c to be so). The adjoint DE (12.4.2) then requires

$$f'(y) = r - \frac{(p - c)'}{p - c} \tag{12.4.5}$$

which determines $y(t)$ as an explicit function of t, which is just $Y_s(t)$ in the Most Rapid Approach solution. Finally, the corresponding control $U_s(t)$ to accomplish all these is

$$U_s(t) = f(Y_s) - Y_s'. \tag{12.4.6}$$

For a normalized logistic growth rate $f(y) = y(1 - y)$ with $f' = (1 - 2y)$, (12.4.5) gives

$$y(t) = \frac{1}{2}\left[(1 - r) + \frac{(p - c)'}{p - c}\right] \equiv Y_s(t) \tag{12.4.7}$$

and

$$u(t) = f(Y_s) - Y_s' \equiv U_s(t). \tag{12.4.8}$$

For the extreme case where p and c are both constants, $Y(t)$ is just a constant

$$Y_s(t) - \frac{1}{2}(1 - r), \qquad U_s(t) = f(Y_s) - \frac{1}{4}(1 - r^2) \qquad (12.4.9)$$

recovering the same extremal by the method of Most Rapid Approach. Whether or not $(p-c)$ is a constant, the *singular solution* Y_s contains no constant of integration and in general does not satisfy the prescribed initial condition.

The lack of flexibility in the singular solution given by (12.4.7) may not pose a problem if we have no restriction on the harvest rate. In reality, the harvest rate is necessarily finite and constrained by an upper and a lower bound. In an ocean fishing ground, $u(t)$ must be nonnegative since re-stocking the fishing ground is not realistic and in any case not practiced. At the other extreme, a fishery has only a limited capacity to harvest for a typical planning period. As such the harvest rate is subject to the inequality constraints (9.5.10), we simply cannot get from Y_0 to $Y_s(0)$ instantaneously if they should be different. While all these observations pose serious issues for an optimal solution, they were resolved by the method of Most Rapid Approach for the case when the size of the available growing fish stock is prescribed at both ends of the planning period. The same approach however cannot be justified in the same way for the present modified problem when the terminal population is not prescribed.

Without a prescribed terminal condition, the modified fishery problem is similar in form to the minimum cost investment problem. It seems reasonable to attempt a solution by the Do Your Best Principle in the subsections below.

12.4.3 Upper corner solution near the end

To illustrate the Do Your Best approach, we limit consideration of the modified problem to the case of constant unit price and cost with $p - c = p_0 - c_0 > 0$ (so that profit is possible). For the modified problem, the Euler BC $\lambda(T) = 0$ (12.4.3) requires

$$\sigma(t) \equiv e^{-rt}\sigma_0(t) = e^{-rt}[(p - c) - \lambda] > 0 \qquad (12.4.10)$$

at $t = T$ and, by continuity of the adjoint function, also for at least some small interval $(t_s, T]$ adjacent to the terminal time for some $t_s \geq 0$. It follows that the stationary condition cannot be met in that interval so that i) an interior control is not determined, and ii) the singular solution does not exist there. On the other hand, by the Do Your Best Principle, a positive sign function, $\sigma(t) > 0$ (at least in $(t_s, T]$), requires the optimal control to be the *upper corner control* u_{max}:

$$U(t) = u_{max} \qquad (t_s < t \leq T) \qquad (12.4.11)$$

and determines the corresponding state and adjoint functions for $t_s < t \leq T$ by

$$y' = f(y) - u_{max}, \qquad \lambda' + \{f'(y) - r\}\lambda = 0, \qquad \lambda(T) = 0. \qquad (12.4.12)$$

If $t_s \leq 0$, then we have $i)$ $\sigma(t) > 0$ in $(0, T]$ so that (12.4.11) holds for $0 < t \leq T$, and $ii)$ the initial condition $y(0) = Y_0$ provides the missing auxiliary condition for the two ODE in (12.4.12) to form a two-point BVP that can be solved to determine the extremal $\{Y(t), \Lambda(t)\}$. It remains to show that the unique extremal for constant p_0 and c_0 renders J maximum (see [34]).

If $t_s > 0$, we denote by $\{y_g(t), \lambda_g(t)\}$ the solution of (12.4.12) augmented by the continuity requirement

$$y_g(t_s) = y_\ell(t_s) \tag{12.4.13}$$

where $y_\ell(t_s)$ is the (yet to be determined) solution of the problem for $t < t_s$. As such $y_g(t)$ and $\lambda_g(t)$ are not completely known until we have information about the solution in the complementary range $t < t_s$ to be found below.

12.4.4   Optimal control for $t < t_s$

For $t_s > 0$, we have $\sigma(t_s) = 0$ (for otherwise we would have a smaller t_s) so that $\lambda(t_s) = p_0 - c_0 > 0$. We consider separately three possible scenarios for $\lambda'(t_s-)$:

12.4.4.1 $\lambda'(t_s-) > 0$

In this case, $0 < \lambda(t) < \lambda(t_s) = p_0 - c_0$ (so that $\sigma(t) = p_0 - c_0 - \lambda > 0$) for $t < t_s$ with the adjoint function continuous at t_s. It follows that the upper corner control is optimal (and the lower corner control is not) so that the upper corner solution $\{y_g(t), \lambda_g(t)\}$ from (12.4.12)–(12.4.13) is extended to a larger interval $(\bar{t}_s, T)$ with $\bar{t}_s < t_s$ taking the place of t_s with $\lambda'(\bar{t}_s-) \leq 0$ (for the development above repeats and extends the range of $\{y_g(t), \lambda_g(t)\}$ to a new switch point for which $\lambda'(\bar{t}_s-) \leq 0$ applies).

12.4.4.2 $\lambda'(t_s-) < 0$

For this more likely scenario, we have $\lambda(t) > \lambda(t_s) = p_0 - c_0 > 0$ for $t < t_s$ so that $\sigma(t) < 0$ for t in some interval (t_0, t_s). By the Do Your Best Principle, the optimal control is necessarily $U(t) = 0$, $t_0 < t < t_s$, with the corresponding state and adjoint functions, to be denoted by $y_\ell(t)$ and $\lambda_\ell(t)$, respectively, determined by

$$y' = f(y), \qquad \lambda' + \{f'(y) - r\}\lambda = 0, \qquad \lambda(t_s) = p_0 - c_0 \tag{12.4.14}$$

in the interval (t_0, t_s) where $\sigma(t) < 0$.

If $t_0 = 0$, then the optimal fishing policy for the modified fishery problem (without a prescribed terminal condition) would be the bang-bang control

$$U(t) = \begin{cases} 0 & (0 \leq t < t_s) \\ u_{max} & (t_s < t \leq T) \end{cases} \tag{12.4.15}$$

with

$$\{Y(t), \Lambda(t)\} = \begin{cases} \{y_\ell(t), \lambda_\ell(t)\} & (0 \leq t < t_s) \\ \{y_g(t), \lambda_g(t)\} & (t_s < t \leq T) \end{cases}$$

where the upper corner solution $\{y_g(t), \lambda_g(t)\}$ is determined by (12.4.12)–(12.4.13) and the lower corner solution is determined by (12.4.14) and the initial condition

$$y_\ell(0) = Y_0.$$

The solution maximizes $J[u]$ since the bang-bang control (12.4.15) is the only control that renders $J[u]$ larger than any other admissible control.

If $t_0 > 0$, we examine the sign if $\lambda'(t_0-)$ is greater or equal to zero and proceed accordingly. (If $\lambda'(t_0-) < 0$, the lower corner control continues to apply to $t < t_0$.)

12.4.4.3 $\lambda'(t_s) = 0$

From the adjoint ODE

$$\lambda'(t_s) + \{f'(y(t_s)) - r\}\lambda(t_s) = 0$$

we have

$$f'(y(t_s)) = r.$$

It follows after repeated differentiation of the sign function and the adjoint ODE that $\sigma'(t_s) = \sigma''(t_s) = \cdots = 0$ to result in $\sigma(t) = e^{-rt}[(p_0 - c_0) - \lambda] = 0$ or the singular solution

$$\lambda_S(t) = p_0 - c_0 \qquad (t \le t_s)$$

and

$$f'(y(t)) = r \qquad \text{or} \qquad Y_S(t) = Y_S^{(k)}(r) \tag{12.4.16}$$

where $Y_S^{(k)}(r)$ is a zero of $f'(y) - r = 0$ that does not depend on t. With singular solution for the fish population $y(t) = Y_S^{(k)}(r)$ independent of time, we have from the equation of state $y' = f(y) - u$ the singular control

$$u_S(t) = f(Y_S^{(k)}(r))$$

also independent of time.

To the extent that the singular fish stock (12.4.16) corresponds to a stationary solution, it is preferred for optimality as long as it does not violate any of the constraints for the problem. Unless $Y_0 = Y_S^{(k)}(r)$, the singular solution generally cannot be made to match the prescribed initial stock and we need to switch to a corner control at some earlier time $t_0 < t_s$ in order to bridge Y_0 and $Y_s(t_0)$. For the case $Y_0 > Y_s(0) = Y_s^{(k)}(r)$, we take

$$U(t) = \begin{cases} u_{\max} & (0 \le t < t_0) \\ f(Y_s^{(k)}(r)) & (t_0 < t < t_s) \\ u_{\max} & (t_s < t \le T) \end{cases} \tag{12.4.17}$$

with t_0 determined by $z(t_0) = Y_s^{(k)}(r)$ where $z(t)$ is the solution of the IVP

$$z' = f(z) - u_{\max}, \qquad z(0) = Y_0.$$

If $z(t) > Y_s^{(k)}(r)$ for all $t < T$, the optimal control is the feasible solution $U(t) = u_{\max}$ with $Y(t) = z(t)$ keeping in mind that there is no prescribed terminal stock for the problem. If we wish to have it, the adjoint variable is to be computed by the terminal value problem

$$\lambda' + \{f'(z) - r\}\lambda = 0, \qquad \lambda(T) = 0 \qquad (0 \le t < T).$$

Exercise 35 *Determine the optimal control and the ODE problem for determining the corresponding state and adjoint function $Y_0 < Y_s(0) = Y_s^{(k)}(r)$.*

12.5 Optimal Allocation of Stem Cells for Medical Applications

Denote by $y(t)$ the total amount of stem cells (in units of biomass) available at time t. Suppose a fraction $u(t)$, $0 \le u(t) \le 1$, of this total is to be used for producing more stem cells according to the simple growth dynamics of

$$y' = \alpha u y, \qquad y(0) = Y_0, \qquad (12.5.1)$$

where α is the growth rate constant. The remaining stem cells, $(1 - u(t))y(t)$, is available for medical applications. The planning goal is to maximize the total stem cells available for medical applications over a planning period $[0, T]$. Some stem cells should be available at the end of the planning period for production of more stem cells in the future. One option would be to prescribe $y(T) = Y_T > 0$. We consider here instead the incorporation of a salvage value for what is left of the stem cells at the end of the planning period with

$$J = \int_0^T (1 - u(t))y(t)dt + \beta y(T)$$

for some constant β. In particular, a positive β corresponds to a benefit accrued to leaving some stem cells for the next planning period. Doing so obviously reduces the benefit associated with cells available for medical applications during the current planning period $[0, T]$; hence there is a trade-off between consumption for current use and saving for future benefit to be adjudicated by the optimization process. We note in passing that the salvage value $\beta y(T)$ may be replaced by a more general value function $G(T, y(T))$ if it is appropriate to do so.

Since $u(t)$ is a fraction of the whole in the present problem, we have the natural constraint that the control function must be restricted by

$$0 \le u(t) \le 1. \qquad (12.5.2)$$

Exercise 36 *a) Use the growth dynamics of stem cell population in the form $uy = y'/\alpha$ to transform this medical use of stem cell problem into a calculus of variations problem and show how it fails to determine an admissible extremal for the problem.*

b) Apply the Do Your Best Principle to determine the optimal control for the special case of the problem with $\alpha = 3/2$, $\beta = 1/2$ and $T = 1$ to obtain

$$U(t) = \begin{cases} 1 & (0 \le t < 5/6) \\ 0 & (5/6 < t \le 1) \end{cases}, \qquad Y(t) = \begin{cases} Y_0 e^{3t/2} & (0 \le t < 5/6) \\ Y_0 e^{5/4} & (5/6 < t \le 1) \end{cases}.$$

Exercise 37 *As satisfying as it may be to be able to determine the optimal use of stem cells with our new approach to constrained optimization problems, the mathematical model is rather unrealistic as it stands.*

a) To see this, compute y(t) for the special case of u(t) = 0 (corresponding to using all available stem cells for medical application and none for reproduction of more stem cells).

b) How are the stem cells to be used physically in order to result in the population y(t) for all t > 0 obtained in part a)?

Exercise 38 *a) Modify the present model so that it is more consistent with reality.*
b) Determine the optimal fraction u(t) for your model.

12.6 The Maximum Principle

While the Do Your Best approach has enabled us to solve optimal control problems with constraints and without a terminal end condition, the solution process can be stream-lined and made less cumbersome. The more efficient approach that is applicable to a broader class of problems is the Maximum Principle of Pontryagin [19, 34]. In this section, we distill the essence of the Do Your Best solution process of Section 12.3 to form the basic elements of the Maximum Principle. In addition to reworking the solution of several problems already solved by the heuristic Do Your Best approach, this powerful principle will then be applied to a constrained optimization problem pertaining to the infectious disease Chlamydia Trachomatis in the next chapter. Its application to a more complex problem on genetic instability and carcinogenesis will be reported in Chapter 14.

12.6.1 The Hamiltonian

While setting up the solution process to implement the Do Your Best Principle is straightforward, the steps that take the original problem to the sign function $\sigma(t)$ is unnecessarily repetitive. Up to now, these steps consist of

- introducing the adjoint function to incorporate the growth dynamics into the performance index,
- obtaining its first variation of the augmented performance index,
- generating the adjoint ODE and Euler BC to reduce the first variation to one involving only the sign function.

If we examine the second line of the expression for $I[u]$ in (12.2.1) and the corresponding δI, we would see that only a part of the integrand is critical to the solution process. To excerpt that essential part, we introduce the quantity

$$H(y, u, \lambda, t) = u^2 + 4y + \lambda u$$

for that part of the integrand of (12.2.1). It is not difficult to see (or verify) the ODE for growth dynamics and the adjoint ODE are generated by

$$y' = \frac{\partial H}{\partial \lambda} = u, \quad \lambda' = -\frac{\partial H}{\partial y} = -4.$$

Furthermore, the relevant sign function is given by

$$\sigma(t) = \frac{\partial H}{\partial u} = (\lambda + 2u).$$

These relations demonstrate the central role of the quantity $H(y, u, \lambda, t)$ occupies in the final solution process for the present class of constrained optimization problem and leads to its designation as the *Hamiltonian* of the problem.

12.6.2   The general problem

Consider now the generic problem of maximizing

$$J[u] = \int_0^T F(t, y, u)dt + G(T, y(T)) \tag{12.6.1}$$

subject to the IVP

$$y' = f(t, y, u), \quad y(0) = Y_0, \quad (\;)' = \frac{d(\;)}{dt}, \tag{12.6.2}$$

and the inequality constraints (12.3.3). The Hamiltonian for this problem is given by

$$H(y, u, \lambda, t) = F(t, y, u) + \lambda f(t, y, u) \tag{12.6.3}$$

where the first term is the Lagrangian of the performance index of our problem and the second term is a λ multiple of the growth rate function. It is again straightforward to verify

$$y' = \frac{\partial H}{\partial \lambda} = f(t, y, u), \quad \lambda' = -\frac{\partial H}{\partial y} = -\frac{\partial F}{\partial y} - \lambda \frac{\partial f}{\partial y} \tag{12.6.4}$$

with the second being the adjoint ODE generated by the Do You Best approach.

To simplify the adjoint DE, we rewrite the second ODE in (12.6.4) the subscript notations for partial derivative

$$\lambda' = -\lambda(t)f_{,y}(t, y, u) - F_{,y}(t, y, u). \tag{12.6.5}$$

Two other important ingredients for the solution process is the Euler BC

$$\lambda(T) = \frac{\partial G(T, y_T)}{\partial y_T} \tag{12.6.6}$$

which involves the salvage value $G(T, y_T)$ where y_T is $y(T)$ for brevity, and the sign function $\sigma(t)$ which is related to the Hamiltonian by

$$\sigma(t) = \frac{\partial H}{\partial u} = F_{,u}(t, y, u) + \lambda f_{,u}(t, y, u). \tag{12.6.7}$$

It is straightforward to verify that these relations reduce to those for the various examples in this chapter.

12.6.3 Maximization of the Hamiltonian

Having generated all the auxiliary relations relevant to the solution process for our constrained maximization problem, there remains the critical step of choosing an appropriate control that would maximizes $J[u]$. For the Do Your Best approach, this is accomplished by working with the sign of the sign function leading to the conclusions

$$
\sigma(t) \begin{cases} = 0 & \text{(generate interior or singular control)} \\ > 0 & \text{(increase } u(t) \text{ until it reaches its upper bound)} \\ < 0 & \text{(decrease } u(t) \text{ until it reaches its lower bound)} \end{cases} \tag{12.6.8}
$$

With $\sigma(t) = H_{,u}(y, u, \lambda, t)$ and

$$
\delta I = \int_0^T \sigma(t) \delta u \, dt = \int_0^T H_{,u}(y, u, \lambda, t) \delta u \, dt,
$$

we may rewrite (12.6.8) in terms of the Hamiltonian to get

$$
H_{,u}(y, u, \lambda, t) \begin{cases} = 0 & \text{(generate interior or singular control)} \\ > 0 & \text{(increase } u(t) \text{ until it reaches its upper bound)} \\ < 0 & \text{(decrease } u(t) \text{ until it reaches its lower bound)} \end{cases}
$$
$$\tag{12.6.9}$$

Given (12.6.9), we may state the results for the Do Your Best Principle in terms of the Hamiltonian (almost) exclusively:

- Whenever possible and feasible, use the stationary condition of the Hamiltonian with respect to the control u to determine the *interior control* in terms of y, λ and t and solve the two-point BVP for the Hamiltonian system (12.6.4) for one or more extremal pairs $y(t)$ and $\lambda(t)$ as candidates for our constrained optimization problem.
- When an interior or singular solution does not exist or is not feasible (with respect to the various constraints), seek the optimal solution $Y(t)$ and $\Lambda(t)$ with an appropriate extreme value of the constrained control.
- With the simple inequality constraints (12.5.2) rendering interior or singular solution infeasible beyond the allowable range, the optimal control $U(t)$ should be the *upper corner control* $u_{\max}$ if $H_{,u}(y(t), u(t), \lambda(t), t) > 0$ and the *lower corner control* 0 if $H_{,u}(y(t), u(t), \lambda(t), t) < 0$.

As such, the Do Your Best solution process is transformed into a problem for the relevant Hamiltonian constructed from the Lagrangian and the growth rate function of the constrained optimization problem. The transformation renders unnecessary any consideration of the first variation of an augmented performance index, including any integration by parts on the first variations. Furthermore, the determination of extremal(s) for the optimization problem is reduced to finding the extrema of the Hamiltonian as an ordinary function of u with y, λ and t as parameters. The solution process in terms of the Hamiltonian (evolved from the Do Your Best Principle herein) is a restricted version of what is known as the *Maximum Principle* [19].

The expression for the control obtained from the stationary condition may be used to eliminate u from the state and adjoint ODE leaving us with a two-point BVP for the state and adjoint functions. Its solution supplies the extremal(s), giving y and λ as functions of t, corresponding to the interior solution with the interior control $u_i(t)$ calculated from the stationary condition.

The stationarity of the Hamiltonian may fail to be useful in at least two ways. The resulting interior solution may violate one or more constraints when checked by (12.5.2) for feasibility. When a bound on u should be violated for a range of t in the solution domain, the optimal control must be a boundary value of the admissible range of u as in the case of (ordinary) function optimization.

The stationarity of the Hamiltonian may also fail to be useful when the condition $\sigma(t) = H_{,u}(y, u, \lambda, t) = 0$ does not determine u in terms of the other parameters as it happens for the problem of medical use of stem cell (and the fishery problem). To handle this case without returning to the expression for the first variation for the present problem (and by working only with the Hamiltonian), we note that for the case when the interior control is feasible, the solution process described above may be cast as the following problem of choosing $U(t)$ to maximize the Hamiltonian:

$$H(Y(t), \Lambda(t), U(t), t) = \max_{u_{min} \leq u \leq u_{max}} [H(Y(t), \Lambda(t), u, t)] \qquad (12.6.10)$$

where $Y(t)$ and $\Lambda(t)$ are the state and adjoint function corresponding to the optimal control $U(t)$. The optimality condition (12.6.10) leads to the stationary condition for an interior control when it determines one or more admissible controls in terms of $Y(t)$ and $\Lambda(t)$. When the stationary condition does not define an interior control, the requirement (12.6.10) still applies; the question is how to make use of it to get an optimal control (with the answer to the question translated from the Do You Best Principle and given in (12.6.9)).

Evidently, the requirement (12.6.10) makes explicit the simplification underlying the Do Your Best Principle in the search for an interior control. Since $y(t)$ and $\lambda(t)$ generally vary with the choice of $u(t)$, the Maximum Principle as defined by (12.6.3), (12.6.4), (12.6.6) and (12.6.10) above clearly asserts that we do not have to be concerned with such changes but consider them to be at their optimal value in the process of finding the optimal control.

As in the extremization of an ordinary function, an optimal control $U(t)$ and the corresponding extremal $\{Y(t), \Lambda(t)\}$ found from the stationary condition

$$H_{,u}(Y(t), U(t), \Lambda(t), t) = 0$$

may be a maximum, a minimum or an "inflection point" and must be further analyzed (by higher order conditions) before being accepted as the solution of the problem. This is particularly important when there are more than one extremals to the problem. It is therefore important to keep in mind that (an optimal control for) maximizing the Hamiltonian is not synonymous with maximizing the performance index J.

To illustrate the applications of the Maximum Principle, we re-work in this and the next subsection some problems already solved by the Do Your Best approach. In the next two subsections, we first revisit a previous exercise on the medical use of stem cells. We then apply the Maximum Principle approach to the fishery problem previously solved by the Do Your Best Principle. The last two chapters of this volume are devoted to some recent research on two rather complex life science phenomena that require intricate application of the same principle.

12.6.4 *The stem cell problem*

To add some new elements and generality to the previously solved Exercise 36 on the optimal allocation of stem cells for medical applications, we consider here the same problem but with $\beta = 0$. By the Maximum Principle, we are to find the optimal control $U(t)$ by the requirement

$$[\{1 + u\,[\alpha\Lambda(t) - 1]\}\,Y(t)]_{0\,\le\,u\,\le\,1} \le \{1 + U(t)\,[\alpha\Lambda(t) - 1]\}\,Y(t) \qquad (12.6.11)$$

with the stationary condition $H_{,u} = \sigma(t)$ not yielding an expression for u, only two singular solutions from

$$\sigma_s(t) \equiv \{\alpha\Lambda_s(t) - 1\}\,Y_s(t) = 0$$

instead. To see how (12.6.10) may be used to find the optimal solution, we begin by making use of what we know about the adjoint variable.

12.6.4.1 *Singular solution not feasible*

From the Euler BC, we have $\Lambda(T) = \beta = 0$ and by continuity $\Lambda(t) < 1$ in some interval $(t_s, T]$. Hence, neither singular solution is feasible adjacent to T since $\Lambda_s(t) = 1/\alpha > 0$ and $Y(t) > 0$ for any $Y_0 > 0$. In fact, we have the following more definitive negative result on the singular solutions:

Proposition 21 *There is no admissible control that sustains a singular solution in any sub-interval of* $(0, T)$.

Proof We already indicated the inadmissibility of $Y_s(t) = 0$. If $\Lambda_s(t) = 1/\alpha$ should apply away from the terminal time T, the adjoint DE would reduce to $0 = -1$ which cannot be met by any admissible control $u(t)$. $\qquad\square$

12.6.4.2 *Lower corner control adjacent to T*

With $[H]_{t=T} = \{(1 - u(T))\}\,Y(T)$, $[H]_{t=T}$ is maximized by $U(T) = 0$ and, by continuity of state and adjoint variables, the lower corner control $U(t) = 0$ is optimal for a sub-interval $(t_s, T]$. For that interval, the Hamiltonian system of DE for state and adjoint functions is

$$y'_\ell = 0, \quad \lambda'_\ell = -1, \quad \lambda_\ell(T) = 0$$

or

$$\lambda_\ell(t) = T - t, \quad Y(t) = \bar{Y}_s$$

for some constant of integration $\bar{Y}_s = Y(t_s-)$ for continuity of the state variable. The situation changes only if $\alpha\Lambda(t) - 1$ should turn positive and the lower corner control no longer maximizes the Hamiltonian. The change takes place at the instant t_s when $\alpha\Lambda(t_s) - 1 = 0$ or

$$t_s = T - \frac{1}{\alpha}.$$

For $t < t_s$, we have

$$\sigma(t) = \alpha\lambda_\ell(t) - 1 = \alpha\,(T - t) - 1 > \alpha\,(T - t_s) - 1 = 0$$

so that $u_\ell(t) = 0$ is not optimal there. The optimal control must switch into another feasible control for $t < t_s$.

12.6.4.3   Upper corner control for $t < t_s$

With $u_\ell(t) = 0$ not optimal and the singular solution not feasible for $t < t_s$, we must examine the upper corner control for that range of time. For $u(t) = u_{\max} = 1$ and with the adjoint function denoted by λ_g, the adjoint problem becomes

$$\lambda_g' = -\alpha\lambda_g, \quad \lambda_g(t_s) = \frac{1}{\alpha},$$

where continuity provides the boundary condition for $\lambda_g(t_s)$. The exact solution is

$$\lambda_g(t) = \frac{1}{\alpha}e^{-\alpha(t-t_s)}$$

with $\alpha\lambda_g(t) = e^{-\alpha(t-t_s)} > 1$ for $t < t_s$. Hence the upper corner control is optimal for the period $[0, t_s)$.

12.6.4.4   The optimal solution by Maximum Principle

Putting together all the pieces found above, we have:

Proposition 22 *By the Maximum Principle, the optimal control $U(t)$ for our medical use of stem cells problem with $\beta = 0$ is*

$$U(t) = \begin{cases} 1 & (0 \le t < t_s = T - 1/\alpha) \\ 0 & (T - 1/\alpha = t_s < t \le T) \end{cases}$$

independent of the values for the parameters α and T.

To complete the problem, we obtain $y(t)$ from the IVP (12.5.1)

$$Y(t) = \begin{cases} Y_0 e^{\alpha t} & (0 \le t \le t_s = T - 1/\alpha) \\ Y_0 e^{(\alpha T - 1)} & (T - 1/\alpha = t_s \le t \le 1) \end{cases} \qquad (12.6.12)$$

where we have required the stem cell population to be continuous at t_s. The corresponding adjoint function can also be obtained by solve the relevant adjoint problem to get

$$\Lambda(t) = \begin{cases} \frac{1}{\alpha}e^{-\alpha(t-t_s)} & (0 \leq t \leq t_s = T - 1/\alpha) \\ T - t & (T - 1/\alpha = t_s \leq t \leq 1) \end{cases}. \qquad (12.6.13)$$

12.6.5   The fishery problem with constant unit harvest cost

12.6.5.1   The problem with $c(h) = c_0$

For another illustration of the solution process by the Maximum Principle, we revisit here the fishery problem where the unit harvesting cost is a constant $c(u) = c_0$ (replacing the previously used variable h for harvest rate by the more generic control variable u). For this problem, we wish to maximize the present value

$$J[u] = \int_0^T e^{-rt}\,[p - c(u)]\,u\,dt \qquad (12.6.14)$$

of the profit stream total over the planning period T from harvesting fish in the high sea at the rate $u(t)$ ton of fish per unit time. As before, we consider the planner to be a price taker with unit fish price $p(t)$ being given. For what appears to be the simplest looking problem, we treat here the case $c(u) = c_0$. The optimization problem is subject to the growth dynamics of the fish population $y(t)$:

$$y' = f(y) - u, \qquad y(0) = Y_0 \qquad (12.6.15)$$

and the inequality constraints on the irreversible harvesting rate

$$0 \leq u(t) \leq u_{\max}. \qquad (12.6.16)$$

To apply the Maximum Principle, we first construct the Hamiltonian from the Lagrangian and the growth dynamics to get

$$\begin{aligned} H(y, \lambda, u, t) &= e^{-rt}\,[p - c(u)]\,u + \lambda\,[f(y) - u] \\ &= e^{-rt}\,\{[p - c_0 - \bar{\lambda}]\,u + \bar{\lambda}f(y)\} \end{aligned}$$

after setting

$$\lambda(t) = \bar{\lambda}(t)e^{-rt}$$

and deduce from H the associated adjoint problem

$$\lambda' = -\frac{\partial H}{\partial y} = -\lambda\frac{df}{dy}, \quad \lambda(T) = 0.$$

There remains only the task of determining the optimal control from the optimality condition

$$\begin{aligned} H[Y(t), \Lambda(t), U(t), t] &= \max_{0 \leq u \leq u_{\max}}\ H[Y(t), \Lambda(t), u, t] \\ &= \max_{0 \leq u \leq u_{\max}}\ [e^{-rt}\,(p - c_0 - \bar{\Lambda})\,u + \bar{\Lambda}f(Y)]. \quad (12.6.17) \end{aligned}$$

12.6.5.2 Stationary and singular solution

From elementary calculus, an obvious candidate for optimal solution is from the stationary condition

$$\frac{\partial H[y, \lambda, u, t]}{\partial u} = e^{-rt}[p - c_0] - \lambda = 0.$$

However, this condition does not contain the unknown control variable u and therefore cannot use to determine candidates for the optimal control. Furthermore, while the condition can be satisfied by choosing

$$\lambda_S(t) = e^{-rt}[p(t) - c_0], \tag{12.6.18}$$

this choice of adjoint function $\lambda(t)$ generally does not satisfy the Euler BC $\lambda(T) = 0$. (This is certainly the case for a constant unit price $p(t) = p_0$.) It is known as the *singular solution* of the problem that may play a role in the final solution. For the range $(t_s, T]$ where the singular solution does not apply, we need to return to the optimality condition (12.6.17) to find other alternative to a stationary solution. For the purpose of illustrating the solution process, we limit further discussion to the case of constant unit fish price so that $p(t) = p_0$.

For the optimization problem to be realistic, two restrictions or assumptions will be made. As it is the case for most fishing grounds in sea or ocean, the fish population of commercial interest are nowhere near the carrying capacity of the region (if not already overfished). Therefore, only growth rates f with the property $df/dy > 0$ are of interest and will be so assumed. The other is that the discount rate r is expected to be small compared to the growth rate of the fish population so that $r - f'(y) < 0$. It follows that $\bar{\lambda}(t)$ is a decreasing function of time in some interval $(t_s, T]$ adjacent to the terminal time of the planning period. These assumptions will contribute to our application of the Maximum Principle for the solution of the problem. (If the restrictions are not met, the Maximum Principle still applies to yield an appropriate solution for the problem.)

12.6.5.3 Upper corner control adjacent to terminal time

With

$$\left[e^{-rt}\left(p - c_0 - \bar{\lambda}\right)\right]_{t=T} = e^{-rT}\left(p_0 - c_0\right) > 0,$$

the Hamiltonian is maximized by the upper corner control $u_{\max}$ at terminal time. By continuity, we have

$$0 < \bar{\lambda}(t) < p_0 - c_0 \qquad (t_s < t < T)$$

for some interval $(t_s, T]$ so that $u(t) = u_{\max}$ in that interval. We denote the solution for the upper corner control $u_{\max}$ by $\{y_g(t), \lambda_g(t)\}$ with

$$y'_g = f(y_g) - u_{\max} \tag{12.6.19}$$

and the corresponding adjoint function $\bar{\lambda}_g(t)$ determined by

$$\bar{\lambda}'_g = (r - f'(y_g))\bar{\lambda}_g, \qquad \bar{\lambda}_g(T) = 0. \tag{12.6.20}$$

The system (12.6.19)–(12.6.20) is augmented by the continuity condition for $y(t)$ at t_s:

$$y_g(t_s+) = y(t_s-).$$ (12.6.21)

Given $(r - f'(y_g)) < 0$, $\bar{\lambda}_g(t)$ is monotone decreasing in $(t_s, T]$ and the upper corner control may cease to maximize the Hamiltonian when

$$\bar{\lambda}_g(t_s) = p_0 - c_0.$$ (12.6.22)

This condition determines the value of t_s as the root of (12.6.22) nearest to T. However, it can be found only after we have specified $f(y)$ and solved the terminal value problem (12.6.20). The latter in turn requires the optimal solution of the problem for $t < t_s$.

12.6.5.4 *The stationary solution phase*

Since a stationary solution that maximizes the Hamiltonian is superior to the corner controls, we consider the possibility of a singular solution for the problem for $t < t_s$ to be denoted by $\{u_S(t), y_S(t), \lambda_S(t)\}$. For that solution, we have from (12.6.18) $\bar{\lambda}_S(t) = p_0 - c_0$ so that

$$\bar{\lambda}'_S(t) = (r - f'(y_S))\bar{\lambda}_S = 0.$$

The singular solution for $y(t)$ is a positive root $y_S^{(k)}(r)$ of

$$f'(y_S) = r.$$ (12.6.23)

For a normalized logistic growth rate for example, we have $f(y) = y(1 - y)$ so that $1 - 2Y_S = r$ or $Y_S(r) = (1 - r)/2$. We limit discussion here to (12.6.23) having only a single positive root as in the case of logistic growth. The growth rate equation (12.6.15) then requires

$$u_S(t) = f(y_S(r))$$

which would be $(1 - r^2)/4$ for a logistic growth rate. In general, the singular (stationary or interior) solution $\{u_S(t), y_S(t), \lambda_S(t)\}$ is the triplet of constants $\{f(y_S(r)), y_S(r), e^{-rt_s}(p_0 - c_0)\}$ where $y_S(r)$ is a positive root of (12.6.23) and t_s is the positive root of (12.6.22) closest to T.

Since we are interested in a phase of singular solution to be joined to the upper corner solution that satisfies the Euler BC, continuity of the state variable requires

$$y_g(t_s) = y_S(t_s)$$ (12.6.24)

providing the initial condition (12.6.21) for the complete determination of $y_g(t)$.

12.6.5.5   *The optimal control near the starting time*

(i) $Y_0 = y_S^{(k)}(r)$ The fish population at the start of the planning period is generally not the singular solution $y_S^{(k)}(r)$. But if Y_0 should equal $y_S^{(k)}(r)$, then the singular solution persists for all t in $[0, t_s]$.

Proposition 23 *If $Y_0 = y_S^{(k)}(r)$, the optimal fishing strategy for maximum discount profit over the planning period $[0, T]$ is*

$$U(t) = \begin{cases} f(y_S^{(k)}(r)) & (0 \le t < t_s) \\ u_{\max} & (t_s < t \le T) \end{cases}$$

with

$$\{Y(t), \Lambda(t)\} = \begin{cases} \left\{ y_S^{(k)}(r), e^{-rt_s}(p_0 - c_0) \right\} & (0 \le t \le t_s) \\ \left\{ y_g(t), e^{-rt_s} \bar{\lambda}_g(t) \right\} & (t_s \le t \le T) \end{cases}$$

where $y_g(t)$ and $\bar{\lambda}_g(t)$ are as previously defined.

(ii) $Y_0 > y_S^{(k)}(r)$ If the initial fish population is above the singular solution, it needs to be driven from Y_0 down to $y_S^{(k)}(r)$. This is possible only if $u_{\max} > f(Y_0)$ and $Y_\infty = f^{-1}(u_{\max}) < y_S^{(k)}(r)$. If so, the corresponding upper corner solution, denoted by $\{y_m, \bar{\lambda}_m\}$

$$y'_m = f(y_m) - u_{\max}, \qquad y_m(0) = Y_0 \tag{12.6.25}$$
$$\bar{\lambda}'_m = (r - f'(y_m))\bar{\lambda}_m, \qquad \bar{\lambda}_m(t_0) = p_0 - c_0 \tag{12.6.26}$$

with t_0 determined by

$$y_m(t_0) = y_S^{(k)}(r).$$

Proposition 24 *If $u_{\max} > f(Y_0)$, $Y_\infty = f^{-1}(u_{\max}) < y_S^{(k)}(r)$ and $t_0 < t_s$, the optimal fishing strategy is given by*

$$U(t) = \begin{cases} u_{\max} & (0 \le t < t_0) \\ f(y_S^{(k)}(r)) & (t_0 < t < t_s) \\ u_{\max} & (t_s < t \le T) \end{cases} .$$

Corollary 8 *If any one of the three conditions of the previous proposition is not met, then the optimal fishing strategy is with $U(t) = u_{\max}$ for all t in $[0, T]$ with the corresponding state and adjoint variable determined by*

$$y'_0 = f(y_0) - u_{\max}, \qquad y_0(0) = Y_0 \tag{12.6.27}$$
$$\bar{\lambda}'_0 = (r - f'(y_0))\bar{\lambda}_0, \qquad \bar{\lambda}_0(T) = 0. \tag{12.6.28}$$

Proof A maximum of $\bar{\lambda}$ occurs at some t_m with $\bar{\lambda}'_0(t_m) = 0$ so that $y_0(t_m)$ is a root of $r = f'(y_0)$. In other words, $y_0(t_m) = y_S^{(k)}(r)$ is at the singular solution with $\bar{\lambda}_0(t_m) = p_0 - c_0$. It follows that $\bar{\lambda}_0(t) < p_0 - c_0$ so that $U(t) = u_{\max}$ is optimal for all t in $[0, T]$. $\qquad \square$

Remark 9 *While $U(t)$ is optimal by the Maximum Principle, we still need to prove that it maximizes the discounted profit stream total $J[u]$ since that principle imposes only a set of necessary conditions for a maximum and $U(t)$ is one candidate that meets them. It is not the only one. Another is $u(t) = u_{max}$ in the corollary of the proposition. That $U(t)$ actually maximizes $J[u]$ can be shown by the method of most rapid approach but the proof will be postponed until Chapter 13 (see Proposition 34 in that chapter).*

(iii) $Y_0 < y_S^{(k)}(r)$ The initial fish population is below the singular solution, it needs to be driven from Y_0 up to $y_S^{(k)}(r)$. Since $u(t)$ is bounded below by 0, the best we can do to get there is by not fishing. The corresponding lower corner solution, denoted by $\{y_\ell, \bar{\lambda}_\ell\}$

$$y_\ell' = f(y_\ell), \quad y_\ell(0) = Y_0 \tag{12.6.29}$$
$$\bar{\lambda}_\ell' = (r - f'(y_\ell))\bar{\lambda}_\ell, \quad \bar{\lambda}_\ell(t_\ell) = p_0 - c_0. \tag{12.6.30}$$

with t_ℓ determined by

$$y_\ell(t_\ell) = y_S^{(k)}(r). \tag{12.6.31}$$

In order to have a subinterval of stationary solution, we need $t_\ell < t_s$.

Proposition 25 *If $Y_0 < y_S^{(k)}(r)$ and $t_\ell < t_s$, the optimal fishing strategy is given by the two-switch optimal control*

$$U(t) = \begin{cases} 0 & (0 \leq t < t_\ell) \\ f(Y_S(r)) & (t_\ell < t < t_s) \\ u_{max} & (t_s < t \leq T) \end{cases}. \tag{12.6.32}$$

Corollary 9 *If $Y_0 < y_S^{(k)}(r)$ and $t_\ell > t_s$, the optimal fishing strategy is given by the bang-bang control*

$$U(t) = \begin{cases} 0 & (0 \leq t < t_s) \\ u_{max} & (t_s < t \leq T) \end{cases}. \tag{12.6.33}$$

The single switch t_s is the zero of (12.6.22) nearest to T where the upper corner solution $\{y_g(t), \bar{\lambda}_g(t)\}$ for $(t_s, T]$ is determined by

$$y_g' = f(y_g) - u_{max}, \quad y_g(t_s) = y_\ell(t_s) \tag{12.6.34}$$
$$\bar{\lambda}_g' = (r - f'(y_g))\bar{\lambda}_g, \quad \bar{\lambda}_g(T) = 0. \tag{12.6.35}$$

Remark 10 *The proofs of these two results will be left as exercises.*

12.6.6 On the Maximum Principle

As illustrated by the re-working of two previously solved problems in the last two subsections, we have effectively re-configured the Do Your Best solution process in terms of the relevant Hamiltonian for constrained optimization problems. Though

motivated by, and evolved from the Do Your Best Principle expressing the solution
process in terms of the Hamiltonian has enabled us to streamline its details. For
the sake of completeness, we summarize below the essential steps in the Maximum
Principle of Pontryagin as a replacement and generalization of the Do Your Best
Principle:

(1) Given the general performance index $J[u]$ as defined in (12.6.1) and the general
    growth dynamics (12.6.2), we construct the related Hamiltonian $H(y, u, \lambda, t)$
    shown in (12.6.3).
(2) Use the Hamiltonian to generate the adjoint DE (12.6.5) (and to recover the
    growth dynamics (12.6.2)).
(3) The adjoint DE is to be augmented by a Euler BC (12.6.6) whenever an end
    condition is not prescribed.
(4) Choose the optimal control $U(t)$ by finding $u(t)$ to maximize the Hamiltonian:

$$H(Y(t), \Lambda(t), U(t), t) = \max_{u \, \varepsilon \, \Omega}[H(Y(t), \Lambda(t), u, t)].$$

While the simple models in this chapter have enabled us to illustrate the implementation or execution of the Maximum Principle, the results show that some of
the models require refinements to capture more features of the phenomena being
modeled that are essential to what we want to know about the phenomena.

Chapter 13

Chlamydia Trachomatis

13.1 A Proof of Concept Model of C. Trachomatis

13.1.1 *Life cycle of Chlamydia Trachomatis*

Chlamydial bacteria cause several diseases in humans, most are not life threatening and in some cases not even noticeable by physicians for a long period. One strain of the bacteria called Chlamydia Trachomatis is an exception as it was for some time the principal cause of eye infection that leads to blindness (and possibly death). Until the turn of the century, it was the leading cause of blindness worldwide but has dropped from 15% to less than 4% of such cases in recent years due to an on-going organized effort for its eradication. For the final phase of the eradication process, it is useful to gain some insight on the reproductive aspects of this strain of the bacteria.

For this purpose, we consider Chlamydia Trachomatis to be in one of two forms: a state RB (reticulate bodies) that can divide and multiply inside the host mammalian cell as well as convert to a state EB (elementary bodies) that is resistant to environmental changes but cannot divide. When the host cell dies, it releases the EB 'spores', but importantly, all RB cells die as well because they cannot survive in the new environment. The EB *cells* are then free to invade another host cell, turn into an RB cell, and begin dividing. What is the optimal conversion rate of RB population, $R(t)$, at time t into EB type if we want to maximize the EB *population* $E(t)$ *when the host cell dies at time T.*

13.1.2 *A linear growth rate model*

To introduce the mathematical techniques for the type of optimization problems involved in the spread of the infectious bacteria, we consider here a simple proof-of-concept model for its proliferation. With an initial size $R_0 > 0$, the population $R(t)$ of the reticulate bodies at a later time t (for this proof-of-concept model) is to grow at a linear natural growth rate (when not being converted to EB) with a rate constant α. When a fraction of the RB population is converted continuously

261

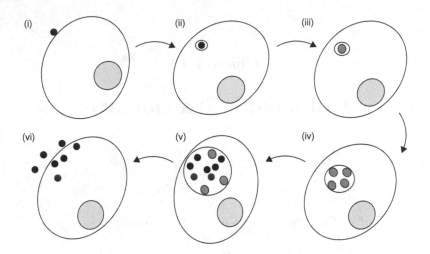

Fig. 13.1   Life cycle of Chlamydia Trachomatis.

in time at a rate uR, we have the linear growth dynamics

$$R' = (\alpha - u)R, \qquad R(0) = R_0, \tag{13.1.1}$$

where $(\;)' = d(\;)/dt$ and the component αR is the natural linear growth rate when $u = 0$. In the absence of apoptosis over the life span T of the host cell, the growth rate of the population $E(t)$ of the elementary bodies is equal to the rate of conversion uR with

$$E' = uR, \qquad E(0) = E_0, \tag{13.1.2}$$

where E_0 is the initial EB population which we take to be zero at the start. The EB population at the time T when the host cell dies is

$$E(T) = \int_0^T uR\,dt. \tag{13.1.3}$$

The simplest optimization problem is to maximize the total EB population $E(T)$ generated by $E' = u(t)R$, $E(0) = 0$ over the host cell life span T. The conversion is irreversible so that the conversion "rate constant" $u(t)$ is nonnegative. It is also bounded above by some constant $u_{\max}$ since there is a finite capacity to convert. Hence the control mechanism for maximization $u(t)$ is subject to the inequality constraints

$$0 \le u \le u_{\max}. \tag{13.1.4}$$

Our proof-of-concept optimization problem is to find an optimal control $U(t)$ for

$$\max_{0 \,\le\, u(t) \,\le\, u_{\max}} \left[J[u] = \int_0^T uR\,dt \right] \tag{13.1.5}$$

subject to the growth dynamics

$$R' = (\alpha - u)R, \qquad R(0) = R_0. \tag{13.1.6}$$

13.1.3 Hamiltonian and Maximum Principle

To apply the method of the Maximum Principle, we form the Hamiltonian

$$H[u] = uR + \lambda(\alpha - u)R = uR(1 - \lambda) + \lambda\alpha R \qquad (13.1.7)$$

and generate the adjoint DE and Euler (adjoint) BC

$$\lambda' = u(\lambda - 1) - \lambda\alpha, \qquad \lambda(T) = 0. \qquad (13.1.8)$$

The critical step is to choose u to maximize $H[u]$ subject to the IVP (13.1.6), the terminal value problem (13.1.8) and the inequality constraints (13.1.4).

Candidates for the optimal control is normally determined by the zeros $u_i(t)$ of

$$\frac{\partial H}{\partial u} = R(1 - \lambda). \qquad (13.1.9)$$

In our case however, the expression (13.1.9) does not involve u. Furthermore, $R(t) > 0$ for all $t > 0$ and $\lambda(t) - 1$ does not vanish for any finite subinterval of $[0, T]$. (If $\lambda(t_0) = 1$ for some t_0 in $(0, T)$, then $\lambda'(t_0) = -\alpha < 0$ by the adjoint DE. Hence, t_0 can only be an isolated zero.) By the Maximum Principle, the solution of our simple optimization problem is expected to come from the corner controls.

With $\lambda(T) = 0$, we have by continuity $(1 - \lambda) > 0$ for $t_s < t < T$ so that $U(t) = u_{\max}$ in that interval. In that case, we have for $t_s < t \le T$

$$R' = (\alpha - u_{\max})R, \qquad \lambda' = (u_{\max} - \alpha)\lambda - u_{\max}, \qquad \lambda(T) = 0$$

or

$$R = C_s e^{(\alpha - u_{\max})t}, \qquad \Lambda(t) = \frac{u_{\max}}{u_{\max} - \alpha}\left[1 - e^{(u_{\max} - \alpha)(t - T)}\right] \qquad (13.1.10)$$

where C_s is some constant of integration to be specified. Two different cases will have to be considered separately.

13.1.3.1 $u_{\max} > \alpha$

For this case, $\lambda(t)$ is a monotone decreasing function of t (that increases as t decreases from T). At some instant $t_s < T$, we have $\lambda(t_s) = 1$. This condition determines

$$t_s = \frac{1}{u_{\max} - \alpha} \ln\left(\frac{\alpha}{u_{\max}}\right) + T < T \qquad (13.1.11)$$

since $u_{\max} > \alpha$. At $t = t_s$, we have

$$\lambda'(t_s) = -\alpha < 0$$

independent of the choice of control $u(t)$! Hence, $\lambda(t) > 1$ (and therewith $1 - \lambda(t) < 0$) in (t_0, t_s) for some $t_0 < t_s$ and $U(t) = 0$ there.

Exercise 39 *Show that the model does not admit (a "singular solution") $\lambda(t) = 1$ over any finite interval (t_1, t_s).*

In that same interval (t_0, t_s), the relevant Hamiltonian system (13.1.6)–(14.3.1) becomes

$$R' = \alpha R, \quad R(0) = R_0; \qquad \lambda' = -\alpha\lambda, \qquad \lambda(t_s) = 1 \tag{13.1.12}$$

where the end condition $\lambda(t_s) = 1$ is required by the continuity of the adjoint function. The solution of the BVP above is

$$R(t) = R_0 e^{\alpha t}, \qquad \lambda(t) = e^{\alpha(t_s - t)}. \tag{13.1.13}$$

With $\lambda(t)$ monotone increasing as t decreases, $U(t) = 0$ continues to maximize the Hamiltonian for smaller and smaller t. Altogether we have

$$U(t) = \begin{cases} 0 & (0 \leq t < t_s) \\ u_{\max} & (t_s < t \leq T) \end{cases}, \tag{13.1.14}$$

with t_s determined by (13.1.11), and correspondingly

$$R(t) = \begin{cases} R_0 e^{\alpha t} & (0 \leq t < t_s) \\ R_0 e^{(\alpha - u_{\max})t + u_{\max}t_s} & (t_s < t \leq T) \end{cases} \tag{13.1.15}$$

$$\lambda(t) = \begin{cases} e^{\alpha(t_s - t)} & (0 \leq t < t_s) \\ \frac{u_{\max}}{u_{\max} - \alpha}\left[1 - e^{(u_{\max} - \alpha)(t - T)}\right] & (t_s < t \leq T) \end{cases}. \tag{13.1.16}$$

From (13.1.11), we see that t_s could be negative, say for sufficiently small T for example. If so, the optimal conversion strategy would convert at maximum rate from the start and throughout the life of the host cell.

13.1.3.2 $u_{\max} < \alpha$

For this case, we rewrite (13.1.10) as

$$\lambda(t) = \frac{u_{\max}}{\alpha - u_{\max}}\left[e^{(\alpha - u_{\max})(T - t)} - 1\right]. \tag{13.1.17}$$

Evidently, $\lambda(t)$ is also monotone decreasing (and increasing from 0 as t decreases from T). At some instant $t_s < T$, we have $\lambda(t_s) = 1$. This condition determines

$$t_s = T - \frac{1}{u_{\max} - \alpha}\ln\left(\frac{u_{\max}}{\alpha}\right) < T \tag{13.1.18}$$

since $u_{\max} < \alpha$. At $t = t_s$, we have again from (13.1.8)

$$\lambda'(t_s) = -\alpha < 0$$

so that $\lambda(t) > 1$ (and therewith $1 - \lambda(t) < 0$) in some interval (t_0, t_s) for some $t_0 < t_s$. The situation is analogous to the $u_{\max} > \alpha$ case so that $U(t) = 0$ in $[0, t_s)$ and the optimal solution for the life of the host cell for the $u_{\max} < \alpha$ case is again given by (13.1.14), (13.1.15) and (13.1.16) with t_s (equivalently) given by (13.1.18).

From (13.1.18), we also have $t_s \leq 0$ for a sufficiently short cell life span. Consequently, an optimal strategy for a sufficiently small T would also be to convert at maximum rate from the start and throughout the life of the host cell. In that case, the optimal strategy would correspond to $u_{op}(t) = u_{\max}$, i.e., converting at $u_{\max}$ from the start. In any case, with $\lambda_g(t) < 1$ for all $t \geq 0$, the upper corner control for the entire interval $[0, T]$ satisfies the optimality condition (12.6.11) and the corresponding $E[T]$ is superior to that by any other admissible control. More specifically, we have

Proposition 26 *Let $T_0 = \ln[u_{\max}/\alpha]/u_\alpha$ (> 0 independent of the sign of u_α), the optimal conversion strategy is to convert from the start at $u_{\max}$ if $T \leq T_0$, i.e., $u_{op}(t) = u_{\max}$ for all t in $[0, T]$.*

Exercise 40 *Experimental evidence shows that the natural growth rate parameter α for RB decreases as conversion rate constant increases. Suppose $\alpha = \alpha_0(1 - u(t))/u_{\max}$. Show that for sufficiently large $T(> T_0)$, the optimal control $U(t)$ is again bang-bang. Determine the switch time t_s and the threshold value T_0 for $t_s > 0$.*

While the optimal controls are mathematically the same for the two different relative magnitudes of $u_{\max}$ and α, the corresponding optimal solutions are different with the RB population increasing with t beyond t_s in the case of $\alpha > u_{\max}$ but begins to decrease for $t > t_s$ in the case $\alpha < u_{\max}$. To the extent that only the relative magnitude of α and $u_{\max}$ is relevant to the conversion activities for this simple model, we may simplify the model in two ways. One is to measure time in unit of α. The second is to recognize that it is $u_\alpha = u_{\max} - 1$ (or $u_\alpha = u_{\max} - \alpha$ before we change the unit of measurement for time) that plays a principal role in the optimal conversion rate and thus the relevant parameter instead of $u_{\max}$ itself.

We adopt both of these changes here and for the rest of this chapter. To implement the new unit of measurement of time, we set $\hat{t} = \alpha t$, $\hat{u} = u/\alpha$ and $\hat{u}_{\max} = u_{\max}/\alpha$. We may then write, for example, the threshold condition $T \leq T_0$ in the normalized form $\hat{T} \leq \hat{T}_0$ where

$$\hat{T}_0 = \frac{1}{\hat{u}_\alpha} \ln[\hat{u}_{\max}] = \frac{1}{\hat{u}_{\max} - 1} \ln[\hat{u}_{\max}]$$

where $\hat{u}_\alpha = \hat{u}_{\max} - 1$. In the normalized form, the condition is now unencumbered by the growth rate parameter α. The "^" notation will be dropped henceforth with the understanding that we are working with the new unit of time

Remark 11 *With $U(t) = u_{\max}$ for all $t \geq 0$ when $u_{\max} < 1$ for the present simple exponential growth of the RB population, we need only to focus on the range $u_{\max} > 1$ for this model. As we shall see, this may not be appropriate for more complex models of the proliferation of the same bacteria when the singular solution plays a role in the optimal conversion strategy.*

13.1.4 Biological mechanisms for optimal strategy

As an infectious disease, C. Trachomatis faces an important choice during its infection of a mammalian host. Each RB bacterium can decide to proliferate (by division) in order to increase the RB population size, or it can convert into a sturdy EB form that survives cell lysis and is ready to infect other cells. If too many RB convert to EB early on, the total C. trachomatis population size would remain moderate before the death of the host cell. Conversely, if too few cells convert into EB at the early stage, a good portion of the resulting larger RB population would remain unconverted at the end (because of the upper bound on the conversion rate) and not survive upon host cell lysis. This would leave only a smaller EB population than it could have had if conversion takes place sooner and/or at a higher rate if possible. In either case, it would seem that the terminal EB population may be made larger by adjusting the proportion of replication to conversion rate ratio.

Working under the assumption that evolutionary forces drive C. trachomatis infection to spread fast by maximizing the terminal EB population, the mathematics of optimal control theory has led to an optimal strategy for RB to only divide and proliferate first, and then eventually start converting. What biological mechanisms may there be for implementing such a strategy? One prevailing hypothesis is that conversion is inhibited while infectious bacteria are in contact with the thin protective inclusion membrane surrounding an individual RB or a few of the initial batch of RB [37]. Early in the infection process the inclusion over a few RB is small, allowing most or all reticulate bodies to be in contact with their own inclusion membrane. As the bacterial population increases gradually, some RB form bacteria will eventually not be in contact with their inclusion membrane and these can then begin to convert. This mechanism would implement a conversion strategy of early replication and proliferation followed by late conversion.

Another mechanism recently discovered in the lab of Ming Tan and Christine Suetterlin [14] involves the size of the descendant RB. It was observed that when an RB divides, the offspring RB do not grow to the original size of their progenitor as many other bacteria do, but instead become progressively smaller. This suggests that cell size can itself be used as a regulator for conversion. During the first five or so cell divisions after infection, the reticulate bodies are still too large to convert. If RB size does in fact regulates conversion, it would also lead to a delay before conversion, during which the population would only proliferate through division.

While these two and other possible mechanisms for implementing the proliferate-then-convert strategy seem plausible, the strategy itself had never been substantiated with mathematical rigor as we managed to do in [36] and summarized above. To the extent that fast spreading of C. trachomatis infection requires a large terminal EB population at the time of host cell lysis, it would seem that the decision-making mechanism regulating when to convert is likely to have been shaped by evolution to maximize the terminal EB population. This latter perspective played a decisive role in the development of the different optimal control type models in this chapter. It should be noted that most of the resulting optimal conversion strategies are compatible with the possible biological mechanisms for their implementation.

13.2   Chlamydia with Finite Carrying Capacity

13.2.1   The model

In the proof-of-concept model, the natural growth rate of the reticulate body population $R(t)$ was taken to be proportional to the size of the population. This limits its applicability to an initial RB population R_0 that is small compared to the carrying capacity of the host cell, a moderate growth rate constant α, and a relatively short host cell life span T. For a more realistic model, we consider here a new differentiation and proliferation model characterized by the IVP

$$R' = (\alpha - u - \beta R)R, \qquad R(0) = R_0. \tag{13.2.1}$$

The new model includes the effect of a finite carrying capacity appropriate for larger R_0 and longer T and reduces to the linear growth rate case if $\beta = 0$ (that corresponds to the limiting case of an infinite carrying capacity). The growth rate of the EB population remains equal to the rate of conversion uR as given (13.2.1):

$$E' = uR, \qquad E(0) = E_0 \; (= 0). \tag{13.2.2}$$

We wish to maximize the EB population size at the terminal time T of the host cell:

$$\max_{0 \, \leq \, u \, \leq \, u_{\max}} E(T) = \int_0^T uR\,dt. \tag{13.2.3}$$

The optimization problem as stated is to choose a control $u_{op}(t)$ to maximize the total EB population $E(T)$ at the time of the host cell's demise. Again, E_0 plays no role in our model and has been set to 0.

The choice of conversion strategy is limited by the obvious lower bound $u = 0$ (given that the conversion is not reversible) and a natural maximum rate $u_{\max}$ imposed by a limited conversion capacity. Hence, $u(t)$ is subject to the constraints (13.1.4). The optimization problem is then given by (13.2.3) subject to the new growth dynamics (13.2.1).

To reduce the number of parameters in the problem, we normalize the RB and EB populations by the carrying capacity of the RB population by setting $\beta R/\alpha = \hat{R}$ and $\beta E/\alpha = \hat{E}$, respectively. We also measure time in units of the growth rate constant for RB, effectively working with $\hat{t} = \alpha t$ (and $\hat{u} = u/\alpha$) instead of t (and u). We henceforth take the growth dynamics in the normalized form

$$R' = R(1 - u - R), \qquad R(0) = R_0, \tag{13.2.4}$$

instead of (13.2.1) (after dropping the "$\char`\^$" from the normalized quantities).

For our previous proof-of-concept investigation by way of the linear natural RB growth rate model, the actual (normalized) growth rate "constant" $1 - u$ does not depend on the RB population size and correspondingly the optimal conversion strategy for maximum $E(T)$ does not depend on the initial RB population size R_0. For the new model, the role of $1 - u$ is replaced by $1 - u - R$ so that growth (or decline) is with diminishing return. In particular, the optimal conversion strategy, now dependent on the RB population and the "conversion rate constant" $u(t)$, will be seen to depend on initial RB population R_0 and the maximum conversion rate constant $u_{\max}$ in a rather complicated way. For example, reticulate bodies in excess of the carrying capacity would be lost due to rate limiting attrition and therefore do not contribute to RB growth and the terminal EB population. Whether this should take place would depend on R_0 and $u_{\max}$. We will delineate in the subsequent development how optimal conversion strategy varies with different ranges of R_0. Before that, we need to establish a special solution of our problem, known as the *singular solution*, that had no role in the simple proof-of-concept model of the previous section.

13.2.2 *The singular solutions*

To apply the Maximum Principle, we again begin with the *Hamiltonian* for the new optimal control problem:

$$H[u] = uR + \lambda R(1 - u - R) = uR(1 - \lambda) + \lambda R(1 - R). \tag{13.2.5}$$

From this expression, we generate the *adjoint differential equation* and *adjoint (Euler) end condition*

$$\lambda' = -\frac{\partial H}{\partial R} = -\lambda(1 - u - 2R) - u, \quad \lambda(T) = 0 \tag{13.2.6}$$

for the *adjoint function* $\lambda(t)$. The Maximum Principle then requires that we choose u to maximize H:

$$\max_{0 \leq u \leq u_{max}} [H[u] = uR(1 - \lambda) + \lambda R(1 - R)] \tag{13.2.7}$$

subject to the initial value problem (IVP) for $R(t)$ in (13.2.4), the terminal value problem (13.2.6), the constraints on the control $u(t)$ (13.1.4) and the nonnegativity constraints on the state function $R(t)$.

To maximize H, we note that the stationary condition

$$\frac{\partial H}{\partial u} = R(1 - \lambda) = 0 \tag{13.2.8}$$

again does not determine the stationary point or *interior control* $u_i(t)$ that would normally induce an extremal $R_i(t)$ by the IVP (13.2.4) and an associated adjoint function $\lambda_i(t)$ by (13.2.6). However, the condition may be met by *singular solutions* that are combinations of $R(t)$ and $\lambda(t)$ that satisfy (13.2.8) as well as their Hamiltonian system of ODE. For our model, $R(t) \equiv 0$ satisfies the stationary condition (13.2.8) but is not a singular solution since (13.2.4) requires $R(t) > 0$ for all t in $[0, T]$ for any prescribed $R_0 > 0$. The other way of satisfying (13.2.8) is $\lambda(t) = 1$; but this does not apply near the terminal time because the Euler BC requires $\lambda(T) = 0$. However (13.2.8) may be satisfied away from the terminal time (as we shall see) in which case

$$\lambda(t) = 1 \equiv \lambda_S(t) \tag{13.2.9}$$

may constitute a singular solution in some sub-interval away from T. For this possible singular solution $\lambda_S(t)$, the adjoint DE requires the corresponding singular RB population, denoted by $R_S(t)$, to be

$$R_S(t) = \frac{1}{2}. \tag{13.2.10}$$

When (13.2.10) is satisfied, possibly away from the initial time (since R_0 is generally different from $R_S = 1/2$), the growth rate equation for the RB population requires the corresponding singular control to be

$$u_S(t) = \frac{1}{2}. \qquad (13.2.11)$$

Whether or not the singular solution (13.2.9)–(13.2.10) applies from the start or in some subinterval(s) of $[0, T]$, the optimal strategy must switch to another control prior to the terminal time T in order to satisfy the Euler BC $\lambda(T) = 0$. In the next subsection, we show the switch is to the upper corner control $u_g(t) = u_{\max}$.

Whatever appropriate combination of control components that forms the optimal control $u_{op}(t)$ may be, it must solve the IVP (13.2.4) and the terminal value problem (13.2.6), it must also satisfy the remaining requirement of the Maximum Principle, namely, the optimality condition (13.2.7). Together, they constitute a set of necessary conditions for the solution of the optimization problem of maximizing $E(T)$.

13.2.3 *The upper corner control is optimal adjacent to T*

Given $\lambda(T) = 0$, we have by continuity $1 - \lambda > 0$ for $t_s < t < T$ for some $t_s < T$. The optimality condition (13.2.7) requires that we convert as much as possible to maximize H. With the constraints imposed by (13.1.4), the optimal choice of conversion rate is by the *upper corner control* $u_{op}(t) = u_{\max}$ in $(t_s, T]$. The adjoint DE now involves $R(t)$; we need to solve the following two ODE simultaneously for R and λ for $u = u_{\max}$ (with the solution denoted by $R_g(t)$ and $\lambda_g(t)$, respectively):

$$R'_g = R_g(1 - u_{\max} - R_g), \qquad R_g(t_s) = R_s \qquad (13.2.12)$$
$$\lambda'_g = -(1 - u_{\max} - 2R_g)\lambda_g - u_{\max}, \qquad \lambda_g(T) = 0. \qquad (13.2.13)$$

The auxiliary condition $R_g(t_s) = R_s$ for the ODE for $R_g(t)$ is to come from the continuity condition at the point t_s.

Note that $\lambda'_g(T) = -u_{\max} < 0$ so that $\lambda_g(t)$ is a positive decreasing function of time in some interval (t_s, T) adjacent to T. It follows that $t_s < T$ is the instant nearest T when λ_g attains the value 1, i.e.,

$$\lambda_g(t_s) = 1. \qquad (13.2.14)$$

The solutions for $R_g(t)$ and $\lambda_g(t)$ for $t > t_s$ are

$$R_g(t) = \frac{u_\alpha R_s e^{-u_\alpha(t-t_s)}}{u_\alpha + R_s - R_s e^{-u_\alpha(t-t_s)}}, \quad \lambda_g(t) = u_{\max} e^{-I(t)} \int_t^T e^{I(\tau)} d\tau, \qquad (13.2.15)$$

where $u_\alpha (= u_{\max} - \alpha) = u_{\max} - 1$, R_s is the yet unknown value $R_g(t_s)$, and

$$e^{I(t)} = \frac{u_\alpha^2 e^{u_\alpha(t_s-t)}}{\left[u_\alpha + R_s - R_s e^{u_\alpha(t_s-t)}\right]^2} \qquad (13.2.16)$$

with $e^{I(t_s)} = 1$. From (13.2.16), we obtain after carrying out the integration in the expression for $\lambda_g(t)$ in (13.2.15)

$$\lambda_g(t) = \frac{u_{\max}}{u_\alpha} \frac{u_\alpha + R_s - R_s e^{u_\alpha(t_s-t)}}{u_\alpha + R_s - R_s e^{u_\alpha(t_s-T)}} \left[1 - e^{u_\alpha(t-T)} \right] \qquad (13.2.17)$$

for all t in $[t_s, T]$.

With (13.2.17), the switch condition (13.2.14) becomes

$$\frac{u_{\max} \left[1 - e^{u_\alpha(t_s-T)} \right]}{u_\alpha + R_s - R_s e^{u_\alpha(t_s-T)}} = 1 \qquad (13.2.18)$$

which may be solved for t_s to get

$$t_s = T - \frac{1}{u_\alpha} \ln \left(\frac{u_{\max} - R_s}{1 - R_s} \right). \qquad (13.2.19)$$

Observe that R_s will come from the solution in the complementary time interval $[0, t_s]$ and may be a function of t_s. The condition for determining the switch point is then a complicated nonlinear equation for t_s. Similar to the simpler proof-of-concept model, the switch point t_s could be negative (if T is sufficiently small for example) or undefined (when $u_{\max} < R_s < 1$ for example). In either case, the upper corner control $u_{\max}$ should be examined for its optimality from the start. However, unlike the simpler model analyzed earlier, whether $u_{\max}$ should be optimal from the start or a switch should take place for some positive t_s cannot be decided for the present improved model without investigating the optimal control in the complementary interval $[0, t_s)$ given that we do not yet know what R_s should be. That uncertainty notwithstanding, the Maximum Principle unequivocally singles out the upper corner control $u_{\max}$ as the control that maximizes the Hamiltonian for an interval of time adjacent to the terminal time T:

Proposition 27 *The upper corner control maximizes the Hamiltonian of the optimization problem (13.2.3) in the interval $(t_s, T]$.*

Before turning to the optimal conversion strategy for the interval $[0, t_s)$, we note that the solution for the special case of $u_{\max} = 1$ requires a little more care but poses no obstacle to the solution of the problem. It will be addressed for the different situations below. Here we observe instead that if $\lambda_g(t) < 1$ for all $t > 0$, then the Hamiltonian is maximized by the upper corner control for all $t \geq 0$ without switching to another admissible control (corresponding to the case $t_s < 0$). To the extent that a maximized Hamiltonian is not synonymous with a maximum $E(T)$, we will address the issue of a maximum $E(T)$ in a later section possibly by the method of most rapid approach [3, 33] when it is necessary to do so.

In the next few sections, we analyze and determine the different optimal conversion strategies for all t in $[0, T]$ that maximize $E(T)$ for different ranges of R_0. Since it is most likely that only a few EB cells enter and infect a host cell initially so that $R_0 \ll 1$, i.e., well below the carrying capacity of the host cell, we consider first in the next section the range $R_0 \leq 1/2$ and analyze briefly the range $R_0 > 1/2$ in the section after. Before leaving this section, we elucidate in the last subsection the reason for a separate treatment of these two ranges of R_0.

13.2.4 *Lower corner control optimal only for* $R_0 \leq 1/2$

If $1 - \lambda_g(t)$ should vanish at some $t_s > 0$ **and** becomes non-positive for $t < t_s$, then the Maximum Principle would require a switch to another more appropriate conversion strategy for $t < t_s$. Continuity of the adjoint function $\lambda(t)$ and the RB population $R(t)$ requires $\lambda(t_s) = 1$ for any switch from u_{max} to another admissible control. It follows that

$$[\lambda']_{t=t_s} = [-\lambda(1 - u - 2R) - u]_{t=t_s} = -[1 - 2R_s]$$

for any control $u(t)$ (including the two corner controls) with $R_s = R(t_s)$ for that control. Depending on the sign of $1 - 2R_s$, $\lambda'(t_s) = \lambda_g'(t_s) = \lambda_\ell'(t_s)$ may be positive, zero or negative. As such $\lambda(t)$ (for any control $u(t)$) may be increasing, stationary or decreasing for $t < t_s$. Evidently, the sign of $\lambda(t) - 1$ for $t < t_s$ determines the appropriate control there. With that in mind, we have the following lemma that eliminates a switch to the lower corner control if $R_0 > 1/2$.

Proposition 28 *The lower corner control is not optimal in* $[0, t_s)$ *if* $R_0 > 1/2$ *(or, for the unnormalized form of the problem, when the initial RB population is not less than half of the carrying capacity of the host cell).*

Proof For $R_0 > 1$, we have $R_\ell' < 0$ so that $R_\ell(t)$ is monotone decreasing asymptotic to 1 from above and $R_\ell(t) > 1$ for all finite $t > 0$. With $u = 0$, the adjoint DE (13.2.6) becomes

$$\lambda_\ell' = -\lambda_\ell(1 - 2R_\ell), \qquad \lambda_\ell(t_s) = 1 \tag{13.2.20}$$

so that $\lambda_\ell(t) < 1$ for all $t < t_s$ (since $\lambda_\ell(t_s) = 1$). With $1 - \lambda_\ell > 0$, the lower corner control is not optimal in $[0, t_s)$ if $R_0 > 1$.

For $R_0 = 1$, we have $R_\ell(t) = 1$ (with $R_\ell'(t) = 0$) and $\lambda_\ell' = \lambda_\ell$ so that $\lambda_\ell(t) = e^{t-t_s} < 1$ for $t < t_s$. With $\lambda_\ell(t) < 1$ for $t < t_s$, the lower corner control is again not optimal in $[0, t_s)$ if $R_0 = 1$.

For $1/2 < R_0 < 1$, we have from (13.3.1) $R_\ell'(0) > 0$ and $\lambda_\ell'(0) > 0$. Hence, both $R_\ell(t)$ and $\lambda_\ell(t)$ are increasing function of time with $\lambda_\ell(t_s) = 1$ and therewith $\lambda_\ell(t) < 1$ for $t < t_s$. The lower corner control is also not optimal in $[0, t_s)$ if $1/2 < R_0 < 1$. $\qquad\square$

By Proposition 28, the lower corner control may be appropriate for the interval $[0, t_s)$ (or a subinterval thereof) only if $R_0 \leq 1/2$. Since the infection of a healthy cell usually starts with the endocytosis of a few EB spores (and hence $R_0 \ll 1$), we will first focus on the range $R_0 \leq 1/2$ of the initial RB population and determine the optimal control for maximum $E(T)$ for this range of R_0 before turning to examine the optimal conversion strategy for the complementary range $R_0 > 1/2$.

13.3   Optimal Conversion Strategy for $R_0 \leq 1/2$

13.3.1   *The optimal control is bang-bang for $u_{\max} < 1/2$*

With the lower corner control eliminated for the range $R_0 > 1/2$, we can focus our discussion of $u_\ell(t) = 0$ for the range $R_0 \leq 1/2$. In that range, we have from (13.2.4) for $u(t) = 0$

$$R'_\ell = (1 - R_\ell)R_\ell, \quad R_\ell(0) = R_0 \tag{13.3.1}$$

with the explicit solution

$$R_\ell(t) = \frac{R_0}{R_0 + (1 - R_0)\, e^{-t}} \tag{13.3.2}$$

monotone increasing and asymptotic to 1 as $t \to \infty$.

If $R_0 = 1/2$, we have $R_\ell(t) > 1/2$ for all $t > 0$ and a singular solution phase of $R_S(t_0) = 1/2$ is not possible for $t > 0$. We need only to consider $R_0 < 1/2$. For that case, we let $R_\ell(t_0) = 1/2$ so that

$$t_0 = \ln\left(\frac{1 - R_0}{R_0}\right) \tag{13.3.3}$$

with t_0 increasing from 0 to ∞ as R_0 decreases from $1/2$ to 0. We may now explore the possibility of a singular solution phase, $\{R_S(t), \lambda_S(t)\} = \{1/2, 1\}$, in some interval (t_0, t_S). Given $R_S(t) = 1/2$, we have from the ODE in (13.2.4) $0 = R'_S(t) = (1 - u - 1/2)/2$ so that we need $u(t) = u_S(t) = 1/2$ to sustain this singular solution phase. But this is not possible since we have restricted $u_{\max} < 1/2$ in this subsection. As a consequence, we have the following optimal conversion strategy for $u_{\max} < 1/2$ (when $R_0 \leq 1/2$):

Proposition 29 *For $R_0 \leq 1/2$, the optimal conversion strategy is the bang-bang control*

$$u_{op}(t) = \begin{cases} 0 & (0 \leq t < t_s) \\ u_{\max} & (t_s < t \leq T) \end{cases} \tag{13.3.4}$$

if $u_{\max} < 1/2$ with the switch point t_s being the unique positive root of

$$t_s = T - \frac{1}{u_\alpha} \ln\left(\frac{u_{\max}(1 - R_0) + u_\alpha R_0 e^{t_s}}{1 - R_0}\right). \tag{13.3.5}$$

The corresponding RB population and adjoint function are given by (13.3.2) and

$$\lambda_\ell(t) = e^{t_s - t}\left(\frac{1 - R_0 + R_0 e^t}{1 - R_0 + R_0 e^{t_s}}\right)^2, \tag{13.3.6}$$

respectively.

Proof Since a singular control is not possible, then there are only two possible scenarios for the optimal conversion strategy. The first is $\lambda_g(t) \leq 1$ for all t in $[0, T]$ so that $u_{op}(t) = u_{\max}$ in $[0, T]$ corresponding to the case $t_s \leq 0$. The second scenario is $\lambda_g(t_s) = 1$ and $\lambda_g(t) > 1$ for some interval $t_\ell < t < t_s$ adjacent to t_s. In the range (t_ℓ, t_s), the upper corner control is not optimal and a switch in control at t_s can only be to the lower corner control. The corresponding adjoint function $\lambda_\ell(t)$ is continuous at t_s and $\lambda'_\ell(t)$ is necessarily < 0 in (t_ℓ, t_s) since

$$\lambda'_\ell(t_s) = -[1 - 2R_\ell(t_s)] = -[1 - 2R_g(t_s)] = \lambda'_g(t_s) < 0.$$

This requires $R_\ell(t_s) = R_g(t_s) < 1/2$ and, by continuity, $R_\ell(t) < 1/2$, $\lambda'_\ell(t) < 0$ and $\lambda_\ell(t) > 1$ in some interval (t_c, t_s). Hence the lower corner control is optimal in (t_c, t_s). Since $R'_\ell(t) > 0$ for $R_0 \leq 1/2$, we must have $R_\ell(t) < 1/2$ in $[0, t_s)$. It follows that lower corner control is optimal in $[0, t_s)$.

The bang-bang control (13.3.4) is the only control that satisfies all the necessary conditions imposed by the Maximum Principle and $[E(T)]_{u=u_{op}(t)} > [E(T)]_{u=0} = 0$; hence, it maximizes $E(T)$. □

Corollary 10 *For $R_0 \leq 1/2$ and $u_{\max} < 1/2$, the optimal control $u_{op}(t) - u_{\max}$ for all t in $[0, T]$ maximizes $E(T)$ if the unique root t_s of (13.3.5) is negative. The corresponding RB population $R_g(t)$ and adjoint function $\lambda_g(t)$ are the solutions of the IVP*

$$R'_g = R_g(1 - u_{\max} - R_g), \quad R_g(0) = R_0$$

and terminal value problem

$$\lambda'_g = (1 - u_{nax} - 2R_g)\lambda_g - u_{\max}, \quad \lambda_g(T) = 0,$$

respectively.

Proof The unique root t_s of (13.3.5) is the instant nearest to T when $\lambda_g(t_s) = 1$ with $\lambda_g(t) < 1$ for $t > t_s$ and $\lambda'_g(t_s) = (1 - 2R_g(t_s)) < 0$. Hence $u_{op}(t) = u_{\max}$ maximizes $H[u]$ and is optimal by the Maximum Principle. It is the only admissible control by that principle. It maximizes $E(T)$ since $[E(T)]_{u=u_{op}(t)} > [E(T)]_{u=0} = 0$. □

It is evident from (13.3.5) that t_s should be positive if T is sufficiently large. More specifically, t_s should be positive if

$$T > T_0 \equiv \frac{1}{u_\alpha} \ln \left(\frac{u_{\max}(1 - R_0) + u_\alpha R_0}{1 - R_0} \right).$$

Table 13.1 provides values of T_0 for a range of $u_{\max}$ and R_0.

13.3.2 *Sub-interval of singular solution possible for* $u_{\max} \geq 1/2$

13.3.2.1 $R_0 < 1/2$

For $u_{\max} \geq 1/2$ and $R_0 < 1/2$, the lower corner solution $R_\ell(t)$ again reaches the singular solution by t_0 as given by (13.3.3); it is now possible to sustain the singular solution (13.2.10) and (13.2.9) for $t > t_0$ by taking $u_{op}(t) = u_S(t) = 1/2$ given

Table 13.1 $T_0(u_{max}, R_0)$.

$u_{max}\backslash R_0$	0.01	0.05	0.10	0.15	0.20	0.25	0.4	0.5
0.10	2.66	3.27	∞	×	×	×	×	×
0.15	2.30	2/65	3.40	∞	×	×	×	×
0.20	2.06	2.31	2.75	3.54	∞	×	×	×
0.25	1.89	2,08	2.39	2,85	3,70	∞	×	×
0.30	1.75	1.91	2.15	2.48	2.97	3.88	×	×
0.35	1.64	1.77	1.97	2.23	2.58	3.10	×	×
0.40	1.55	1.66	1.83	2.04	2.31	2.68	∞	×
0.45	1.47	1.57	1.72	1.89	2.12	2.40	4.52	×
0.495	1.41	1.50	1.63	1.79	1.98	2.22	3.65	×

$u_{max} \geq 1/2$. As the Maximum Principle requires the Euler BC $\lambda(T) = 0$ to be satisfied, the optimal solution must leave the singular solution (if it should contain such a solution segment) and switch to an upper corner solution at a second switch point t_S to conform with Proposition 27. With $R_s = 1/2 = R_S(t_S)$, this upper corner solution, to be denoted by $R_{gS}(t)$ and $\lambda_{gS}(t)$, is determined by

$$R'_{gS} = R_{gS}(1 - u_{max} - R_{gS}), \qquad\qquad R_{gs}(t_S) = 1/2 \qquad (13.3.7)$$
$$\lambda'_{gS} = -\lambda_{gS}(1 - u_{max} - 2R_{gS}) - u_{max}, \quad \lambda_{gS}(T) = 0 \qquad (13.3.8)$$

with t_S determined by the continuity of the adjoint function at the switch point:

$$\lambda_{gS}(t_S) = 1. \qquad (13.3.9)$$

The new upper corner solution differs from $\{R_{g0}(t), \lambda_{g0}(t)\}$ with the latter satisfying the continuity condition $R_s = R_\ell(t_s)$. The explicit solution of the BVP (13.3.7)–(13.3.9) for $u_{max} \neq 1$ is

$$R_{gS}(t) = \frac{u_\alpha e^{-u_\alpha(t-t_S)}}{2u_\alpha + 1 - e^{-u_\alpha(t-t_S)}} \qquad (13.3.10)$$

$$\lambda_{gS}(t) = \frac{u_{max}}{u_\alpha} \frac{2u_\alpha + 1 - e^{u_\alpha(t_S - t)}}{2u_\alpha + 1 - e^{u_\alpha(t_S - T)}} \left[1 - e^{u_\alpha(t-T)}\right] \qquad (13.3.11)$$

with

$$t_S = T - \frac{1}{u_{max} - 1} \ln\left(2u_{max} - 1\right). \qquad (13.3.12)$$

For the special case $u_{max} = 1$, we get by taking the limit as $u_\alpha \to 0$

$$R_{gS}(t) = \frac{1}{2 + t - t_S} \qquad (13.3.13)$$

$$\lambda_{gS}(t) = \frac{2 + t - t_S}{2 + T - t_S}(T - t) \qquad (13.3.14)$$

with

$$t_S = T - 2. \qquad (13.3.15)$$

To delineate the relation between t_s and t_S, we first observe that $t_S(T, u_{max})$ as given by (13.3.12) does not depend on R_0 while t_0 as found in (13.3.3) depends only on R_0. As such, t_0 may be smaller than, equal to, or greater than t_S (for a prescribed combination of T and $u_{\max} \geq 1/2$). When $t_0 < t_S$, the gap (t_0, t_S) allows for a singular solution phase in the optimal solution of the problem; this is not true if $t_0 > t_S$. The threshold value of R_0 that separates the two optimal solution types, denoted by R_c, is when $t_0(R_0 = R_c) = t_S$. Since t_0 is a monotone decreasing function of R_0 that ranges from 0 to ∞ as R_0 decreases from $1/2$ to 0, there is a unique root R_c of the equation

$$t_0(R_0) = \ln\left(\frac{1 - R_0}{R_0}\right) = t_S(T, u_{\max}).$$

With $t_S(T, u_{\max})$ known from (13.3.12), we have

$$R_c(T, u_{\max}) = \frac{1}{1 + e^T (2u_\alpha + 1)^{-1/u_\alpha}} \qquad (13.3.16)$$

with $0 < R_c \leq 1/2$ for the range $R_0 \leq 1/2$ relevant to our discussion.

In terms of R_c, we may now differentiate the optimal control for different ranges of $R_0 < 1/2$ (for $u_{\max} \geq 1/2$) in the following proposition (with $R_0 = 1/2$ to be discussed in the next section):

Proposition 30 *For $R_0 < 1/2$ and $u_{\max} \geq 1/2$, $E(T)$ is maximized by the two-switch control*

$$u_{op}(t) = \begin{cases} 0 & (0 \leq t < t_0) \\ 1/2 & (t_0 < t < t_S) \\ u_{\max} & (t_S < t \leq T) \end{cases} \qquad (13.3.17)$$

if $R_c < R_0 \ (< 1/2)$ and by the bang-bang control (13.3.4) if $R_0 < R_c$, where the only switch from lower to upper corner control at t_s as t increases is given by the unique root t_{s0} of (13.3.5) inside the interval (t_S, t_0). The notation t_{s0} denotes a switch points from the lower corner control in $[0, t_s)$ to the upper corner control in $(t_s, T]$.

Proof For $R_0 \leq R_c$, the bang-bang control (13.1.14) that satisfies all the necessary conditions imposed by the Maximum Principle is unique with $[E(T)]_{u=u_{op}(t)} > [E(T)]_{u=0}$. Hence, $[E(T)]_{u=u_{op}(t)}$ is maximized. In contrast, there are also admissible controls other than the two-switch control (13.3.17) in the case of $R_c < R_0 < 1/2$ that meet all the requirements of the Maximum Principle. That (13.3.17) maximizes $E(T)$ can be proved by the method of the most rapid approach (see proof of Proposition 34). At this time, we only note that for $t_0 < t_S$ and the continuity requirements being met at the switch point t_0, the inclusion of a singular solution phase in $u_{op}(t)$ should be considered since it (and no other) renders the Hamiltonian stationary in (t_0, t_S):

$$\frac{\partial H}{\partial u} = R_S(t)[1 - \lambda_S(t)] = 0, \qquad (t_0 < t < t_S). \qquad (13.3.18)$$

That $H[u]$ is maximized can be seen from its stationary value. For the optimal solution in the interval (t_0, t_S), we have

$$H[u_{op}(t)] = \frac{1}{4}.$$

Given $\lambda_S(t) = 1$, the value $1/4$ is in fact the maximum of the remaining right-hand member of $H[u]$ $(= \lambda_S R(1 - R) = R(1 - R))$.   □

To emphasize the importance of the threshold initial population R_c, we have for $R_c < R_0$ $(< 1/2)$ and $u_{max} \geq 1/2$ an optimal conversion strategy that involves a three-segment optimal control with two switches at t_0 and t_S. The first switch point t_0 is as given by (13.3.3) and the second t_S by (13.3.12). For the first segment $[0, t_0)$, the lower corner solution, (13.3.2) and (13.3.6) with $t_s = t_0$, applies to raise $R(t)$ to $R_S(t) = 1/2$. At the other end adjacent to T, we have the upper corner solution $\{R_{gS}, \lambda_{gS}\}$ as given by (13.3.10) and (13.3.11) for the interval $(t_S, T]$ in order to satisfy the Euler BC. These two solution segments are bridged by the singular solution $\{R_{gS}(t), \lambda_{gS}(t)\} = \{1/2, 1\}$ in the interval (t_0, t_S) that renders $H[u]$ maximum in that interval.

On the other hand, the optimal conversion strategy is given by the bang-bang control (13.3.4) if R_0 $(< 1/2) \leq R_c$ (for which $t_0 > t_s$). The lower corner solution is now given by (13.3.2) and (13.3.6) and the corresponding upper corner solution after switching is given by (13.1.10) and (13.2.17), respectively. The single switch point t_s which appears in both solutions is the unique positive root t_{s0} of (13.3.5). If t_s should turn out to be negative, then Corollary 10 applies with $\lambda_g(t) < 1$ for all $t \geq 0$.

At this point, we should call attention to the fact that the expression (13.3.12) for the switch point t_S (from the singular control to the upper corner control) is no longer real-valued for $u_{max} < 1/2$. However $u_{max} < 1/2$ is outside the range of interest for the present discussion (as it has already been treated in a previous subsection). As $u_{max} \downarrow 1/2$ from above, the switch point $t_S \downarrow -\infty$ monotonically. It follows that for any prescribed terminal time T, there is a threshold value for u_{max} for which $t_S = 0$ so that the conversion should be at maximum rate from the start.

13.3.2.2 $R_0 = 1/2$

For the special case of $R_0 = 1/2$ and $u_{max} = 1/2$, the preferred singular (stationary) solution $\{R_S(t) = 1/2, \lambda_S(t) = 1, u_S(t) = 1/2\}$ satisfies the initial conditions and the constraints (13.1.4) and is optimal from the start. However, it must switch to the upper corner control at the switch point t_S given by (13.3.12) to satisfy the Euler BC as required by the Maximum Principle. Thus, we have for this special case

$$u_{op}(t) = \begin{cases} 1/2 & (0 \leq t < t_S) \\ u_{max} & (t_S < t \leq T) \end{cases} \qquad (13.3.19)$$

with the upper corner solution for $t_S < t \leq T$ again given by (13.3.10), (13.3.11) and (13.3.12).

For $R_0 - 1/2$ but $u_{max} > 1/2$, we have $1 - u_{max} < 1/2$ so that $R'_{g0}(0) = (1 - u_{max} - 1/2)/2 < 0$ and $R'_{g0} = R_{g0}(1 - u_{max} - R_{g0}) < 0$ with $R_{g0}(t) \downarrow 1 - u_{max} < 1/2$. Since a stationary solution that maximizes is expected to be superior (and will be proved to be so by the method of most rapid approach later) and $u_{max} > 1/2$, we should take $u_{op}(t) = u_S(t) = 1/2 \ (< u_{max})$ for a singular solution from the start that persists prior to the enforcement of Proposition 27. The switching would again take place at t_S, the last instant the required Euler BC can still be met.

In the second case (for $R_0 = 1/2$ and $u_{max} > 1/2$), the optimal control (13.3.19) may not be the only admissible control by the Maximum Principle. Until proven that it is not optimal, an example would be for $u_{op}(t) = u_{max}$ for the entire host cell life span (as it also gives a positive $E(T)$).

13.3.3 On various threshold values

In Table 13.2, we report the values of the four quantities $\{t_0, t_{s0}, t_S, R_s\}$ for $u_{max} = 2, T = 2$ and a range of values for R_0 in the interval $(0, 0.5)$ with the values for t_{s0} obtained numerically as the root of the switching condition (13.3.5) (obtained from (13.2.14) with $R_s = R_\ell(t_{s0})$ as given by (13.3.2)). The results in the table confirm the following:

1) The existence of a unique $R_c < 1/2$,
2) $t_S < t_0$ (thereby eliminating the possibility of a singular solution phase and leading to a bang-bang optimal control) for $R_0 < R_c$.
3) The existence of an interval (t_0, t_S) for a singular solution phase leading to the two-switch optimal control (13.3.17) for $R_0 > R_c$.

Table 13.2 Variation of switch points and R_s with R_0 ($u_{max} = 2, T = 2$).

R_0	0.1	0.15	0.20	0.25	0.289	0.30	0.35	0.40	0.45
t_0	2.197	1.735	1.386	1.099	0.901	0.847	0.619	0.406	0.201
t_{s0}	1.146	1.077	1.011	0.949	0.901	0.888	0.827	0.766	0.704
t_S	0.901	0.901	0.901	0.901	0.901	0.901	0.901	0.901	0.901
R_s	0.259	0.341	0.407	0.463	0.500	0.510	0.552	0.589	0.623

Similar calculations have also been done for other values of u_{max} and T with similar results. Table 13.3 reports the results for $u_{max} = 2.5$ and $T = 2$. For this case, except for a smaller threshold value $R_c \ (= 0.255\ldots)$, how the optimal conversion strategy varies with $R_0 \ (< 1/2)$ is similar to the previous case.

It should be evident from (13.3.12) that t_S is linear in T. For the two values u_{max} of Tables 13.2 and 13.3, the value of t_S for other T values are just a translational adjustment (of adding $T - 2$ to the values reported). This is shown numerically in Table 13.4 by comparing t_S for $T = 2$ and $T = 4$ for different u_{max}. The nonlinear dependence of R_c on T and u_{max} can be read off the exact solution (13.3.16) and is

Table 13.3　Variation of switch points and R_s with R_0 ($u_{\max} = 2.5, T = 2$).

R_0	0.1	0.15	0.20	0.255	0.30	0.35	0.40	0.45
t_0	2.197	1.735	1.386	1.076	0.847	0.619	0.406	0.201
t_{s0}	1.448	1.384	1.322	1.076	1.205	1.147	1.088	1.028
t_S	1.076	1.076	1.076	1.076	1.076	1.076	1.076	1.076
R_s	0.320	0.413	0.484	0.500	0.588	0.629	0.669	0.696

Table 13.4　Variation of t_S and R_c with $u_{\max}$ and T.

$T \setminus u_{\max}$		0.6	0.70	0.85	1.00	1.25	1.50	2.00
2	t_S	−2.02	−1.05	−0.38	0	0.38	0.61	0.90
	R_c	0.88	0.74	0.59	0.50	0.41	0.35	0.29
4	t_S	−0.02	0.95	1.62	2	2.38	2.61	2.90
	R_c	0.51	0.28	0.17	0.12	0.09	0.07	0.05

shown numerically for $T = 2$ and 4 and for a range of $u_{\max}$ in Table 13.4. Note that only the portion of the results with $t_S \geq 0$ (and $R_c \leq 1/2$) in the table is relevant to the problem of this section.

We do not have an explicit solution for the switch point t_{s0} (for $R_0 < R_c$), but the following lemma provides some information about its dependence on T:

Lemma 2 *For $R_0 < 1/2$, the switch point t_{s0} is an increasing function of the terminal time T.*

Proof Upon differentiating (13.3.5) with respect to T, we obtain

$$\frac{dt_{s0}}{dT} = \frac{(1 - R_0)\, e^{-u_\alpha (T - t_{s0})}}{u_{\max}\,(1 - R_0 + R_0 e^{t_{s0}})} > 0.$$

　　□

How t_{s0} varies with $u_{\max}$ is a little more complicated. Upon differentiating both sides of (13.3.5) with respect to $u_\alpha = u_{\max} - 1$, we obtain the following lemma on t_{s0} being an increasing function of $u_{\max}$ independent of the sign of $u_\alpha = u_{\max} - 1$:

Lemma 3 *For $R_0 < 1/2$ (and hence $u_{\max} > 1/2$ for the lower corner control to be optimal in $[0, t_s)$), the switch point $t_s = t_{s0}$ is an increasing function of $u_{\max}$ if T is sufficiently large.*

Proof For this case, the switch point t_s is the root t_{s0} of the equation (13.3.5) defining t_s. Routine calculations from differentiating both sides of (13.3.5) with respect to $u_{\max}$ leads to

$$\frac{dt_{s0}}{du_{\max}} = \frac{1}{u_\alpha} \frac{(1 - R_0)\,(T - t_{s0})\, e^{u_\alpha (T - t_{s0})} - (1 - R_0) - R_0 e^{t_{s0}}}{(1 - R_0)\, e^{u_\alpha (T - t_{s0})} + R_0 e^{t_{s0}}}$$

$$\sim \begin{cases} \frac{1}{u_\alpha}\,(T - t_{s0}) > 0 & (u_\alpha > 0) \\ -\frac{1}{u_\alpha R_0}\,\{(1 - R_0)\, e^{-t_{s0}} + R_0\} > 0 & (u_\alpha < 0) \end{cases}.$$

　　□

That t_{s0} is an increasing function of $u_{\max}$ is already evident from the values of t_{s0} in Tables 13.2 and 13.3 with a rather modest T magnitude of 2 (for the range

$R_0 < 1/2$ where the lower corner control is optimal). Tables 13.5 and 13.6 provide additional support for this behavior for $T = 2$ and $T = 4$, respectively, for a range of $R_0 < 1/2$. It should be noted that the results for t_{s0} are relevant only to the problem with $R_0 < R_c$ with R_c given in terms of u_{max} and T by (13.3.16) (see also Table 13.4). We highlighted in boldface values of t_{s0} in these tables that are not relevant since a singular solution phase exist in a finite subinterval in these cases. Even when there is no singular solution phase for $R_0 < R_c$, the switch point t_{s0} may also not be relevant for the problem. Cases with $t_{s0} < 0$ correspond to $\lambda_{g0}(t) < 1$ for all t in $[0, T]$ so that Corollary 10 applies throughout the host cell's life span.

Table 13.5   Variation of t_{s0} with u_{max} and R_0 ($T = 2$).

$u_{max} \backslash R_0$	0.15	0.25	0.30	0.40	0.45	0.495
0.25	−0.419	−0.725	−0.874	−1.161	−1.306	−1.440
0.45	0.070	−0.195	−0.321	−0.572	−0.700	−0.819
0.60	0.294	0.054	−0.060	−0.289	−0.406	−0.515
0.85	0.548	0.341	0.242	0.043	−0.059	−0.154
1.00	0.659	0.408	0.376	0.192	0.098	**0.010**
1.25	0.804	0.634	0.552	0.389	**0.305**	**0.227**
1.50	0.915	0.762	0.689	**0.542**	**0.467**	**0.397**
2.00	1.077	0.949	**0.888**	**0.788**	**0.704**	**0.647**
2.50	1.190	1.080	**1.028**	**0.924**	**0.872**	**0.823**

Table 13.6   Variation of t_{s0} with u_{max} and R_0 ($T = 4$).

$u_{max} \backslash R_0$	0.15	0.25	0.30	0.40	0.45	0.495
0.25	0.339	−0.189	−0.450	−0.808	−0.997	−1.165
0.45	0.984	0.505	0.303	−0.066	−0.243	−0.401
0.60	1.304	0.860	0.672	0.327	0.162	0.013
0.70	1.474	1.045	**0.875**	**0.546**	**0.377**	**0.234**
0.85	1.688	**1.298**	**1.133**	**0.827**	**0.680**	**0.548**
1.00	**1.863**	**1.504**	**1.349**	**1.065**	**0.929**	**0.806**
1.25	**2.097**	**1.778**	**1.642**	**1.391**	**1.276**	**1.162**
1.50	**2.280**	**1.995**	**1.874**	**1.651**	**1,544**	**1.447**
2.00	**2.551**	**2.317**	**2.218**	**2.038**	**1.951**	**1.874**
2.50	**2.741**	**2.545**	**2.462**	**2.312**	**2.240**	**2.175**

13.3.4   Summary for $R_0 \leq 1/2$

The lower corner solution $\{R_\ell(t), \lambda_\ell(t)\}$ induced by the lower corner control $u_\ell(t) = 0$ has the distinct ability to increase the RB population from $R_0 \leq 1/2$ to the level of the stationary solution $\{R_S(t), \lambda_S(t)\} = \{1/2, 1\}$ at some instant $t_0 > 0$ (and even larger $R(t)$ for $t > t_0$). Once the stationary RB population $1/2$ is reached, a stationary solution that maximizes is to be sustained since it is superior to a corner solution. As a consequence, the optimal conversion strategy for $R_0 \leq 1/2$ will be dictated by whether the singular control can be sustained. A simple case of (the host cell's) inability to have a segment of singular control is $u_{max} < 1/2$ for which $R_S(t) = 1/2$ cannot be reached.

By Proposition 27, the optimal control $u_{op}(t)$ cannot remain a singular control and must switch to the upper corner control at some instant $t_S > t_0$ prior to the terminal time T in order to satisfy the required Euler BC. To the extent that the stationary solution is preferred, the switch should be as late as possible but still enable the adjoint function to vanish at terminal time. This upper corner solution has been designated as $\{R_{gS}(t), \lambda_{gS}(t)\}$. Continuity of $R(t)$ and $\lambda(t)$ requires $\{R_{gS}(t_S), \lambda_{gS}(t_S)\} = \{1/2, 1\}$ (in addition to the Euler BC $\lambda_{gS}(T) = 0$). This BVP and its solution are given by (13.3.7)–(13.3.9) and (13.3.10)–(13.3.12), respectively.

A more complete classification of possible optimal conversion strategies for $R_0 \leq 1/2$ can now be given in terms of a threshold initial RB population $R_c(T, u_{\max})$ defined by the conditions $t_S = t_0$ and $R_\ell(t_0) = R_{gS}(t_S) = 1/2$. Depending on the magnitude of the initial RB population R_0 ($\leq 1/2$) relative to R_c, we have the following two qualitatively different optimal conversion strategies:

13.3.5   Case I — $0 < R_0 \leq R_c$

For this range of R_0, we have $t_S \leq t_0$. It is not possible for $u_{op}(t)$ to have a finite time interval of singular control. The optimal conversion strategy is given by the bang-bang control (13.3.4) of Proposition 29 where the switch point t_s is the unique positive root of (13.3.5) and $R_g(t_s) = (R_s =)R_\ell(t_s) < 1/2$. The corresponding RB population and adjoint function are given by (13.3.2) and (13.2.20) for $0 \leq t \leq t_s$ and by (13.2.15) and (13.2.17) for $t_s \leq t \leq T$. There would be no switch if the root t_s is negative and upper corner control applies throughout $[0, T]$ (see Corollary 10).

13.3.6   Case II — $R_c < R_0 < 1/2$

For this range of R_0, we have $t_0 < t_S$ and the optimal conversion strategy is given by Proposition 30 with a two-switch control (13.3.17): starting with the lower corner control $u_\ell(t) = 0$, switching to the singular control $u_S(t) = 1/2$ at t_0 as given by (13.3.3) and then to the upper corner control $u_g(t) = u_{\max}$ at the switch point t_S given by (13.3.12). The corresponding RB populations and adjoint functions are given by (13.3.2) and (13.3.6) in $[0, t_0)$, by the stationary solution $\{R_S(t), \lambda_S(t)\} = \{1/2, 1\}$ in (t_0, t_S) and by (13.3.10)–(13.3.12) in $(t_S, T]$.

13.4   Optimal Conversion Strategy for $R_0 > 1/2$

13.4.1   Upper corner control for $1 - u_{\max} \geq R_0$ ($> 1/2$)

We know from Proposition 28 that the lower corner control is not appropriate near the starting time for $R_0 > 1/2$. Neither is the singular solution since $R_0 > 1/2 \neq R_S(t)$. We must then examine the only other option of an upper corner control at the early stage of the host cell's life span, the interval $[0, t_s)$ or some subinterval of it.

Let $R_{gm}(t)$ and $\lambda_{gm}(t)$ be $R(t)$ and $\lambda(t)$ generated by $u = u_{\max}$ for t adjacent to the starting time with the change of the RB population in time governed by the IVP

$$R'_{gm} = R_{gm}(1 - u_{\max} - R_{gm}), \qquad R_{gm}(0) = R_0 \ (> 1/2). \qquad (13.4.1)$$

Its exact solution is

$$R_{gm}(t) = \frac{u_\alpha R_0}{(u_\alpha + R_0)\, e^{u_\alpha t} - R_0}. \qquad (13.4.2)$$

In this section, we focus on the range $1 - u_{\max} \geq R_0$ and show that the optimal strategy is to convert at maximum rate from the start.

13.4.1.1 $1 - u_{\max} = R_0$

For the very special value of $u_{\max} = 1 - R_0 < 1/2$, we see from (13.4.1) that $R_{gm}(t) = R_0$ for all t. The corresponding adjoint function that satisfies the Euler BC is determined by

$$\lambda'_{gm} = \lambda_{gm} R_0 - u_{\max}, \qquad \lambda_{gm}(T) = 0$$

or

$$\lambda_{gm}(t) = u_{\max} \left\{ 1 - e^{R_0(t-T)} \right\}.$$

With $u_{\max} = 1 - R_0 < 1/2$, we have $0 < \lambda_{gm}(t) < 1$ in $[0, T)$. Hence, the optimal control for this special case is $u_{op}(t) = u_{\max}$ for all t in $[0, T]$.

13.4.1.2 $1 - u_{\max} > R_0$

For the broader range $u_{\max} < 1 - R_0 < 1/2$, we see from (13.4.1) that $R'_{gm}(t) > 0$ so that $R_{gm}(t) > R_0 > 1/2$ and tends to $1 - u_{\max} > R_0 > 1/2$ from below. The corresponding adjoint function that satisfies the Euler BC is determined by

$$\lambda'_{gm} = -(1 - u_{\max} - 2R_{gm})\lambda_{gm} - u_{\max}, \qquad \lambda_{gm}(T) = 0. \qquad (13.4.3)$$

We need only to ensure that $\lambda_{gm}(t) < 1$ for $0 \leq t < T$ for $u_{op}(t) = u_{\max}$ to be optimal in in $[0, T]$. For this purpose, suppose the maximum of $\lambda_{gm}(t)$ is attained at t_m with $\lambda'_{gm}(t_m) = 0$. It follows from the adjoint ODE

$$\lambda_{gm}(t_m) = \frac{u_{\max}}{2R_{gm}(t_m) - 1 + u_{\max}} < 1$$

since $R_{gm}(t_m) > 1/2$.

Altogether, these observations lead to the following optimal solution for the range $1 - u_{\max} \geq R_0 \ (> 1/2)$:

Proposition 31 *For $1 - u_{\max} \geq R_0 > 1/2$ (or $u_{\max} \leq 1 - R_0 < 1/2$), the control $u_{op}(t) = u_{\max} \ (< 1/2)$ for all t in $[0, T]$ maximizes $E(T)$.*

Proof Since $u_{op}(t) = u_{\max}$ is the only control that satisfies all necessary conditions for optimality required by the Maximum Principle and is superior to the lower corner control, it maximizes $E(T)$. □

13.4.2 Optimal conversion strategies for $1 - u_{\max} < R_0$

For the complementary range $u_{\max} > 1 - R_0 \; (> 1/2)$, we have from (13.4.1) $R'_{gm} < 0$ so that $R_{gm}(t) < R_0$ is a monotone decreasing function that tends to $1 - u_{\max} < R_0$ from above. As the limit population $1 - u_{\max}$ may be below $R_S(t) = 1/2$ we need to examine separately the two different possibilities of having and not having a stationary solution component for the optimal control $u_{op}(t)$.

13.4.2.1 $1 - R_0 < u_{\max} \le 1/2$

For this range of $u_{\max}$ (which is a part of the range $1 - u_{\max} < R_0$), we have ($R'_{gm} < 0$ as noted above) $R_{gm}(t) \downarrow 1 - u_{\max} \ge 1/2$ from above so that $R_{gm}(t) > 1/2$ for any finite t. The RB population does not reach the singular solution $R_S(t) = 1/2$. The only conversion rate not ruled out is with $u(t) = u_{\max}$ for all t in $[0, T]$. To see if it is optimal, we again examine the maximum of $\lambda_{gm}(t)$ attained at t_m with $\lambda'_{gm}(t_m) = 0$. It follows from the adjoint ODE

$$\lambda_{gm}(t_m) = \frac{u_{\max}}{2R_{gm}(t_m) - 1 + u_{\max}}.$$

Since $R_{gm}(t)$ tends to $1 - u_{\max}$ from above, we get from the exact expression above

$$\lambda_{gm}(t_m) < \frac{u_{\max}}{1 - u_{\max}} \le 1$$

given $u_{\max} \le 1/2$. Hence $u_{op}(t) = u_{\max}$ is optimal by Maximum Principle and the only one that meets all the necessary conditions required by the principle. With $[E(T)]_{u_{op}(t) = u_{\max}} > [E(T)]_{u_\ell(t) = 0} = 0$, we have

Proposition 32 *For $u_{\max} \le 1/2$ (and $R_0 > 1/2$), the control $u_{op}(t) = u_{\max}$ is optimal in $[0, T]$ and maximizes $E(T)$.*

13.4.2.2 $u_{\max} > 1/2$

For this complementary range of $u_{\max}$, we have again $R'_{gm} < 0$ but now the monotone decreasing function $R_{gm}(t)$ tends to $1 - u_{\max} < 1/2$ from above. Hence it is possible to have a sub-interval of stationary solution, depending on the relative magnitude of the instant t_x when $R_{gm}(t)$ (see (13.4.1) and (13.4.2)) reaches $R_S(t) = 1/2$ and the instant t_S when $R_{gS}(t)$ (as previously defined by the BVP (13.3.7)–(13.3.9)) is equal to $1/2$. With the exact expression (13.4.2) for $R_{gm}(t)$, the condition

$$R_{gm}(t_x) = \frac{1}{2} \tag{13.4.4}$$

defining t_x gives

$$t_x = \frac{1}{u_\alpha} \ln\left(\frac{R_0 \, (2u_\alpha + 1)}{u_\alpha + R_0} \right).$$

Similar to the case of $R_c < R_0 \le 1/2$ where there would be no stationary solution component if $t_0 > t_S$, we have the following analogous proposition for $u_{\max} > 1/2$ (and $R_0 > 1/2$):

Proposition 33 For $u_{\max} > 1/2$ (and $R_0 > 1/2$), $E(T)$ is maximized by the optimal control $u_{op}(t) = u_{\max}$ for all t in $[0, T]$ if $t_x \geq t_S$.

Proof When $t_S \leq t_x$, the maximum allowable control $u_{\max}$ is insufficient for reducing the initial RB population fast enough to get to the level of the singular solution $R_S(t) = 1/2$ before the adjoint function should leave its singular value of $\lambda_S(t) = 1$ to head toward the Euler BC $\lambda(T) = 0$ required by the Maximum Principle. In that case, the upper corner control would apply throughout the host cell's life span.

To show that it is optimal, i.e., it maximizes the Hamiltonian, we note that $R_{gm}(t)$, with $u_{op}(t) = u_{\max}$ for all t in $[0, T]$, is monotone decreasing (since $u_{\max} > 1/2$) and tends to $1 - u_{\max} < 1/2$ from above as $t \to \infty$. Concurrently, the corresponding adjoint function λ_{gm} (required to satisfy the Euler BC) is determined by (13.4.3) to be monotone decreasing in some interval adjacent to T. Suppose again $\lambda_{gm}(t)$ attains a maximum at t_m prior to declining to zero at T. In that case $\lambda'_{gm}(t_m) = 0$ and

$$\lambda_{gm}(t_m) = \frac{u_{\max}}{2R_{gm}(t_m) - (1 - u_{\max})}.$$

Since $t_m < t_x$ (otherwise we would have a solution for $\{R_{gS}, \lambda_{gS}\}$ with $\lambda_{gS}(t_S) = 1$ for some $t_S > t_x$ given $\lambda_{gm}(t_m) > 1$), we must have $R_{gm}(t_m) > R_{gm}(t_x) = 1/2$. It follows that

$$\lambda_{gm}(t_m) < \frac{u_{\max}}{1 - (1 - u_{\max})} = 1$$

and $u_{op}(t) = u_{\max}$ for all t in $[0, T]$ is optimal by the Maximum Principle.

Now $u_{op}(t) = u_{\max}$ is the only available optimal control. With $[E(T)]_{u_{op}(t) = u_{\max}} > [E(T)]_{u_\ell(t) = 0} = 0$, $u_{op}(t) = u_{\max}$ maximizes $E(T)$.   □

On the other hand, the optimal conversion strategy would be quite different if $t_x < t_S$.

Proposition 34 For $u_{\max} > 1/2$ (and $R_0 > 1/2$), the two-switch control

$$u_{op}(t) = \begin{cases} u_{\max} & (0 \leq t < t_x) \\ 1/2 & (t_x < t < t_S) \\ u_{\max} & (t_S < t \leq T) \end{cases} \tag{13.4.5}$$

maximizes $E(T)$ if $t_x < t_S$.

Proof With t_x being the first instant for $R_{gm}(t)$ to reach $R_S(t) = 1/2$ and t_S ($> t_x$) the latest time for $\lambda_{gS}(t)$ to leave $\lambda_S(t) = 1$ and still satisfy the Euler BC, we take $u_{op}(t)$ to be the two-switch control (13.4.5) and prove below that it maximizes $E(T)$ by the method of the most rapid approach. Our goal is to show that

$$[E(T)]_{u = u_{op}(t)} - [E(T)]_{u = u_c(t)}$$

$$= \int_0^T u_{op}(t) R_{op}(t)\, dt - \int_0^T u_c(t) R_c(t)\, dt$$

$$= \int_0^T \left\{ [(1 - R)R - R']_{u_{op}(t)} - [(1 - R)R - R']_{u_c(t)} \right\} dt > 0$$

where $u_C(t)$ is any admissible comparison control for which the corresponding $R_C(t)$
has the same initial and terminal values as shown in the figure below. Note that
starting from the same initial population R_0, the $R_C(t)$ resulting from any comparison control (decreases with time and) can only reach the singular value $1/2$ at some
instance $t_D > t_x$ (and not sooner) since $R_{op}(t)$ approaches the singular solution
most rapidly with $u_{op}(t) = u_{max} > u_C(t)$ for the interval $[0, t_x]$. Similarly, $R_C(t)$
can only leave the singular solution (and descend toward the terminal value $R(T)$)
no later than $R_{op}(t)$ so that $t_D < t_{Sx}$ and not later.

The key to our proof is to visualize the two graphs $R_{op}(t)$ and $R_C(t)$ as trajectories in the $R - t$ plane and the integral as line integral written as

$$[E(T)]_{u=u_{op}(t)} - [E(T)]_{u=u_C(t)}$$
$$= \int_{C_{op}} [(1-R)Rdt - dR] - \int_{C_u} [(1-R)Rdt - dR]$$
$$= \int_{C_{op}} [G(t,R)dt + H(t,R)dR] - \int_{C_u} [G(t,R)dt + H(t,R)dR]$$

where $G = (1 - R)R$ and $H = -1$. We now consider each of the two closed sub-
contours C_1 and C_2 separately: the loop integral around the closed contour C_1
traversed counterclockwise from point A, to point B to point D and back to point
A and a second integral C_2 traversed counterclockwise from point D, to E to F and
back to D (see Figure 13.2).

The first loop integral around C_1 may be written with the help of Green's
theorem for the $t - R$ plane as

$$\oint_{C_1} [G(t,R)dt + H(t,R)dR] = \iint_{A_1} \left[\frac{\partial H}{\partial t} - \frac{\partial G}{\partial R} \right] dtdR$$
$$= \iint_{A_1} (2R - 1)dtdR > 0$$

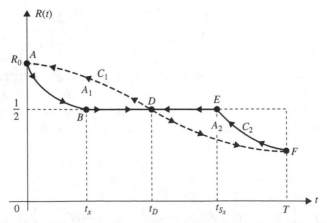

Fig. 13.2   Contour for the optimal solution and an alternative path.

since we have $R(t) > 1/2$ for the region enclosed by C_1. Similarly, the second loop integral around C_2

$$\oint_{C_2} [G(t,R)dt + H(t,R)dR] = \iint_{A_2} \left[\frac{\partial H}{\partial t} - \frac{\partial G}{\partial R} \right] dtdR$$

$$= \iint_{A_2} (2R-1)dtdR < 0$$

since we have $R(t) < 1/2$, for the region enclosed by C_2. With

$$\int_{C_{op}} [Gdt + HdR] - \int_{C_u} [Gdt + HdR]$$

$$= \oint_{C_1} [Gdt + HdR] - \oint_{C_2} [Gdt + HdR]$$

$$= \iint_{A_1} (2R-1)dtdR - \iint_{A_2} (2R-1)dtdR > 0$$

we have

$$[E(T)]_{u=u_{op}(t)} - [E(T)]_{u=u_C(t)} > 0.$$

The same method of proof applies to any other admissible control $u_C(t)$ with the corresponding RB population having the same two end values. Hence, $E(T)$ is maximized by $u_{op}(t)$.   □

Remark 12 *That any admissible control must start with the same initial RB population R_0 is dictated by reality. At the terminal point where $R(T)$ is not prescribed, any admissible control $u_C(t)$ should also end with the terminal RB population corresponding to the optimal control $u_{op}(t)$. Obviously the comparison control $u_C(t)$ should not leave behind more RB than $u_{op}(t)$. At the same time, $R_C(T)$ should not be less than $R_{op}(T)$ as it would leave the singular (stationary) solution earlier than necessary in order to get to a lower end population. A control that leads to a shorter interval of stationary solution is inferior to one with a longer interval assuming all other requirements of the Maximum Principle are met by both.*

13.4.3   *The relative magnitude of* t_x *and* t_S

Given the importance of the relative magnitude of t_x and t_S in determining the optimal conversion strategy for $u_{max} > 1/2$ (and $R_0 > 1/2$), we establish in this subsection some quantitative measures to elucidate this relative magnitude. As in the case of relative magnitude of t_0 and t_S for $R_c < R_0 \leq 1/2$, an important threshold relation for the present problem is the combination of parameter values that leads to $t_x = t_S$.

For $u(t) = u_{max}$ $(> 1/2)$, the exact solution of the IVP (13.4.1) is

$$R_{gm}(t) = \begin{cases} u_\alpha R_0 e^{-u_\alpha t}/(u_\alpha + R_0 - R_0 e^{-u_\alpha t}) & (u_{max} \neq 1) \\ R_0/(1 + R_0 t) & (u_{max} = 1) \end{cases}. \tag{13.4.6}$$

The condition $R_{gm}(t_x) = 1/2$ gives the following exact dependence of t_x on R_0 and u_{max}:

$$t_x = \begin{cases} \ln\left(R_0(2u_\alpha + 1)/(u_\alpha + R_0)\right)/u_\alpha & (u_{max} \neq 1) \\ (2R_0 - 1)/R_0 & (u_{max} = 1) \end{cases}. \qquad (13.4.7)$$

Note that t_x is well defined and nonnegative for all $R_0 > 1/2$ (and $2u_{max} > 1$) with

$$[t_x]_{R_0=1/2} = 0,$$

as it should be, while

$$[t_x]_{R_0=1} = \begin{cases} \ln\left((2u_{max} - 1)/u_{max}\right)/u_\alpha > 0 & (u_{max} \neq 1) \\ 1 & (u_{max} = 1) \end{cases}.$$

The existence of a finite positive t_x for a range of $R_0 > 1/2$ when $u_{max} > 1/2$ gives rise to the possibility of a switch of control to the singular control $u_S(t) = 1/2$ associated with the stationary solution $\{R_S(t), \lambda_S(t)\} = \{1/2, 1\}$. To ensure the switching is to the singular solution when it is possible to do so at t_x (in the range $u_{max} > 1/2$), we need the corresponding adjoint function, denoted by $\lambda_{gx}(t)$, determined by the terminal value problem

$$\lambda'_{gx} = -(1 - u_{max} - 2R_{gm})\lambda_{gx} - u_{max}, \qquad \lambda_{gx}(t_x) = 1 \qquad (13.4.8)$$

or

$$\lambda_{gx}(t) = \frac{u_\alpha + R_0 - R_0 e^{-u_\alpha t}}{u_\alpha + R_0 - R_0 e^{-u_\alpha t_x}} e^{u_\alpha(t - t_x)}. \qquad (13.4.9)$$

While the singular (stationary) solution may be preferred, the Euler BC $\lambda(T) = 0$ requires the optimal strategy to switch to the upper corner control at an opportune time t_S in order to convert as much RB as possible thereafter and reduce the (shadow) worth (measured by the adjoint function λ) of any remaining RB population to zero by the time the host cell dies. For that requirement, we again work with $R_{gS}(t)$ and $\lambda_{gS}(t)$ previously defined by (13.3.10), (13.3.11) and (13.3.12). In particular, the second switch point t_S is given by (13.3.12) to be

$$t_S = T - \frac{1}{u_\alpha} \ln(2u_\alpha + 1). \qquad (13.4.10)$$

The critical relation $t_S > t_x$ needed for the two switch control (13.4.5) to be optimal is met by

$$T > \frac{1}{u_\alpha} \ln\left(\frac{R_0(2u_\alpha + 1)^2}{u_\alpha + R_0}\right) \equiv T_0. \qquad (13.4.11)$$

Proposition 35 *For a prescribed combination of R_0 and u_{max}, the two-switch control (13.4.5) maximizes $E(T)$ if $T > T_0$. Proposition 32 applies if $T \leq T_0$ (so that $t_x \geq t_S$).*

For the special case of $u_{max} = 1$, the value of T_0 may be obtained by taking the limit of (13.4.11) as $u_\alpha \to 0$ to get

$$T_0 = 2 - \frac{1}{R_0} \quad (u_{max} = 1).$$

Results of some sample calculations for t_x (and previously calculated t_S) reported in Table 13.7 show that $E(T)$ is maximized by the two-switch control (13.4.5) only for the boldfaced cases in the table.

Note that t_x as given by (13.4.4) is independent of T while the dependence of t_S on T as given by (13.3.12) is linear and additive so that its value for a different T is obtained simply by adding $T - 2$ to the relevant value in Table 13.7. Graphs of $t_S - t_x$ are plotted as functions of $u_{max} - 1$ for several values of R_0 in Figure 13.3.

Table 13.7 t_x vs. t_S for $T = 2$.

$R_0 \backslash u_{max}$		0.505	0.55	0.75	1.1	1.5	2
0.55	t_x	4.652	1.329	0.348	**0.153**	**0.093**	**0.063**
	t_S	−7.303	−3.117	−0.773	**0.177**	**0.614**	**0.901**
0.75	t_x	7.124	3.081	1.151	0.572	**0.365**	**0.251**
	t_S	−7.303	−3.117	−0.773	0.177	**0.614**	**0.901**
1.25	t_x	8.285	4.125	1.880	1.054	0.713	**0.511**
	t_S	−7.303	−3.117	−0.773	0.177	0.614	**0.901**
1.5	t_x	8.494	4,324	2.043	1.178	0.811	**0.588**
	t_S	−7.303	−3.117	−0.773	0.177	0.614	**0.901**
2.0	t_x	8.729	4.550	2.239	1.335	0.940	**0.693**
	t_S	−7.303	−3.117	−0.773	0.177	0.614	**0.901**

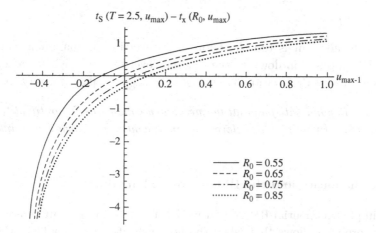

Fig. 13.3   Graph of $t_S(T = 2.5, u_{max}) - t_x(R_0, u_{max})$ vs. $u_{max} - 1$ for several values of R_0.

13.4.4   Summary of the optimal conversion strategy for $R_0 > 1/2$

For a fixed pair of $u_{\max}$ and T, the optimal conversion strategy for $R_0 > 1/2$ again depends on whether the constitution of the host cell can support a period of stationary solution that maximizes the Hamiltonian. But unlike $R_0 \leq 1/2$, the appropriate control depends principally on the magnitude of $u_{\max}$.

13.4.5   Case III — $u_{\max} \leq 1/2$

For this range of $u_{\max}$, $R_{gm}(t) > 1/2$ does not reach the stationary solution for any finite t. Proposition 31 applies with the optimal control $u_{op}(t) = u_{\max}$ for all t in $[0, T]$.

13.4.6   Case IV — $u_{\max} > 1/2$

For this complementary range of $u_{\max}$, the optimal conversion strategy depends on the relative magnitude of t_x and t_S in a way similar to the relative magnitude of t_0 and t_S.

13.4.6.1 $t_x < t_S$

A two-switch control $(u_{\max} - u_S - u_{\max})$ given by (13.4.5) with a segment of stationary solution is proved to maximize $E(T)$ by the method of most rapid approach (see Proposition 34). The corresponding RB population and adjoint function defined by (13.4.6) and (13.4.9) for $0 \leq t \leq t_x$ and by (13.3.10)–(13.3.11) for $t_S \leq t \leq T$. The two switch points t_x and t_S are determined by (13.4.7) and (13.3.12), respectively. The two end segments are bridged by an interval (t_x, t_S) of stationary solution.

13.4.6.2 $t_S \leq t_x$

The host cell life span is too short for the conversion at maximum capacity to drive the RB population down to the singular level early enough for a segment of stationary solution. Proposition 33 applies with $u_{op}(t) = u_{\max}$ for all t in $[0, T]$.

Remark 13 *Beyond satisfying all the necessary conditions for optimality imposed by the Maximum Principle, the relevant optimal control is shown to maximize $E(T)$ in all cases some less trivially by the method of most rapid approach.*

13.5   Mathematics and the Biology of Chlamydia

For the simple exponential RB growth model, the solution for our constrained optimization problem shows that when the host cell life span is sufficiently long, it would be in the Chlamydia bacteria's best interest (for maximum spread of infection) to delay conversion. This theoretical conclusion is consistent with intuition and conventional expectation and provides us with their mathematical justification.

A more refined model with finite carrying capacity shows that the optimal strategy for achieving the same objective may not always be bang-bang of the form (13.1.14). For a relatively large initial Chlamydia population of the order of the carrying capacity (a rather rare event in practice), the optimal strategy would be to convert immediately at the maximum allowable rate.

How do the two possible biological mechanisms that seem so plausible for implementing the bang-bang control of the simpler model (see Section 1 of this chapter) play out for this alternative strategy? If only those RB not in contact with the inclusion membrane should convert, the new strategy (of converting from the start) may be implemented immediately when the initial infecting RB population is (in rare cases) sufficiently large to have a portion of them away from the inclusion membrane (which is consistent with the theoretical requirement of a large R_0 for that strategy). In contrast, the other mechanism (that converts only after several rounds of division) would not be consistent with an immediate conversion strategy. However, it is possible that the second mechanism may be operating concurrently with the first.

Under the heading of other possible biological mechanisms not discussed explicitly earlier at the end of Section 1, one is based on the substantial difference in size between the RB and EB form of C. trachomatis. When a large RB population consumes all available space inside an inclusion, one way to create more space for further division would be for some RB to convert to considerably smaller EB to free up space for the remaining RB to proliferate. This "crowding" induced conversion can take place either from the start when R_0 is large or at a later stage after many rounds of division. It thus offers still another possible biological mechanism for implementing either theoretically optimal strategy.

As long as our models for the constrained maximization of the terminal EB population are linear in the "conversion rate constant" $u(t)$, there appears to be one unrealistic feature common among all possible optimal strategies from these models. When there is a switch from one temporally uniform control to another, the switch would be instantaneous and discontinuous (or bang-bang for brevity). Now an abrupt (discontinuous) change can take place when the switches are implemented by the biological mechanisms described previously and not inconsistent with the data from the experiments of [14]. For example, if conversion should take place only after many divisions of the initial infecting RB population and a new generation of RB becomes sufficiently small, then the conversion rate constant (the control) $u(t)$ would be zero during the first several divisions and switch to some finite value (instantaneously) only when a (sizable) group of later generation RB reaches a critically small size. That is certainly a bang-bang switch since a whole generation of RB would divide and switch pretty much at the same time. In other words, the theoretical occurrence of a bang-bang control is not inconsistent with the biology of C. trachomatis.

What may seem problematic however is the fact that the various finite jump switches in the theoretical optimal strategies from these models can only be between the two corner controls (0 to u_{max}) or between a corner control and the singular control (0 to 1/2 or 1/2 to u_{max}). A switch of 0 to 1/2 (the singular control normalized by the RB growth rate constant α) may not be realistic even if non-smoothness in the change in conversion rate is not an issue. A casual examination of the experimental data reported in [14] seem not to support such "large" abrupt changes. In other words, the issue is not whether a bang-bang strategy is consistent with the experimental data. Not only it could be, a bang-bang strategy is also compatible with the known plausible biological processes of conversion of C. trachomatis discussed in earlier sections of this paper. As delineated, the issue arises from the apparent "large" magnitude (relative to the experimental data) of the conversion rate change at the switch, not its discontinuity or abruptness. In particular, a switch from 0 to 1/2 does not appear to be supported by the experimental evidence reported in [14]. For that perceived incompatibility, a possible resolution depends on the actual biological process(es) underlying the C. trachomatis life cycle.

Since the models investigated do not specify the value of u_{max}, they implicitly leave its specification (hence its magnitude) to the investigators based on what is known about the bacteria. The experimental data obtained in [14] are precisely what is needed for estimating u_{max}. There remains the question: What if u_{max} estimated from the data should be smaller than 1/2 (the normalized singular control)? This situation has already arisen earlier in Section 5.1 and has been addressed: it would eliminate the option of a singular solution phase in the optimal strategy in that case. Altogether then, a bang-bang control with u_{max} determined from available data should not be incompatible with the findings reported in [14].

On the other hand, there is still the possibility that the plausible biological processes discussed earlier in this paper should prove not to be responsible for the Chlamydia life cycle. In addition, suppose the actual biological mechanism for implementing the conversion from RB to EB involves a gradual and smooth transition from proliferation to conversion (and the discontinuous changes in the experimental data are due to limitations of measurement techniques). In that case, a bang-bang control would be truly inappropriate and the models that led to such an optimal conversion strategy would require modifications to incorporate features of the actual biological processes responsible for the Chlamydia life cycle (once we know what they entails).

It may also be necessary for the models to adjust for the noisy environment internal to and outside the host cell that creates a natural variability in the timing and/or rate of differentiation. We should also be mindful of mammalian inclusions for the C. trachomatis bacteria changing size over time and consider a *time-dependent* carrying capacity that may provide a vehicle to gradual transition. Still another important factor is the interplay between population size and the time of host cell death as a large Chlamydial population can actively cause the host to lyse in order to spread the infection to other cells. In any case, more realistic (deterministic or stochastic) models may be needed if a more gradual or less abrupt transition between proliferation and conversion is expected for the optimal strategy in maximizing the terminal EB population. In all cases, additional features would have to be added to models examined herein or existing features modified to address the issue raised on the fundamentally and intrinsically important theoretical results deduced from the models herein for maximizing the spread of C. trachomatis infection.

Chapter 14

Genetic Instability and Carcinogenesis

14.1 Genetic Instability is a Two-Edge Sword

In this chapter, we undertake the study of a class of phenomena that involves the intertwining of the issues of evolution, instability, optimization and control. The class of phenomena of interest is the initiation of various types of cancer. An important feature of our problem is the lack of an explicit analytical solution of the model ODE system in terms of elementary or special functions. For this reason, a complete characterization of the solution for these problems requires a different approach to the process of extracting information from the mathematical model. It requires abstract analysis that often makes use of inequalities deduced from the BVP associated with the process of optimization (see [21, 35]).

To discuss carcinogenesis, we need to introduce some relevant biological processes and terminology. *Homeostasis* is a process for maintaining relative constancy in a biological system. In particular, homeostatic control of cells maintains a dynamic equilibrium (or acceptable variations) of cell populations in (organs of) multicellular organisms. Sometimes cells can break out of homeostatic control and enter a phase of abnormal expansion. Cancer is a manifestation of such uninhibited growth, which results when cell lineages lose the ability to maintain a sufficient rate of *apoptosis*, i.e., programmed cell death, to counterbalance the rate of *mitosis*, i.e., cell division. Mechanisms of homeostatic control can be disrupted by *genetic mutations*, to which every cell is prone to some extent. A highly elevated susceptibility to mutations is known in oncology as *genetic (on genomic) instability* and, being closely related to carcinogenesis, is of much interest to cancer researchers (see [10] and references therein).

Genetics instability is known to involve the following two (among other) competing effects on carcinogenesis: an increased frequency of cell deaths resulting from deleterious mutations, and an increased frequency of mutations producing cancerous cells. The former effect impedes cancer growth; the latter accelerates it. This gives rise to the question, *What "amounts" of instability are most favorable to the onset of cancer?* The first mathematical models for this class of problems [9] were based on two conceptual premises, (a) the Darwinian view of a cell colony as a collection of phenotypes struggling for survival and predominance, and (b) the degree of genetic instability in the colony can be (hypothetically) set to any level and remain constant in time. A colony of cells was regarded as a species undergoing a birth-and-death process, interpreted as a Darwinian microevolution, and subject to genetic instability. The degree of genetic instability was defined quantitatively as the probability of genetic mutation of a specified type per cell, per mitosis. The "microevolutionary success" of the species was defined as the ability to produce a cancerous population of a given size M, and was measured as the inverse of the time required to reach the target population size. The question how fast (if ever) the target cancerous population size is reached was thus recast as a problem in optimization. The first results showed that "too much" instability impedes the growth of the colony by increasing the cell death rate (a result of too many deleterious mutations in cells), while "too little" instability impedes the growth by lowering the rate of acquisition of cancerous mutations. In the same paper, an optimal time invariant level of genetic instability that maximizes the rate of progression was found. This level, defined as the probability of chromosomal loss per cell division, agreed well with available in vitro experimental measurements and turned out to be a robust mathematical result: it depended only logarithmically on the relevant parameter values, and its order of magnitude was consistent with the data (see [10] and references therein).

The restriction that the level of genetic instability remain constant in time, however, served only to facilitate the first steps in the mathematical modeling of the problem. There is growing biological evidence that the degree of genetic instability is high at relatively early stages of carcinogenesis, and decreases as cancer progresses. In [10], we removed the restriction of a time independent level of genetic instability and allowed it to be a function $p(t)$ of time within a biologically admissible range. For an appropriate mathematical model of the new problem, we distinguished among several types of cells: normal cells, cancerous cells with homeostatic control impaired completely, and in some cases also cells containing intermediate mutations but still obeying homeostatic control. The entire cell colony would undergo birth-and-death process and the evolution of these different cell populations in the colony was taken to depend on the mutation rate $p(t)$ due to genetic instability that acts as a control for the population growth. Of interest was an *admissible control* $p(t)$ (or its normalized form $u(t)$ as defined in (14.2.6) below) which steers the cancerous cells to a target population M in the shortest possible time.

This shortest time problem was discussed in [10] for two different mechanisms of homeostatic control loss. The dynamics of the first mechanism to activate genetic instability, *activation of an oncogene*, is described and mathematically modeled in this chapter. The control-theoretic analysis leading to the final solution is developed in this chapter for non-convex death rate (as a function of mutation rate) to illustrate the method of solution for this type of problems. It should be noted at this point that our goal here is mainly to understand the experimentally observed behavior of developing cancer. In this sense, it differs from the conventional motivation for undertaking an optimal control analysis in engineering and other areas of biosciences, which is to find a control that can be implemented to steer the outcome toward some optimal objective.

Before we embark on a discussion on the optimal control solution of the shortest time problem with time-varying mutation rate, we first discuss the same problem but restricting the mutation rate to be constant over the period of carcinogenesis. In so doing, we offer an alternative model to the probabilistic approach of [9] to deduce the optimal mutation rate from the ODE system governing the evolution of the cancerous population.

14.2 Activation of an Oncogene

14.2.1 *A one-step system*

Consider a cell colony undergoing a birth and death process with a time-varying mutation rate within a biologically admissible range. Cells go through a sequence of mutations until an advantageous, i.e., cancer-favorable, phenotype is achieved. Mutation rate can have two effects. On the one hand, an increased mutation rate can lead to a faster production of the advantageous mutants, thus accelerating the growth of the colony. On the other hand, a high mutation rate reduces the fitness of the mutated cells and thus leads to a higher death toll in the population. To delineate the net consequence of these two competing effects, we consider in this paper a colony of normal cells in steady state prior to the onset of mutation. Its offspring mutants evolve to overcome selection barrier for division rate by activating an oncogene through one molecular event, genetic or epigenetic. Denote by X_1 the population of cells that have not undergone cancerous mutations, and by X_2 the mutated type. The probability for a cell to acquire an inactivating mutation of a particular gene upon a cell division is denoted by $\bar{\mu}$; this quantity is called the "basic mutation rate". The probability p is an additional transformation rate resulting from genetic instability. This quantity measures the degree of genetic instability in cells. It is low in stable cells (cells without *chromosomal instability* or CIN), but it can be highly elevated in chromosomally unstable cells. Effectively, if $p \ll \bar{\mu}$, then there is no genetic instability; $p \gg \bar{\mu}$ means a genetically unstable cell population. Both probabilities $\bar{\mu}$ and p are measured per gene, per cell division. For the present investigation, p is allowed to vary with time. The goal is to find a strategy $p(t)$ which maximizes the growth of cancer with the mutation rate $p(t)$ due to genetic instability being the control mechanism in our optimal control problem.

The dynamics of the cell populations is modeled as follows. Cells reproduce and die, and the rate of renewal is normalized to be 1 for type X_1. In the absence of cancerous mutants X_2, X_1, is taken to obey the conventional logistic growth law. The mutants X_2 expand at the rate $a > 1$ (and usually $\gg 1$). With $(.)' = d(.)/dt$, these processes translate into the following two ordinary differential equations (ODE) describing the rate of change of the two cell populations,

$$X_1' = -\,(\bar{\mu}+p)\,X_1 + (1-d)X_1 - \phi X_1, \qquad (14.2.1)$$

$$X_2' = (p+\bar{\mu})X_1 + a(1-d)X_2 - \phi X_2 \qquad (14.2.2)$$

where

$$\phi = (1-d)X_1/N, \qquad (14.2.3)$$

subject to the initial conditions

$$X_1(0) = N, \qquad X_2(0) = 0 \qquad (14.2.4)$$

for a prescribed death rate, $d(p)$, that increases with mutation rate $p(t)$.

In both rate of change equations (14.2.1) and (14.2.2) for normal and cancerous cell population, respectively, we note the following:

(1) The first term on the right-hand side is the combined loss/gain rate of the relevant cell population due to mutation of normal cells.
(2) The second term is the linear proliferation rate net apoptosis for the relevant cell population.
(3) The third term with the factor ϕ provides the growth limiting factor leading to logistic growth in the absence of mutation, and accounts for the homeostatic control present in a population of X_1 cells.

Choices of ϕ other than (14.2.3) are also possible (see [10]). With a fraction of normal cell population starting to mutate into cancerous mutants, X_2 cells break out of regulation and enter a phase of abnormal growth (without limit except for failures of physiological functions). Depending on its net growth rate, cancerous mutants may or may not reach the target cancer cell population.

14.2.2   *Dependence of death rate on mutation rate*

In our models, the death rate, d, is taken to be a function of the mutation rate p due to genetic instability. If p is small, then chromosome losses do not happen. If p is large, a cell often loses chromosomes, which results in an increased death rate. Therefore, in general, the function $d(p)$ is an increasing function of p. An example of this dependence of the death rate on p is

$$d(p) = \frac{d_m}{u_m^\alpha}\{u_m^\alpha - (u_m - (p - p_{\min}))^\alpha\} = d_m\,[1 - (1 - u)^\alpha], \quad (\alpha > 0) \quad (14.2.5)$$

where

$$u_m = p_{\max} - p_{\min}, \qquad u = \frac{1}{u_m}(p - p_{\min}) \qquad (14.2.6)$$

with $p_{min} \leq p \leq p_{max}$. The quantities p_{min} and p_{max} define a biologically relevant range of the mutation rate, p, and the nonnegative u is a normalized *gross
chromosomal change (mutation) rate* which satisfies the inequality constraint

$$0 \leq u \leq 1. \tag{14.2.7}$$

The constant d_m defines the magnitude of the death rate, and is taken to be in the
interval $[0, 1]$. The motivation for the particular form of dependency (14.2.5) on
positive α, the exponent in that expression for $d(p)$, was given in [10]. It was found
in [35] that the optimal control is characteristically different for two distinct cases
of $\alpha \geq 1$ and $\alpha < 1$ with the different effects of two types of convexity (non-convex
and strictly convex) carried over to general form of $d(p)$. In this chapter, we limit
the discussion of the method for deducing the optimal solution to the non-convex
case.

Every normalized genetic instability rate $u(t)$ that satisfies the constraints
(14.2.7) is an *admissible control*, determines a growth process for the cell colony.
We shall seek the choice of an admissible control $u(t)$ in the range $0 \leq u \leq 1$ that
allows the cancerous population to reach a given size, M, in the shortest possible
time. The specified terminal population size M is called the *target*. An admissible
control $U(t)$ that steers the cancerous population to the target faster than any other
control is an *optimal control*.

In the simplest case first considered in [9], the class of admissible controls, $u(t)$,
was restricted to constant functions. Then, the result of the optimization problem
is a single normalized mutation rate value, U (or $\bar{p}$ in un-normalized form), which
depends on the parameters of the system. However, shorter growth times can be
achieved if we allow p (and hence u) to be a function of time. It seems intuitive,
and is evident from experimental results, that higher initial and lower subsequent
values of $u(t)$ will better facilitate the growth. We will show presently that our
mathematical model does lead to an optimal mutation rate of this type and how
the variation of $U(t)$ with time depends on the convexity of the death rate function
$d(u)$.

14.2.3 *Dimensionless formulation*

To reduce the number of parameters in our problem, we let the population size of
normal (harmless) cells at time t be normalized by its initial population size N and
denote it by $x_1(t) = X_1(t)/N$. We let $x_2(t) = X_2(t)/M$ be the population size of
cancerous mutant cells normalized by its final target population size M. In terms
of $\{x_k(t)\}$, the two ODEs (14.2.1)–(14.2.2) governing the time evolution of the two
cell populations become:

$$x_1' = -(\mu + u_m u)x_1 + [1 - d(u)](1 - x_1)x_1 \equiv g_1(x_1, x_2, u), \tag{14.2.8}$$

$$x_2' = \frac{1}{\sigma}(\mu + u_m u)x_1 + [1 - d(u)](a - x_1)x_2 \equiv g_2(x_1, x_2, u), \tag{14.2.9}$$

where $(\ )' = d(\ )/dt$, $\sigma = M/N \gg 1$, $a \geq 2$, $10^{-1} \leq u_m \equiv p_{max} - p_{min} \leq 1$ and
$0 < \mu \equiv \bar{\mu} + p_{min} \ll 1$. (Typically, we have $\sigma \geq 10$, $a = 2$, and $\mu = 10^{-1} \ll u_m$.)

For simplicity, we will first focus our discussion herein on the death rate of the form (14.2.5) with $d_m = 1$, i.e.,

$$d(u) = 1 - (1 - u)^\alpha \tag{14.2.10}$$

for some real parameter α. An in-depth discussion for the more complicated (but realistic) case of $0 \leq d_m < 1$ as well as general $d(u)$ distinguished only by its convexity can be found in [10, 35]. As we shall see, the three cases $\alpha > 1$, $\alpha = 1$ and $\alpha < 1$ for the two specific types of $d(u)$ will have to be treated separately.

The two ODEs (14.2.8) and (14.2.9) are subject to the following three auxiliary conditions

$$x_1(0) = 1, \qquad x_2(0) = 0, \qquad x_2(T) = 1 \tag{14.2.11}$$

where T is the (unknown) terminal time when the cancerous mutant cell population reaches the target size. The evolution of the normalized mutation rate $u(t)$ is subject to the inequality constraints

$$0 \leq u \leq 1. \tag{14.2.12}$$

As we shall see, admissible comparison controls are necessarily allowed to be among the class of piecewise continuous functions on the interval $[0, T]$, broader than the class of C^1 functions originally prescribed for the principal unknown $y(t)$ in the basic problem for the calculus of variations. This class of feasible control is denoted by Ω. Needless to say, the two normalized cell populations are also subject to the non-negativity constraints

$$x_1 \geq 0, \qquad x_2 \geq 0. \tag{14.2.13}$$

14.3 Shortest Time by a Constant Mutation Rate

14.3.1 *An alternative description of the evolving populations*

In the general problem (14.2.8), (14.2.9), (14.2.10), (14.2.11) and (14.2.12), the terminal time when the cancerous population reaches the target size is not known in advance and is a part of the solution for the problem. As such we have what is known as a free boundary problem. It is conventional to normalize or scale the time variable by the unknown terminal time T, i.e., by setting $\tau = t/T$, to result in a BVP with fixed end points and with T as a parameter in the ODE system. For the purpose of numerical solutions for the problem, we take the alternative approach to a fixed domain problem by using the cancerous cell population x_2 as the independent variable and both the normal cell x_1 population and the time progression as functions of x_2. In that case, the two ODEs for the two populations, (14.2.8) and (14.2.9), may be rewritten as

$$\frac{dx_1}{dx_2} = \frac{g_1(x_1, x_2, u)}{g_2(x_1, x_2, u)}, \qquad \frac{dt}{dx_2} = \frac{1}{g_2(x_1, x_2, u)}, \tag{14.3.1}$$

for $x_1(x_2; u)$ and $t(x_2; u)$ with the two initial conditions in (14.2.11) written as
$$x_1(0; u) = 1, \qquad t(0; u) = 0. \tag{14.3.2}$$
The terminal condition $x_2(T; u) = 1$ can be used in the form
$$t(x_2 = 1; u) = T(u) \tag{14.3.3}$$
to give the terminal time T as a function of u.

Note that for a given $u(t)$, the two ODEs in (14.3.1) are uncoupled in either formulation. The first ODE and the first initial condition in (14.3.2) can be solved first with the result used in the second ODE with the second initial condition in (14.3.2) for the determination of $t(x_2, u)$. However, as long as there is not an explicit solution of the ODE for x_1, this observation does not offer much of an advantage for a numerical solution for the problem.

14.3.2 Numerical solution for a time-invariant control

14.3.2.1 A brute force method

Mathematical software (such as Mathematica and MatLab) exist for efficient, fast and accurate numerical solutions of IVP such as the one defined by (14.3.1)–(14.3.2). Hence, we may simply compute the solution of the IVP for a set of representative values of the constant mutation rate over some coarse mesh of the interval $[0, 1]$ in order to get an approximate location T_c of the minimum terminal time T and the corresponding value for the optimal time-invariant mutation rate U_c of the optimal mutation rate U. We can then focus our search in a small neighborhood of U_c for a more accurate approximation U_a of U within a prescribed error tolerance (by simple bisections for example).

To illustrate, we apply this brute force method to a system with $\alpha = 2$, $u_m = 1$, $\mu = 0.1$, $a = 2$ and $\sigma = 10$. In the first table below, we obtain the values $\{T_k\}$ for 11 values of the mutation rate $\{u_k\}$ in the interval $[0, 0.250]$ which shows a local minimum of $T_c = 4.1566\ldots$ around $U_c = 0.125\ldots$ (Additional computation with the same increment of the mutation rate $\Delta u = 0.025$ for the interval $[0.250, 1]$ show that $T_c = 4.1566\ldots$ is in fact an approximate global minimum for u in $[0, 1]$.) In the second table, we obtain the values $\{T_k\}$ for 11 values of the mutation rate in the interval $[0.1000, 0.1250]$. We see from the table a more accurate local minimum terminal time to target population of $T_a = 4.15586\ldots$ (to five significant figures) around $U_a = 0.1175$.

Coarse Mesh with $\Delta u = 0.025$ in the $[0, 0.250]$

k	1	2	3	4	5	6
u_k	0	0.025	0.050	0.075	0.100	0.125
T_k	4.3746	4.2786	4.2162	4.1782	4.1593	4.1566

k	7	8	9	10	11
u_k	0.150	0.175	0.200	0.225	0.250
T_k	4.1683	4.1936	4.2322	4.2843	4.3505

Finer Mesh with $\Delta u = 0.0025$ in the interval $[0.100, \; 0.125]$

k	1	2	3	4	5	6
u_k	0,1000	0.1025	0.1050	0.1075	0.1100	0.1125
T_k	4.1593	4.1584	4.1576	4.1569	4.1564	4.1561

k	7	8	9	10	11
u_k	0.1150	0.1175	0.1200	0.1225	0.1250
T_k	4.15590	4.15586	4.15597	4.1562	4.1566

14.3.2.2 A single pass solution scheme

If a single pass solution scheme is preferred, we note that a requirement for an optimal constant mutation rate U to minimize $T(u)$ is the stationary condition

$$\left[\frac{dT}{du}\right]_{u=U} = 0. \tag{14.3.4}$$

To make use of this condition, we differentiate both ODE (14.3.1) partially with respect to u to get

$$\frac{dv}{dx_2} = \frac{1}{g_2^2(x_1, x_2, u)}\left\{g_2\left(v\frac{\partial g_1}{\partial x_1} + \frac{\partial g_1}{\partial u}\right) - \left(v\frac{\partial g_2}{\partial x_1} + \frac{\partial g_2}{\partial u}\right)\right\}, \tag{14.3.5}$$

$$\frac{dw}{dx_2} = -\frac{1}{g_2^2(x_1, x_2, u)}\left\{v\frac{\partial g_2}{\partial x_1} + \frac{\partial g_2}{\partial u}\right\} \tag{14.3.6}$$

where

$$v = \frac{\partial x_1}{\partial u}, \qquad w = \frac{\partial t}{\partial u}. \tag{14.3.7}$$

The ODEs (14.3.1), (14.3.5) and (14.3.6) form a fourth order system for the four unknowns $\{x_1, t, v, w\}$ and are augmented by the two initial conditions (14.3.2) and two new initial conditions

$$v(0; u) = w(0; u) = 0 \tag{14.3.8}$$

which follow from differentiating partially the conditions of (14.3.2) with respect to u. For a fixed u, the fourth order system is uncoupled as the four equations can be solved sequentially starting with the first ODE of (14.3.1). This observation is of no significant consequences if there is no explicit solution for the system in terms of known functions.

After obtaining the solutions $t(x_2; u)$ and $w(x_2; u)$, we observe

$$t(1; u) = T(u), \qquad w(1; u) = \frac{dT}{du}. \tag{14.3.9}$$

The stationary condition (14.3.4) then requires

$$w(1; U) = 0. \tag{14.3.10}$$

If there is an explicit expression for $w(x_2; u)$, then the time invariant optimal mutation rate U is a zero of $w(1; u)$.

When an explicit expression is not available and the ODE system is to be solved numerically, we have well-known iterative schemes to determine U that satisfies (14.3.10). To formulate a single pass method of solution instead, we introduce artificially the ODE

$$\frac{du}{dx_2} = 0 \qquad (14.3.11)$$

to reflect the fact that u is time invariant. The five ODE, (14.3.11) plus the original four ODE for $\{x_1, t, v, w\}$, augmented by the four initial conditions (14.3.2) and (14.3.8) and the terminal condition $w(1; u) = 0$ constitute a two-point BVP for the five unknowns $\{x_1, t, v, w\}$ and u. Such a BVP can be solved by any existing mathematical software such as Mathematica, Maple or MatLab. Assuming the second order condition

$$\left[\frac{\partial^2 t}{\partial u^2} \right]_{x_2=1} > 0$$

is met, the (constant) solution $u = U$ is the time invariant optimal mutation rate for the problem.

It should be noted that any BVP solver for nonlinear ODE typically involves an iterative solution scheme for a series of linearized problems. As such, the solution scheme developed above may only be single pass in appearance but still involves iterations in the solution process. It differs from our bi-section type iterative scheme of the previous subsection only in the sophistication of the iterative algorithm employed. Furthermore, the ODE (14.3.11), introduced artificially to form a two-point BVP would generally lead to a singular Jacobian in most iterative solution schemes and require some finesse to determine the actual solution for our problem.

14.4   The TOP

14.4.1   *Fastest time to cancer*

Clinical evidence (see [10] and references therein) suggests that a constant mutation rate first investigated in [9] and assumed in the previous section is not consistent with reality for several types of cancer development. For breast, colon and lung cancer, there is generally a high level of mutation rate at the early stage followed by a relatively low rate near the end. It appears that the early high level of genetic instability is to generate a large pool of cancerous cells while a lower rate at later stage allows the cancerous cells to fast proliferate by their natural growth rate. As such, we investigate in the rest of this chapter the consequences of permitting the mutation rate u to vary with time. Of interest is the determination of an optimal mutation rate $U(t)$ to enable the cancerous mutated cells to reach the target population in the fastest time possible. In other words, we wish to find $U(t)$ for which the terminal time T is a minimum subject to the cell dynamics (14.2.8)–(14.2.9), the two initial conditions (14.2.11), the terminal condition $x_2(T) = 1$ and the inequality constraints (14.2.12) and (14.2.13).

This kind of shortest time problem is known as the *time optimal problem* (TOP) in optimal control theory. To cast the problem in the context of this theory, we make use of the Maximum Principle and introduce the rather artificial performance index

$$J = \int_0^T 1 \, dt, \tag{14.4.1}$$

so that the problem becomes one of minimizing J subject to the same set of growth dynamics and auxiliary conditions with $T > 0$ to be determined as a part of the solution.

14.4.2　The Maximum Principle

We wish now to determine the optimal mutation rate $U(t)$ among the admissible (feasible) time-varying functions of class Ω (previously defined to be PWS and subject to the constraining inequality (14.2.12)). We do this by introducing an appropriate Hamiltonian function

$$H = 1 + \lambda_1(t) g_1(x_1, x_2, u) + \lambda_2(t) g_2(x_1, x_2, u) \tag{14.4.2}$$

$$= 1 + \frac{u_m u + \mu}{\sigma} R(t) + \{1 - d(u)\} D(t)$$

where

$$R(t) = x_1(\lambda_2 - \sigma\lambda_1), \qquad P = \lambda_1 x_1 + \lambda_2 x_2, \tag{14.4.3}$$

$$D(t) = (1 - x_1)P + \lambda_2 x_2(a - 1).$$

In the expression for H, λ_1 and λ_2 are the two PWS *adjoint* (or co-state) functions for the problem chosen to satisfy the two *adjoint differential equations*,

$$\lambda_1' = -\left(\lambda_1 \frac{\partial g_1}{\partial x_1} + \lambda_2 \frac{\partial g_2}{\partial x_1}\right) \tag{14.4.4}$$

$$= -\frac{(u_m u + \mu)}{\sigma}(\lambda_2 - \sigma\lambda_1) + \{1 - d(u)\} \{P - \lambda_1(1 - x_1)\},$$

$$\lambda_2' = -\left(\lambda_1 \frac{\partial g_1}{\partial x_2} + \lambda_2 \frac{\partial g_2}{\partial x_2}\right) = -\lambda_2 \{1 - d(u)\} (a - x_1), \tag{14.4.5}$$

and (for the given auxiliary conditions on the state variable x_1 and x_2) one *adjoint* (*Euler*) *boundary condition*:

$$\lambda_1(T) = 0 \tag{14.4.6}$$

for our problem [2, 33].

With the admissible controls restricted to be piecewise continuous and to satisfy the inequality constraint (14.2.12), any *optimal mutation rate* $U(t)$ *(known as optimal control* in the literature in this field), has at worst finite jump discontinuities in $[0, T]$. It follows from the way $u(t)$ appears in the state and adjoint ODE, (14.2.8), (14.2.9), (14.4.4) and (14.4.5), that the state and adjoint variables are continuous throughout $(0, T)$. For a solution by the Maximum Principle introduced in the previous chapter, every *optimal mutation rate function* $U(t)$ of the shortest time problem is to satisfy the following necessary conditions for optimality:

(1) Four continuous functions $\{X_1(t), X_2(t), \Lambda_1(t), \Lambda_2(t)\}$ exist and satisfy the Hamiltonian system of four differential equations (14.2.8), (14.2.9), (14.4.4) and (14.4.5) and four auxiliary conditions in (2.1.7), (9.1.6) and (14.4.6) for $u = U(t)$.

(2) With $u = U(t)$, the terminal time T satisfies a free end condition (10.4.5) of Chapter 10 which in terms of the Hamiltonian (after a Legendre transformation) becomes

$$[H]_{t=T} = [1 + \Lambda_2 \bar{g}_2]_{t=T} = 0, \qquad (14.4.7)$$

with $\bar{g}_2 = g_2(X_1, X_2, \Lambda_1, \Lambda_2, U)$ after simplification by the Euler BC (14.4.6).

(3) For all t in $[0, T]$, the Hamiltonian achieves its minimum with $u = U(t)$, i.e.,

$$H(X_1(t), X_2(t), \Lambda_1(t), \Lambda_2(t), U(t)) = \inf_v [H(X_1(t), X_2(t), \Lambda_1(t), \Lambda_2(t), v)],$$
$$(14.4.8)$$

over all v in the set of admissible controls Ω restricted by (14.2.12).

(4) If there should be a finite jump discontinuity in the optimal control $U(t)$ at the instant T_s, the Hamiltonian is required to be continuous at T_s (known also as the "*switch condition*"):

$$[H]_{t=T_s-}^{t=T_s+} = 0. \qquad (14.4.9)$$

(5) If an interior control $U_i(t)$ determined by the necessary (stationary) condition

$$\left[\frac{\partial H}{\partial u}\right]_{u=U_i} = \left[\frac{u_m}{\sigma} R(t) - D(t) d^{\cdot}(u)\right]_{u=U_i} = 0, \qquad (14.4.10)$$

where $(\ )^{\cdot} = d(\ )/du$, does not violate the inequality constraints $0 \le u(t) \le 1$, the optimality condition (14.4.8) for H (as a function of u) would have $U_i(t)$ as the optimal control,

$$U(t) = U_i(t).$$

In principle, the stationary condition (14.4.10) can be solved for $U_i(t)$ in terms of $\{x_1, x_2, \lambda_1, \lambda_2\} = \{X_1(t), X_2(t), \Lambda_1(t), \Lambda_2(t)\}$ so that the four state and adjoint ODE (14.2.8), (14.2.9), (14.4.4) and (14.4.5) along with the four auxiliary conditions (14.2.11) and (14.4.6) constitute a two-point BVP for the state and adjoint functions. Such a problem may be solved numerically if necessary by available mathematical software such as MatLab, Mathematica or Maple. Together with the expression from (14.4.10) for $U_i(t)$, we have determined the solution of our TOP, numerically if necessary.

(6) If the interior control violates the inequality constraints $0 \le u(t) \le 1$, the optimality condition (14.4.8) for H is to be used to determine an appropriate corner control (the upper corner control $u_g(t) = 1$ or the lower corner control $u_\ell(t) = 0$ for the present problem) where the interior control is not applicable.

(7) If the stationary condition (14.4.10) does not involve the control u, the condition still determines a singular solution for the problem. The optimality condition (14.4.8) for H is to be used to determine the applicability of the singular solution for the entire or a part of the solution domain and helps to determine an appropriate corner solution where the singular solution is not applicable.

14.4.3   *Some preliminary results on adjoint functions*

As in classical optimization of functions, the (most desirable) stationary solution $U_i(t)$ may not be appropriate for some problem. One such occasion would be when it violates the constraining inequality (14.2.12). Another would be when it maximizes instead of minimizes the Hamiltonian as it is required for our problem. On such occasions, our experience with classical optimization of functions tells us that we should choose the optimal control from the boundary of the control set Ω. For the present problem, this amounts to choosing between the two extreme values of the allowable mutation rate: the *upper corner control* $u = 1$ and the *lower corner control* $u = 0$. The instrument at our disposal for a proper selection is the optimality condition (14.4.8). For this condition to produce the result sought, we need to develop some preliminary tools in this and the next subsection.

To help decide the appropriate corner control near the terminal time, we establish the following properties for the adjoint function $\lambda_2(t)$:

Lemma 4 $\lambda_2(t) < 0$ and $\lambda_2'(t) \geq 0$ $(0 \leq t \leq T)$ *for any admissible control $u(t)$.*

Proof Given $0 \leq d(u) \leq 1$, we have from (14.4.7) and (14.4.6)

$$\lambda_2(T) = -\left[\frac{1}{g_2(x_1, x_2, u)}\right]_{t=T}$$

$$= -\left[\frac{\sigma}{(\mu + u_m u)x_1 + \sigma x_2(1 - d(u))(a - x_1)}\right]_{t=T} < 0 \quad (14.4.11)$$

(since both numerator and denominator of (14.4.11) are positive) with

$$[\lambda_2']_{t=T} = -[\lambda_2(a - x_1)\{1 - d(\bar{u})\}]_{t=T} \geq 0 \quad (14.4.12)$$

by (14.4.5). The two conditions (14.4.11) and (14.4.12) implies $\lambda_2(t) < 0$ in some neighborhood of $t = T$. The lemma follows from these local results and the ODE (14.4.5) which requires $\lambda_2' \geq 0$ for $0 < t < T$. $\square$

The result above precludes the optimal mutation rate $U(t)$ to be the upper corner control near the end as shown by the following proposition:

Proposition 36 *An optimal control is not an upper corner control for all t in an interval $(T_0, T]$ for some $T_0 < T$, i.e., $U(t) < 1$ for $T_0 < t \leq T$.*

Proof At the terminal time T, we have $\lambda_1(T) = 0$ and therewith

$$H(t = T) = \begin{cases} 1 + \frac{1}{\sigma}\lambda_2(T)\{\mu x_1(T) + \sigma[a - x_1(T)]\} & (u(T) = 0) \\ 1 + \frac{1}{\sigma}\lambda_2(T)x_1(T)(\mu + u_m) & (u(T) = 1) \end{cases}. \quad (14.4.13)$$

Relevant biological parameter value ranges are $a \geq 2$, $\sigma \gg 1$, $u_m \lesssim 1$ and $\mu \ll 1$. With $a - x_1(T) > a - 1 > 1$, we have $\sigma(a - 1) > u_m \geq u_m x_1(t)$ for any admissible solution so that

$$[H(t = T)]_{u=0} < [H(t = T)]_{u=1} \quad (14.4.14)$$

given $\lambda_2(t) < 0$ in $[0, T]$ by Lemma 4. The proposition follows from (14.4.14) and the continuity of state and adjoint variables.   □

Remark 14 *Note that Proposition 36 applies whether $U_i(t)$ is applicable. However, it is most useful when the interior control is not appropriate for the problem.*

From $\lambda_2(T) < 0$ and $\lambda_1(T) = 0$, we have $\lambda_2 - \sigma\lambda_1 < 0$ at the terminal time. With this observation, we obtain the following result useful in subsequent developments:

Lemma 5 *For some interval (T_c, T) adjacent to the terminal time T, we have: (i) $(\lambda_2 - \sigma\lambda_1) < 0$ for either corner solution, and (ii) $\lambda_1(t) > 0$ for a lower corner control, and (iii) $\lambda_1(t) < 0$ for an upper corner control.*

Proof Property (i) follows immediately from $\lambda_2(T) - \sigma\lambda_1(T) = \lambda_2(T) < 0$ (see (14.4.11)) and the continuity of the adjoint variables. For the remaining two properties, we note that the first adjoint ODE requires

$$\lambda_1'(T) = -\frac{\lambda_2(T)}{\sigma}\{(u_m u + \mu) - \sigma[1 - d(u)]\}_{t=T}$$

which is positive for $u = 1$ and negative for $u = 0$ (given $\sigma \gg (\mu + u_m) = O(1)$). By continuity of the adjoint variables, λ_1 is increasing (toward $\lambda_1(T) = 0$) and hence $\lambda_1(t) < 0$ for $u = 1$ or decreasing (toward $\lambda_1(T) = 0$) and hence $\lambda_1(t) > 0$ for $u = 0$, both at least in a small interval adjacent to the terminal time T. Take (T_c, T) to be the smallest of the relevant intervals for which $\lambda_2 - \sigma\lambda_1 < 0$.   □

Another important property of the state and adjoint functions is that the combination $P(t)$ defined in (14.4.3) is always negative:

Lemma 6 *i)* $P(t) = P(T)E(t, T) = P(T)e^{-\Psi(t)}$, *with*

$$\Psi(t) = \int_t^T [1 - d(u(\tau))]\, x_1(\tau)d\tau.$$

ii) $P(t) < 0$ *for all t in $[0, T]$.*

Proof It is straightforward to use the state and adjoint ODE to show that

$$P'(t) - \lambda_1 x_1' + \lambda_2 x_2' + \lambda_1' x_1 + \lambda_2' x_2 = [1 - d(u)]\, x_1 P$$

or

$$P(t) = P(T)e^{-\Psi(t)}, \qquad \Psi(t) = \int_t^T [1 - d(u(\tau))]\, x_1(\tau)d\tau$$

which proves part *i)* of the lemma. With

$$P(T) = \lambda_1(T)x_1(T) + \lambda_2(T)x_2(T) = \lambda_2(T) < 0,$$

part *ii)* of the lemma follows.   □

14.4.4   Vanishing Hamiltonian for our TOP

The following lemma, related to a well-known result for autonomous systems, provides another important tool for determining the optimal solution.

Proposition 37 *For the optimal growth of the cancerous mutant cell population induced by an optimal control $U(t)$ (that results in the shortest time to the target mutant cell population), the Hamiltonian (14.4.2) of our problem vanishes for all t in $[0,T]$, i.e., $\bar{H} \equiv [H]_{u=U(t)} = 0$, for all t in $[0,T]$.*

Proof Except for locations of simple jump discontinuities of the control $u(t)$, we can differentiate H with respect to time to get

$$\frac{dH}{dt} = \frac{\partial H}{\partial u}\frac{du}{dt}, \qquad (14.4.15)$$

where we have made use of the Hamiltonian structure of the four relevant state and adjoint ODE to eliminate terms involving derivatives of state and adjoint variables. For $u = U(t)$, the right-hand side of (14.4.15) vanishes since either $U(t)$ is an interior control so that $\partial H/\partial u = 0$ or it is a corner control in which case we have $dU(t)/dt = 0$. Hence, $\bar{H} \equiv [H]_{u=U(t)}$ is a constant in any interval where H is continuously differentiable. With the free end condition (14.4.7), we have $\bar{H}(t) = 0$ in the interval $(T_s, T]$ if there should be a simple jump discontinuity in U at some earlier time $T_s < T$ in the interval $(0,T)$. The switching condition (14.4.9) required the constant $\bar{H}(T_s)$ to be the same (and equal to zero by the free end condition) on both sides of the jump in $U(t)$. The observation allows us to extend $\bar{H}(t) = 0$ to the next switch point and finally for the entire interval $[0,T]$.   □

14.5   Strictly Concave Death Rates

14.5.1   Upper corner control near the start

For the death rate (14.2.10) with $d^{\cdot} = \alpha(1-u)^{\alpha-1}$, the stationary condition (14.4.10) may be written as

$$d^{\cdot} = \alpha(1-u_i)^{\alpha-1}$$
$$= \frac{u_m}{\sigma}\frac{R(t)}{D(t)} \equiv \frac{u_m}{\sigma}r(t). \qquad (14.5.1)$$

For $\alpha > 1$, the condition (14.5.1) formally determines an interior control $u_i(t)$ in terms of $\{\lambda_i(t)\}$ and $\{x_j(t)\}$. A corresponding solution of the BVP for $\{x_k(t), \lambda_j(t)\}$ is called an "interior solution" for the shortest time problem.

Lemma 7 $R(t) < 0, x_1 > 0$ *and* $q(t) \equiv \lambda_2 - \sigma\lambda_1 < 0$ *for some interval* $[0,T_1)$ *adjacent to the starting time.*

Proof Given

$$[H]_{t=0} = 1 + \frac{\mu + u_m u(0)}{\sigma}R(0) = 0, \qquad (14.5.2)$$

we must have

$$R(0) = [x_1(\lambda_2 - \sigma\lambda_1)]_{t=0} = \lambda_2(0) - \sigma\lambda_1(0) < 0.$$

By the continuity of the state and adjoint functions, we must have $x_1 > 0$, $R(t) < 0$ and $(\lambda_2 - \sigma\lambda_1) < 0$ for some interval $[0, T_1)$ adjacent to the starting time. $\qquad\square$

Lemma 8 $D(t) = D_1 t + o(t)$ *for sufficiently small* $t > 0$ *where* D_1 *assumes the same sign as* $\lambda_1(0)$.

Proof Given the initial conditions for the two state variables, the conclusion follows from

$$\begin{aligned} D(t) &= (1 - x_1)P + \lambda_2 x_2(a - 1) \\ &= [-x_1'(0)t + o(t)][\lambda_1(0) + O(t)] + O(t) \\ &= \lambda_1(0)\{\mu + u_m u(0)\}t + o(t) \end{aligned}$$

for $0 \le t \ll 1$. $\qquad\square$

Proposition 38 *For* $\alpha > 1$, *any optimal solution for a minimum terminal time requires i) an upper corner control in some finite interval* $[0, T_1)$, *i.e.,* $U(t) = 1$ *for* t *in* $[0, T_1)$ *for some* T_1 *in* $(0, T)$.

Proof With $D(t)$ and $R(t)$ not depending on u explicitly, we make use of (14.2.10) and (14.4.3) and solve the stationary condition (14.4.10) for $u(t)$ to get the interior control

$$U_i(t) = 1 - \left[\frac{u_m}{a\sigma}\frac{R(t)}{D(t)}\right]^\eta \quad \left(\eta = \frac{1}{\alpha - 1} > 0\right). \tag{14.5.3}$$

From Lemmas 7 and 8, we get $D(t)/R(t) \to 0$ as $t \to 0$ with the sign of $D(t)/R(t)$ opposite of that of $\lambda_1(0)$. It follows that the interior control $U_i(t)$ is either negative or imaginary (see also Lemmas 4 and 6) and therefore not admissible at least for some interval $[0, T_1)$ for some positive T_1 and the optimal control must be a corner control there. With $R(0) < 0$, the expression (14.5.2) for the Hamiltonian at $t = 0$ shows that $H(0)$ is minimized by the upper corner control at the start. By continuity, $U(t)$ must be the upper corner control in $[0, T_1)$ for some T_1 in $(0, T)$. $\qquad\square$

We now know that for all $\alpha > 1$, the optimal solution starts with an upper corner control in some interval $[0, T_1)$. Biologically, the population of mutant cells is small initially and further gain by clonal expansion of mutants is small compared to the gain by mutation through genetic instability even at a higher death rate. But as the mutant cell population grows and the normal cell population declines through continual maximum mutation rate, mutants gained by clonal expansion of existing mutants should rise while the new mutants from continued mutation of normal cells falls. After some breakeven point T_s, proliferation of existing mutants would more than offset the loss of potential mutants from lowering or stopping genetic instability. To see what actually takes place beyond the interval $[0, T_1)$ during which upper corner control applies, we begin by proving the following negative result on the interior control $U_i(t)$ for $\alpha > 1$ which is well-defined and feasible for $t > T_1$:

Proposition 39 *For $\alpha > 1$, the interior control $U_i(t)$ is maximizing in $[T_1, T]$, $0 < T_1 < T$.*

Proof Both $D(t)$ and $R(t)$ are continuously differentiable away from the jump discontinuity of $U(t)$. Consequently, we can differentiate the Hamiltonian twice with respect to time where the interior control applies to get

$$\left[\frac{d^2 H}{dt^2}\right]_{u=U_i} = \left[\frac{\partial^2 H}{\partial u^2}\right]_{u=U_i}\left(\frac{dU_i}{dt}\right)^2 = -D(t)d^{\cdot\cdot}(U_i)\left(\frac{dU_i}{dt}\right)^2 \quad (14.5.4)$$

$$= -\left[(1-x_1)P + \lambda_2 x_2(a-1)\right]d^{\cdot\cdot}(U_i)\left(\frac{dU_i}{dt}\right)^2$$

after making use of the Hamiltonian structure of the state and adjoint equations to eliminate terms involving derivatives of the state and adjoint variables. Since $P(t) < 0$ (by Lemma 6), $\lambda_2(t) < 0$ (by Lemma 4) and $d^{\cdot\cdot}(U_i) < 0$ for $\alpha > 1$ (given $0 < U_i < 1$), it follows that

$$\left[\frac{\partial^2 H}{\partial u^2}\right]_{u=U_i} < 0 \quad \text{and} \quad \left[\frac{d^2 H}{dt^2}\right]_{u=U_i} < 0 \quad (14.5.5)$$

wherever U_i is well-defined and differentiable. In other words, the interior control, wherever well-defined and differentiable, is maximizing and therefore is not optimal in any part of the solution domain for the minimum terminal time problem. □

Proposition 40 *For $\alpha > 1$, any optimal control for the 1-step model must end with a lower corner control in some finite interval $(T_0, T]$, i.e., $U(t) = 0$ for t in $(T_0, T]$ for some T_0 in $(0, T)$.*

Proof Given Proposition 39, we can only choose between the two corner controls, Proposition 36 eliminates the upper corner control as a candidate for optimal mutation rate in some interval $(T_0, T]$. It follows that the optimal mutation rate must end with the lower corner control in $(T_0, T]$ for some $T_0 < T$. That $T_0 > 0$ is an immediate consequence of Proposition 38 ruling out the lower corner control for some interval $[0, T_1)$. Hence, T_0 must be $\geq T_1$. □

14.5.2 Optimal mutation rate is bang-bang

With Propositions 40 and 38, we know that the optimal control must be the lower corner control in the interval $(T_0, T]$ and must be the upper corner control in $[0, T_1)$ with $0 < T_1 \leq T_0 < T$. In principle, it is possible to have more (than one) switches between corner controls in T_1 and T_0 (keeping in mind that the interior control is maximizing), we show presently that there can only be one switch in the interval $(0, T)$ (so that $T_1 = T_0 \equiv T_s$ with the only switch point denoted by T_s).

At the switch point, the switch condition requires $H(T_s+) = H(T_s-)$ so that

$$1 + \frac{u_m + \mu}{\sigma} R(T_s) = 1 + \frac{\mu}{\sigma} R(T_s) + D(T_s)$$

or

$$G(T_s) = \frac{u_m}{\sigma} R(T_s) - D(T_s) = 0 \tag{14.5.6}$$

with

$$D(t) = (1 - x_1)P + \lambda_2 x_2 (a - 1).$$

Proposition 41 *The optimal control $U(t)$ starts with the upper corner control and has exactly one switch from upper to lower corner control at some T_s in $(0, T)$ determined by $G(T_s) = 0$ where $G(\cdot)$ is given by (14.5.6).*

Proof At the two ends $t = 0$ and $t = T$, we have from Proposition 38 and Lemma 4

$$G(0) = \frac{u_m}{\sigma} [\lambda_2 - \sigma \lambda_1]_{t=0} < 0, \tag{14.5.7}$$

$$G(T) = -[\lambda_2 \{a - x_1 (1 + u_m/\sigma)\}]_{t=T} > 0, \tag{14.5.8}$$

with the second inequality being a consequence of $a \geq 2 > (1 + u_m/\sigma)$ for the biological realistic ranges of parameter values (including those mentioned earlier in this chapter). These two end values of G ensure the existence of at least one location for the switch point in $(0, T)$.

Next, we establish that $G(\tau) = 0$ in fact has only one root in $(0, T)$ by showing that it is a monotone increasing function of its argument. Hence, the location of the single switch point T_s of $U(t)$ is uniquely determined by $G(\tau) = 0$. Since all quantities in the expression (14.5.6) for G are continuous at T_s, we take them to be the values at T_s- (where the upper corner $u = 1$ control applies) and rewrite it as

$$G(\tau) = -\lambda_2 \left[\frac{x_1}{\sigma} (1 - x_1) + x_2 (a - x_1) \right] + \frac{x_1}{\sigma} (\lambda_2 - \sigma \lambda_1) (1 + u_m - x_1)$$

$$= -\left[\frac{1 + u_m - x_1}{u_m + \mu} + \lambda_2 \left\{ \frac{x_1}{\sigma} (1 - x_1) + x_2 (a - x_1) \right\} \right]_{t=\tau}.$$

By differentiating $G(\tau)$ with respect to its argument τ, we get

$$\frac{dG}{d\tau} = -x_1 \left\{ 1 + \lambda_2 (u_m + \mu) \left(a - x_1 + x_2 + \frac{2x_1 - 1}{\sigma} \right) \right\} \tag{14.5.9}$$

where we have made use of (14.2.8) and (14.2.9) to eliminate the derivatives of x_k. We now make use of (14.4.11) with $u = 1$ to get

$$[(u_m + \mu) \lambda_2(t)]_{u=1} < [(u_m + \mu) \lambda_2(T)]_{u=1} = -\frac{\sigma}{x_1(T)} < -\sigma.$$

Together with the inequality

$$\sigma \gg a - x_1 + x_2 + \frac{2x_1 - 1}{\sigma} > a - 1 - \frac{1}{\sigma} > 0,$$

we have

$$\left[\frac{dG}{d\tau}\right]_{u=1} > -x_1(t)\left[1 - \sigma\left(a - 1 - \frac{1}{\sigma}\right)\right] > -x_1(t)\left(1 - \sigma\right) > 0$$

and the proposition is proved.   □

Altogether, we have the following complete characterization of the optimal mutation rate for a death rate (14.2.10) with $\alpha > 1$.

Theorem 24 *For $\alpha > 1$, the optimal mutation rate for fastest time to cancer for our (1-step) model is the bang-bang control*

$$\bar{u}(t) = \begin{cases} 1 & [0 \le t < T_s) \\ 0 & (T_s < t \le T] \end{cases} \qquad (14.5.10)$$

with its one switch point determined by the switching condition (14.4.9) which may be taken to be the single root of $H(T_s) = 0$ or (14.5.6) in light of Proposition 37.

Proof The proposition follows from Propositions 38, 40, 39 and 41.   □

It remains to determine the location of the unique switch point T_s by finding the unique root of (14.5.6) or some alternative methods to be discussed in Section 14.6.

14.5.3   *General strictly concave death rate*

The conclusions reached above for the death rates (14.2.10) with $\alpha > 1$ also hold for general strictly concave death rates that satisfy the inequality

$$\frac{1 - d(u)}{1 - u} < d'(u) \qquad (14.5.11)$$

The proof of this observation is given in [35].

14.6   Optimal Switch Point for Bang-Bang Mutation

14.6.1   *A brute force scheme on $x_s = x_2(T_s)$*

With the optimal control u known to be bang-bang as given by (14.5.10), it is not necessary to devise any sophisticated numerical scheme for the problem since searching for the principal unknown involves only solving IVP. Furthermore, we have learned from the problem with a time invariant mutation rate that we should compute on a fixed domain by using the mutant population x_2 as the independent variable (instead of time t). For that approach, we think of both t and x_1 as a function of x_2 (which is permissible given $x_2'(t) > 0$) and work with the following two new ODEs

$$\frac{dx_1}{dx_2} = \frac{g_1(x_1, x_2; u)}{g_2(x_1, x_2; u)}, \qquad \frac{dt}{dx_2} = \frac{1}{g_2(x_1, x_2; u)} \qquad (14.6.1)$$

which follow from application of the chain rule with $g_1(x_1, x_2; u)$ and $g_2(x_1, x_2; u)$ as given by (14.2.8) and (14.2.9). The ODEs (14.6.1) are augmented by the initial conditions

$$x_1^{(1)}(0) = 1, \qquad t^{(1)}(0) = 0. \tag{14.6.2}$$

Suppose the time to switch from upper to lower corner control is T_k and the corresponding mutant population is $x_{2k} = x_2(T_k)$. We compute the solution of the IVP (14.6.1)–(14.6.2) with $u = 1$ for the interval $0 \leq x_2 \leq x_{2k}$ and denote the solution by $\left\{x_1^{(1)}(x_2), t^{(1)}(x_2)\right\}$. The same pair of ODE is then solved again but now with $u = 0$ and for the interval $x_{2k} \leq x_2 \leq 1$ augmented by the continuity conditions

$$x_1^{(0)}(x_{2k}) = x_1^{(1)}(x_{2k}), \qquad t^{(0)}(x_{2k}) = t^{(1)}(x_{2k}) \tag{14.6.3}$$

with the solution denoted by $\left\{x_1^{(0)}(x_2), t^{(0)}(x_2)\right\}$. For this last solution, we get the terminal time $T^{(k)}$ for mutant cells reaching the target population

$$T^{(k)} = t^{(0)}(1)$$

when the switch point is x_{2k}.

By carrying out the solution process above for a set of coarse mesh points $\{x_{2k}\}$ of the interval $[0, 1]$, we obtain the value x_{2m} that gives to the shortest approximate terminal time $T^{(m)} = t^{(0)}(1; x_{2m})$ among the collection of terminal times $\{T^{(k)} = t^{(0)}(1; x_{2k})\}$. The corresponding $T_m = t^{(1)}(x_{2m}) = t^{(0)}(x_{2m})$ gives an approximate solution for the optimal switch time T_s for the problem.

Having an approximate location of the optimal switch point x_{2m}, we can refine the search by using a finer mesh for a neighborhood of x_{2m} and continue the refinement until a desired accuracy of the switch point is achieved.

We report here only that for $\sigma = 10$, $\mu = 0.1$, $u_m = 1$ and $a = 2$, the numerical scheme above gives an optimal switch time of $T_s = 1.4125\ldots$ and a fastest terminal time of $T = 5.3986\ldots$ which is substantially lower than the approximate fastest time of $5.68\ldots$ obtained in [10] by an existing code for the discrete SQP algorithm without the benefit of knowing $U(t)$ being bang-bang. (It is noted that the inaccuracy is due to the ad hoc process for discretization of the continuous optimal control problem and not to any inadequacy of the SQP algorithm.) Changing only $a = 2$ to $a = 5$ leads to an optimal switch time of $T_s = 0.41415\ldots$ and an earlier terminal time of $T = 1.9817\ldots$

14.6.2 Some bounds for the optimal switch time

While the brute force solution scheme above requires only seconds of desktop computing, it is still of some interest to avoid unnecessary computation. For this reason, we present below upper and lower bounds, T_g and T_ℓ for T_s (first reported in [21]) to limit our search for the optimal mutation rate to within the interval $[T_\ell, T_g]$.

14.6.2.1   Crossover of corner solutions for x_2

An obvious *upper bound* for the switch time would be the crossover time T_c when the upper and lower corner solutions for the IVP problem for state variables, (14.2.8)–(14.2.9) and (2.1.7), have the same cancerous mutant population, i.e., $x_2^{(0)}(T_c) = x_2^{(1)}(T_c)$. For any later switch time, we have $x_2^{(0)} > x_2^{(1)}$ since there would be less normal cells to mutate and a larger mutant population to multiply and proliferate. Hence it would be inferior to switch later.

While the determination of T_c by the exact solutions or numerical integration is straightforward, the use of T_c as a first approximation for T_s is too conservative. The rapid proliferation of the mutants in the case of a lower corner control, especially for $a \gg 1$, would more than offset the corresponding smaller mutant population at an earlier time.

14.6.2.2   Crossover of corner cancerous growth rates

A less conservative choice would be the time T_g when the two corner growth rates for mutants are equal: or

$$\left[\frac{dx_2^{(1)}}{dt}\right]_{t=T_g} = \left[\frac{u_m + \mu}{\sigma}x_1^{(1)}\right]_{t=T_g} = \left[\frac{\mu}{\sigma}x_1^{(0)} + x_2^{(0)}(a - x_1^{(0)})\right]_{t=T_g} = \left[\frac{dx_2^{(0)}}{dt}\right]_{t=T_g}.$$
$$(14.6.4)$$

For $\mu = 0.1$, $u_m = 1$, $a = 2$ and $\sigma = 10$, we have $T_g = 2.215\ldots$ compared to $T_c = 3.267\ldots$. Evidently, switching at T_g would benefit from proliferating at the same rate but from a larger mutant population for proliferation given $x_2^{(1)}(T_g)$ $(> x_2^{(0)}(T_g))$.

Remark 15 *Switch at T_g may still be later than optimal. By switching earlier, the faster growth rate ax_2 of the mutant cells may still be able to offset the effect of the shortfall $x_2^{(1)}(T_g) - x_2^{(0)}(T_k)$ incurred by an earlier switch time $T_k < T_g$.*

14.6.2.3   A lower bound

The natural proliferation of cancerous mutants is much faster than the normal cells. For that reason, the lower corner solution will be more effective eventually. This suggests consideration of the threshold T_ℓ when the rate of increase of cancerous mutants from the mutation of $x_1^{(1)}(t)$ (with $u = 1$) equals proliferation rate of the mutant cells $x_2^{(1)}(t)$ without mutation ($u = 0$). In other words, take T_ℓ to be the time when

$$\left[\frac{u_m + \mu}{\sigma}x_1^{(1)}\right]_{t=T_\ell} = \left[\frac{\mu}{\sigma}x_1^{(1)} + x_2^{(1)}(a - x_1^{(1)})\right]_{t=T_\ell} \qquad (14.6.5)$$

or

$$\left[x_2^{(1)}(a - x_1^{(1)}) - \frac{u_m}{\sigma}x_1^{(1)}\right]_{t=T_\ell} = 0. \qquad (14.6.6)$$

The condition (14.6.6) determines the threshold T_ℓ using $\left\{x_1^{(1)}, x_2^{(1)}\right\}_{t=T_\ell}$ from the solution of the IVP for the two populations with $u(t) = 1$ in $[0, T_\ell)$. That T_ℓ is a lower bound for the switch point will be explained below and confirmed by a specific example.

For $a = 5$, $\mu = 0.1$, $u_m = 1$ and $\sigma = 10$, the threshold condition (14.6.6) gives $T_\ell = 0.41415\ldots$ with $[T]_{T_\ell} = 1.98177\ldots$ which is indistinguishable (at least for the number of digits calculated) from the (accurate approximate) optimal switch point T_s. We note however that T_ℓ is not the actual switch point or an upper bound. For $a = 2$ (while keeping the other three parameters unchanged), we have $T_\ell = 1.2630\ldots$ with $[T]_{T_\ell} = 5.4060\ldots$. The actual optimal switch time is found to be $T_s = 1.4125\ldots$ which is larger. We note in passing that by using the bounds on the switch time T_s, we would stay with t as the independent variable and work with the original state equations for the first half of the solution interval $[0, T_s]$ and switch to x_2 as the independent variable for the remaining half of the interval $[x_s, 1]$ with $x_s = x_2^{(1)}(T_s)$ from the solution for the first half.

There is a simple explanation why T_ℓ is only a lower bound for the optimal switch point T_s. Though the condition (14.6.6) shows that natural growth rate for the mutant population $x_2^{(1)}$ has caught up with (and equal to) the gain rate from the mutation of the $x_1^{(1)}$ population at T_ℓ, that parity may be lost after switching to the lower corner control. With the switch (to the lower corner control), the population $x_1^{(1)}$ begins to rise more rapidly and tends to $1 - \mu$ because of the addition of the logistic growth term. Correspondingly, the natural proliferation rate of x_2 is reduced somewhat through the factor $a - x_1$ in the second state equation (14.2.9). As such, the parity between the two modes of gaining mutant cells at the threshold T_t may be lost (for a short time after T_ℓ), much less so for $a \gg 1$ and more so if $a = O(1)$. The latter scenario requires a later switch time than T_ℓ to get a larger x_1 at the switch to balance the change in logistic growth rate after the switch to a lower corner control as shown for the $a = 2$ case. The precise condition specifying the optimal switch time is given by (14.4.9).

The observations on the various bounds is summarized in the following proposition useful for bracketing the two starting iterates in the application of the iterative numerical solution scheme:

Proposition 42 *The threshold value T_ℓ defined by the condition (14.6.6) provides a lower bound for the optimal switch point T_s while T_g ($< T_c$) provides an upper bound.*

Remark 16 *Numerical solutions for optimal switch have been obtained for several sets of parameter values. It was found that T_ℓ constitutes a tight lower bound, often very close to the optimal switch point for larger a, i.e., for more aggressive cancerous proliferation rate. Its determination only involves the upper corner control problem (for which we have an explicit exact solution).*

14.7   Other Types of Death Rates

14.7.1   A death rate linear in mutation rate

For $\alpha = 1$, the specific class of death being investigated simplifies to $d(u) = u$. The optimal control for this case is the same bang-bang control for $\alpha > 1$. An additional complication arises in the proof for this case is the possibility of a *singular solution* associated with the stationary condition

$$\left[\frac{\partial H}{\partial u} \right]_{u=u_i} = \frac{u_m}{\sigma} R(t) - D(t) = 0.$$

While this condition does not determine an interior control $U_i(t)$, it may nevertheless hold for a segment of the (and possibly the entire) solution domain. This was shown not to be the case in [35] leaving us with the following result for the oncogene activation model:

Proposition 43 *For $\alpha = 1$ in (14.2.10), the minimum time T to the target mutated cell population, with $x_2(T) = 1$, is attained by the unique optimal (bang-bang) control*

$$\bar{u}(t) = \begin{cases} 1 & (0 \le t \le T_s) \\ 0 & (T_s \le t \le T) \end{cases} \tag{14.7.1}$$

with the terminal time T determined by the transversality condition (14.4.7) and with the only switch point $T_s < T$ determined by the switch condition (14.4.9).

Proof See [35].   □

14.7.2   Strictly convex death rate

For $0 < \alpha < 1$ in (14.2.10), we proved in [35] the following result for our one-step model:

Proposition 44 *For $0 < \alpha < 1$ in the expression (14.2.10) for $d(u)$, the optimal control for our TOP is*

$$\bar{u}(t) = \begin{cases} u_i(t) & (0 \le t \le T_s) \\ 0 & (T_s \le t \le T) \end{cases} \tag{14.7.2}$$

with $u_i(T_s) = 0$. The interior control applies to the entire solution domain if $T_s \ge T$.

Proof See [35].   □

Similar to the case of a strictly concave death rate, Proposition 44 has been extended in [35] to apply to all strictly convex death rates that satisfy the inequality

$$\frac{1 - d(u)}{1 - u} > d'(u). \tag{14.7.3}$$

Chapter 15

Mathematical Modeling Revisited

Through a good number of mathematical models and their analysis worked out in the preceding chapters, the readers have been exposed to many of the commonly used techniques and best practices for mathematical modeling. In principle, aspiring mathematical modelers would have acquired what they need to be successful in mathematical modeling. For those who still need help to distill from these models the effective tools for mathematical modeling and analysis, we comment below on some of the more subtle aspects of these principles, techniques and practices that are not so transparent to new students of mathematical modeling.

15.1   From Simple to Complex

15.1.1   *Simple question*

Mathematical modeling usually starts with a question on a phenomenon or an activity of interest. We may want to know why an observed phenomenon takes place, or whether our expectation of what should happen next is justified. To investigate the growth of human population on Earth can take place at many level. We can start at the level of fertilized eggs that develops into human beings and add to the human population or how the current human population has evolved from single cell organisms and should further evolve in the future. Our specific interest, the state of our knowledge and the available resources often dictate the scope of the investigation and the mathematical model we should try to formulate. Developments in the early chapters of this volume focused on questions such as "When will the human population reach 10 billion?" and "When will it double the 2011 population of 7 billion?" Questions of this kind may be investigated phenomenologically (without dealing with the behavior of individuals or cell biology). They allowed us to seek answers at a level similar to the study of the planetary motion by Newton's laws of motion (without getting into the detailed interaction of individual molecules or atoms of materials that make up the planet). In both cases, the question was relatively simple and led to a mathematical model that generated useful information relevant to the inquiry.

Simplicity is important for a first model in investigating a new phenomenon. There is a natural tendency on the part of a modeler to be comprehensive. But a more comprehensive model is necessarily more complex and more difficult to extract useful informations mathematically or computationally. This is clearly illustrated throughout this volume as we progress from the simple exponential growth model of Chapter 2 to the two-component model for genetic instability and carcinogenesis in Chapter 15. An astute question for developing a first mathematical model is important for getting a good start. It should not be too vague or too specific for a meaningful inquiry. It also should not be too detailed or too comprehensive for a tractable mathematical problem. "What is the future of the human race?" would qualify for the former and "What is the birthday of the 10 billionth individual on Earth?" would be an example of the latter.

15.1.2   *Simple model and upgrading*

While a case can also be made for starting with a comprehensive model and then making it tractable by various levels of simplifications, the modeling cycle described in Chapter 1 and practiced throughout this volume provides the framework for the opposite approach. As those adopted in Chapter 2, the first question(s) posed for a phenomenon not previously examined should lead to a relatively simple first model that enables the modeler to gain insight to the phenomenon under consideration. When appropriate, it is preferable that the question would keep the model at the macroscopic level where more mathematical tools are available for extracting useful information from the model. Even if the first model should lead to results not consistent with observations, the modeling cycle offers the modeler the opportunity to upgrade the model after gaining some insight on the phenomenon. In the early days of physics, we have questions such as "Is the Earth's orbit around the Sun circular?" When observational data seemed not to support this expectation, the discrepancy prompted physicists to change to the right question, "Whether the orbit is elliptical and why?" As the exponential growth model in Chapter 2 found wanting for long elapsed time, that it can be analyzed to yield some useful information about population growth over relatively short time spans constitutes progress. The deficiency of the model for longer elapsed time then led us to incorporate the notion of carrying capacity of the environment.

While there is no definitive argument against starting an investigation of a phenomenon with a comprehensive mathematical model of the phenomenon, there is no assurance that an attempt for comprehensiveness would succeed in formulating a truly comprehensive model. There is always the issue of what is considered truly comprehensive. Even if there should be a general agreement of comprehensiveness such as Newtonian mechanics had been for nearly two centuries, it was eventually found inadequate in addressing the small precession of the perihelion of planet Mercury. A more elaborate theory of relativity eventually provided the theoretical basis for the small precession. (See [33] for more details.)

15.1.3 Instantaneous rate of change

Striving for simplicity goes beyond asking relatively simple questions. To develop a mathematical model for the question posed, we have to know something about what regulates or drives the phenomenon. In physics, the motion of planets or objects in our daily life is governed by Newton's laws of motion. Unfortunately, we do not have a simple universally applicable law for population growth (or most biological phenomena) analogous to Newton's laws of motion. For a specific environment, the growth rate may be estimated from available census data or by examining the various birth and death processes. If we want to avoid plunging immediately into such highly technical activities, we may stay at the phenomenological level by extracting from what is observable a reasonable assumption about the relation between current and future population sizes and then see if its consequences fit the facts. Historically, Newton's second laws also started out as a hypothetical assumption; but its predicted consequences fit the corresponding actual observations so well (including the highly sensitive space capsules in outer space flights today) that it was accepted eventually as the law that governs the dynamics of mass objects (as long as they do not travel near the speed of light).

For population growth, the goal we set for the modeler was to determine the Earth's future population $y(t)$. Since we do not know $y(t)$ for $t > 0$ (where $t = 0$ is now or some fixed reference time), we need to know or reasonably assume something about the human population beyond the past. Drawing on the success of (the relatively simple models for) Newtonian mechanics, the modeler should naturally try to learn (or could reasonably assume) something about the rate of change of the population. Now it is an observed fact that people generally beget more people; the larger the current population, the faster the growth of the population. It is therefore not unreasonable to adopt the following

- *Instantaneous Rate of Change Hypothesis*: The rate of change of a population should depend on its current size (and, for the time being, nothing more).

What is good about this simple *rate of growth* hypothesis (which may or may not be appropriate) is that it can be translated into a mathematical relation on the evolution of the population size with time, thereby giving us a mathematical model for the growth of the human population.

More generally, it is worth emphasizing that for phenomena that evolve with time (the only kind we are concerned with in this volume), their most basic characteristic (beyond the actual evolution of the phenomenon with time) is the rate of change with time. From information available or reasonable assumption, some kind of instantaneous rate of change hypothesis would allow us to make use of our knowledge of calculus to extract information about the phenomenon. Only when such a hypothesis fails to be useful or meaningful do we go to the next level and consider the "rate of change of a rate of change" (of the position of our vehicle for example) our vehicle to answer our question. Historically, Newton's second law of motion relates the rate of change of momentum to the relevant force. When the force depends on the position of the object involved (such as a spring-mass system and a planet in motion), Newton's second law (in the case of constant mass) is taken as a relation between "the rate of change of the rate of change of the position $y(t)$", i.e., the acceleration of the body, and the force experienced by the object in order to have one (vector) equation for one (vector) unknown.

It is not our intention to leave an impression that we should not consider anything complex or complicated. If the information provided by a (relatively simple) model does not meet some of the required validation (such as available data), we would have to modify the first model by refining it to include more features and properties of the phenomenon being modeled. There are no recipe on what more to include or how the model should be modified for improved performance. Fruitful modifications are often model-specific but the models examined in these notes were selected to provide some concrete approaches for *model improvement* that are applicable to other models. Even if the first model should prove to be very representative of the modeled phenomenon, it may still be important to improve further its accurate characterization of the phenomenon. A historically important example is the tiny discrepancy between the prediction of the Newtonian theory of planetary motion and the actual orbit of planet Mercury mentioned above. A much more complex theory of relativity was developed to provide the correct model and the theoretical foundation for that phenomenon.

15.2   Open to Options

15.2.1   *Higher order rates of change?*

To move from a model based on some "rate of change" to one about the "rate of change of some rate of change" is just one example of being flexible in mathematical modeling. J. Kepler spent years of his life trying to show Mars' trajectories around the sun to be an egg-shaped ovoid orbit but the available observed data (including those of his former boss Tycho Brahe) did not support his hypothesis. He eventually switched to an elliptical orbit after more than 40 failed attempts but probably should have done so sooner. (Newton made no assumption about the orbits and let them be consequences of his mathematical model based on his postulated laws of motion. His more ready success rested more on his correct postulate about the attractive force governed by the inverse square law and his development of calculus for his mathematical analysis.) Being flexible allows us to consider alternative approaches and not being tied down by one particular approach toward a mathematical model.

In addition to being flexible and open to options, it also pays to be imaginative. In the usual models for planetary motion, each huge and massive planet is treated as a point mass in space. An example from the biological sciences is the modeling of a nerve axon as a conductive cylindrical cable which has spawned many informative research projects based on such a model (see [28] for references on the pioneering work on cable neuron).

A modeler may be flexible and imaginative in many other ways. Applications of Newton's second law of motion with the relevant force field depending on the location of the object (of constant mass) in question have naturally led to the conventional form of $F = ma$ or, in the form of an ODE,

$$m\frac{d^2y}{dt^2} = F(y). \tag{15.2.1}$$

The motion of a planet around the Sun subject to Newton's inverse square law is a conspicuous example with $y(t)$ being the radial distance $r(t)$ between the planet and the Sun.

The mathematical model (15.2.1) involving an "instantaneous rate of change of some instantaneous rate of change" (also known as a second order derivative or a second order instantaneous rate of change) is less tractable compared to a first order rate of change and therefore less attractive for mathematical analysis and numerical computation. Yet, by an imaginative use of elementary calculus, it is possible to transform the model into one for a first order instantaneous rate of change. With $v(t) = dy/dt$ corresponding to the velocity of the object (planet), we have by the chain rule in calculus

$$m\frac{d^2y}{dt^2} = m\frac{dv}{dt} = m\frac{dv}{dy}\frac{dy}{dt} = mv\frac{dv}{dy}. \tag{15.2.2}$$

This identity transforms Newton's second law (15.2.1) into a first order rate of change relation

$$mv\frac{dv}{dy} = \frac{m}{2}\frac{d}{dy}(v^2) = F(y) \tag{15.2.3}$$

albeit that the rate of change in (15.2.3) is with respect to the location y (and not time t). Significantly, the transformation from a second order rate of change to one for first order rate of change is more than cosmetic. In the form (15.2.3), we have a calculus problem for the exact solution for v, the velocity of the object, as a function of position y. As simple, elegant and substantively rewarding as the transformation (15.2.2) may be, it is not the preferred approach to the final solution for the problem of planetary motion subject to the inverse square law. Readers are referred to Chapter 2 of [33] for an even more surprisingly simple complete solution (known as Kepler's second law of planetary motion) for the problem.

15.2.2 Mathematical effectiveness & computational efficiency

The simplest model for the growth of the human population on Earth formulated in Chapter 2 is characterized by the IVP

$$\frac{dy}{dt} = ay, \qquad y(0) = Y_0.$$

It was shown that the model would have the Earth's population growing exponentially with time

$$y(t; a, Y_0) = Y_0 e^{at}. \tag{15.2.4}$$

There are two parameters a and Y_0 that may be chosen to make the model accurately matching the existing data and give us confidence in its ability to predict the human population in future years. It was shown that the two parameters could be fixed by two pieces of census data available for 1974 and 2011 (or any two pieces of census data). The model resulting from these parameter values generates population sizes that are in reasonably good agreement over nearly a decade of census data.

With a view toward improving the predictive power of (15.2.4), we used the conventional method of least squares as an alternative way for determining the values of the same two parameters. This method minimizes the sum of the squares of the difference between a number model solutions $\{y(t_k; a, Y_0)\}$ at several instants of time and the know data $\{\bar{y}_k\}$ at these instants:

$$\min_{\{a, Y_0\}} \left[E_2(a, Y_0) = \sum_{k=1}^{N} \{y(t_k; a, Y_0) - \bar{y}_k\}^2 \right]. \tag{15.2.5}$$

The conditions of stationarity provide the number of equations needed to determine the two unknowns:

$$\frac{\partial E_2}{\partial a} = 0, \qquad \frac{\partial E_2}{\partial Y_0} = 0. \tag{15.2.6}$$

Applications of this method of parameter estimation were carried out in Chapter 2 using census data for a few years between 1974 and 2011 with better agreement for the same range of census data.

It is evident from the solution process that the least square method led to relatively tedious algebra and numerical calculations. For a least square fit of four data points, the usual solution process requires the roots for a 36 degree polynomial (see (2.3.6) in Chapter 2). While root finding poses no particular obstacle with the computing software currently available for this purpose, the mathematical issue of a unique real root seems unnecessarily complicated. It turns out that the solution process for the same least square problem can be streamlined to avoid such complications.

Instead of (15.2.4), consider working with its equivalent expression

$$z(t) = \ln(y) = \ln(Y_0) + at \equiv b + at.$$

With $b = \ln(Y_0)$, $z(t_k) = \ln(y(t_k))$ and $\bar{z}_k = \ln(\bar{y}_k)$, the least square fit takes the form

$$E_2(a, b) = \sum_{k=1}^{N} \{z(t_k; a, b) - \bar{z}_k\}^2$$

and

$$\frac{\partial E_2}{\partial a} = 2 \sum_{k=1}^{N} t_k \left(b + at_k - \bar{z}_k \right) = 0,$$

$$\frac{\partial E_2}{\partial b} = 2 \sum_{k=1}^{N} \left(b + at_k - \bar{z}_k \right) = 0.$$

The two stationary conditions for E_2 are simply two linear equations for b and a. Its unique solution can be obtained by a "back-of-the-envelop" calculation with no intricate mathematical issues or tedious numerical computation. While the difference is not a game changer in this case, it may be for other analogous improvement in computational requirements.

These and other wonderful examples show the benefits for a modeler willing to explore and open to options. However, they do not tell you when to be flexible or how to be imaginative. The when and how often depend on the particular problem. There is no recipe for the modeler to follow and implement. In that sense, mathematical modeling is like wine, you get better at it with age and experience. A variety of model developments is offered in this volume to illustrate how these recommendations may be implemented.

15.3 Improvement or Alternative

15.3.1 *Weighted parameter estimation*

In the previous section, mathematical transformations, when used innovatively, were seen to lead to a substantial improvement in the solution process for particular mathematical models. However, the improvements were on the effectiveness and efficiency of the solution process and not on contents of the models themselves. Improvements on a method used in the mathematical analysis and computation phase of the modeling cycle can also improve the explanatory and predictive power of the mathematical model being studied. The change of variables in the least squares method for an exponentially growing population (15.2.4) is one such example. The conventional method minimizes the sum of the squares of the difference between a number model solutions $\{y(t_k; a, Y_0)\}$ at different instants of time and the known census data $\{\bar{y}_k\}$ at these instants as previously shown in (15.2.5). The conditions of stationarity (15.2.6) provides the correct number of equations needed to determine the two unknowns, a and Y_0. Agreements with known data not used by the estimation process are good for the recent past but poor for much earlier years, notably for 1807.

Other statistical methods for estimating the parameters a and Y_0 can be found in standard texts in statistics. Some may be used for estimating a and Y_0 for better agreement between data and model solution for a longer period of time. Even without going to these references, one modification of the conventional least squares method naturally suggests itself. Since some data points are known to correspond to periods of abnormal times (such as famine, draught, war or plaque), their role in determining the two parameters should probably be less and their contribution reduced. In that case, a more sensible least squares estimation method would be to assign different weights to the different local error terms to emphasize such reduction. This amounts to introducing a set of weight factors $\{w_k\}$, with $0 \leq w_k \leq 1$, and replacing the expression E_2 by

$$E_w(a, Y_0) = \sum_{k=1}^{N} w_k \{y(t_k; a, Y_0) - \bar{y}_k\}^2.$$

The method now chooses a and Y_0 to minimize $E_w(a, Y_0)$.

By choosing the weight factor appropriately to reduce the effects of abnormality of certain data points, the resulting estimated parameter values are expected to make (15.2.4) more in agreement with the census data for a longer period of elapsed time and more accurately predict future growth. While the modification of the least squares method does not change the mathematical model resulting from the instantaneous rate of change hypothesis, it does change the efficacy of the model, assuming we have chosen the weight factors appropriately. This calls attention to the fact that there are ways to improve the performance of a model without changing the particular instantaneous rate of change hypothesis adopted for the model.

15.3.2 *Rate limiting carrying capacity*

An important part of mathematical modeling is the validation and improvement phase of the modeling cycle. We encountered throughout this volume numerous examples of a need to refine the models investigated. When exponential growth was deemed unrealistic, we developed a more realistic model by taking into account a rate limiting carrying capacity y_c for an improved model that better reflects reality. We did so by introducing the popular logistic growth model:

$$\frac{dy}{dt} = \alpha y \left(1 - \frac{y}{y_c}\right).$$

It is important to mention that this is not the only way to improve the exponential growth model even if our goal is to incorporate the feature of a rate limiting carrying capacity for growth. Leaving out models that allow for additional features such as a threshold for extinction, there are numerous other models that have the same feature of a carrying capacity. The Gompertz model

$$\frac{dy}{dt} = -\alpha y \ln\left(\frac{y}{y_c}\right),$$

with $\alpha > 0$ and $y_c > 0$, is among those that have two critical points, an asymptotically stable one at y_c and an unstable one at the origin. As such, both logistic growth and Gompertz offer the same rate limiting feature that reflects real population growth, whether it is human population on Earth or a population of cultured cells in a petri dish. Without specifying additional characteristics of the growth process, one is not a better improvement than the other. In that sense, they offer alternatives as an improvement for the exponential growth.

In the context of human population growth, it is felt that the logistic growth is the more appropriate model for its near constant "marginal growth rate for small population size":

$$\frac{df}{dy} = \alpha \left(1 - \frac{2y}{y_c} \right) \simeq \alpha$$

for $y/y_c \ll 1$. In contrast, we have for the Gompertz model

$$\frac{df}{dy} = -\alpha - \alpha \ln \left(\frac{y}{y_c} \right) \sim -\alpha \ln \left(\frac{y}{y_c} \right) \to \infty$$

as $y \to 0$ which reflects more closely the behavior of an aggressive form of tumor growth.

In Chapter 3, we introduced the logistic growth model to limit growth rather naturally by retaining one more term in the Taylor series expansion for $f(y)$ and did not give any indication that there are other options to focus our attention on the consequence of that new model. With the Gompertz model, we now see an alternative that is not differentiable at the origin and therefore not a part of any Taylor expansion about the origin. Its existence as an unexpected competing model for the logistic growth model should serve as a warning to modelers to be on the lookout for possible alternatives unknown to, or unexpected by them which may trump the one they know.

Again, there is no recipe on how we may uncover possible alternative models to the one under investigation. The fact that mathematics does not have scientific boundaries is one opportunity for seeing possible alternatives. That cable model of an axon was adopted from an electrical cable model and other examples provide some hope for discovering useful models for one field from another.

15.4   Active Modeling Experience

Through some specific examples in this short concluding chapter, we called attention to some special features in mathematical modeling that are only apparent to those who have had considerable experience in modeling activities. While there are no recipe for addressing them, being aware of them would help to sensitize mathematical modelers to their possible importance for a particular project. No attempt is made to exhaust such features; instead, we turn to address briefly another type of issues pertaining to mathematical modeling of real life phenomena.

For pedagogical purposes, some of the models discussed in this volume may be seen as not sufficiently close to the corresponding real life situations. Some aspiring mathematical modelers and teachers of mathematical modeling may feel that the students should learn through active participation in the modeling of real life phenomenon and not as passive audience in lectures. There is certainly merit to this point of view and case study courses are available in some institutions to serve this purpose. In the life sciences, there is the tradition for students to work in senior researchers' laboratories typically through quarter long "lab rotations." Even mathematics majors have opportunities to participate in active learning of mathematical modeling through summer long immersion in their mentor's research project in one of the many NSF funded REU Programs. The feedback from some students in such a program was that they had found the demand of mathematical modeling overwhelming, principally because of the complexities of the phenomenon to be modeled and the extensive amount of science background to be acquired in order to do meaningful work on the project. Sometimes, a student's role was reduced to taking samples and analyzing their contents by automated instrumentation.

While active learning by working on a real life problem is valuable, it is felt that some student modelers need some organized overview of the mathematical modeling process. More specifically, exposure to some examples of model formulation and analysis would better prepared them for the real life modeling on their own. Several aspects of such first exposure should be beneficial to aspiring mathematical modelers:

- Distilling from the mathematical modeling process a concise description of that process in the form of the *modeling cycle*: This allows aspiring modelers to focus on the relevant activities at different stages of the modeling process.
- Clarifying the mathematical issues of interest for a given modeling problem: This allows the modelers to identify (or at least narrow down) the mathematical and computational tools for analyzing the model.
- Identifying different ways of validating the model formulated and analyzed for the phenomenon of interest.
- Modifying a model for improved characterization of the phenomenon.

Knowledge from such an introductory course would teach modelers not to be too ambitious initially. They should start with the simplest meaningful model and gradually incorporate more features for more complete characterization of the phenomenon being studied (in the model modification phases) as it becomes better understood. These and other benefits of an organized lecture course prior to active modeling should render mathematical modeling of other more complex phenomena less daunting. This volume grew out of an attempt to meet this need for an organized first exposure.

Appendix A

First Order ODE

A.1 Separable ODE

A.1.1 Reduction to a calculus problem

For the ODE (2.1.1), re-arrange the ODE for a general $f(y)$ to get $dy/f(y) = dt$ or

$$\frac{1}{f(y)}\frac{dy}{dt} = \frac{d}{dt}[F(y)] = 1 \qquad (A.1.1)$$

where $F(y)$ is the anti-derivative of $f(y)$ with $f(y) = dF/dy$. By this re-arrangement process, we have reduced the original ODE to a calculus problem of finding the anti-derivative of $f(y)$. The calculus problem (A.1.1) may be solved by simple integration to get

$$\int_0^y \frac{dz}{f(z)} = t + C_0 \qquad (A.1.2)$$

where C_0 is a constant of integration. Whether we can find an anti-derivative of $1/f(y)$, the relation (A.1.2) gives t as a function of y. Under suitable condition, we can solve (A.1.2) for y in terms of t (to be shown below for some special cases). Whether (A.1.2) is invertible, the relation describe how one variable evolves as the other variable changes. Note that we may take a different lower limit of integration instead of 0 since C_0 is arbitrary at this point.

The same method of solution may be extended to the more general first order ODE

$$\frac{dy}{dt} = f(y,t) \qquad (A.1.3)$$

provided that it can be written as

$$\frac{dy}{dt} = g(y)h(t). \qquad (A.1.4)$$

Definition 17 *A first order ODE (A.1.3) of the form (A.1.4) is said to be separable if $f(y,t) = g(y)h(t)$.*

Definition 18 *The first order ODE (A.1.3) is said to be non-autonomous if the function f depends explicitly on the independent variable t. It is an autonomous ODE if f does not depend on t explicitly as in the case of (2.1.1). (The terminology applies whether the unknown y is a scalar or vector function of t.)*

For example, the growth rate of the human population may be naturally seasonal or artificially impose (such as a one-child policy for a fixed period of time). As such the growth rate function f would be different depending on the time period of interest (even if the dependence on y remains unchanged) and the corresponding ODE is non-autonomous.

For the separable non-autonomous ODE (A.1.4), we can reduce the solution process for the equation to a calculus problem by re-arranging it to read

$$\frac{1}{g(y)}\frac{dy}{dt} = h(t).$$

Upon integrating both sides with respect to t, we obtain

$$G(y) = H(t) + C$$

where

$$G'(y) \equiv \frac{dG}{dy} = \frac{1}{g(y)}, \qquad H'(t) \equiv \frac{dH}{dt} = h(t)$$

and C is a constant of integration needed for fitting the initial data.

Example 35 *Solve the ODE $y' = \sin(t)e^{-y}$.*

Re-arrange the ODE to read

$$e^{-y}y' = \sin(t).$$

Upon integrating with respect to t, we obtain

$$-e^{-y} = -\cos(t) + C.$$

The solution may be taken in the form

$$e^{-y} = \cos(t) + C_0$$

or

$$y = -\ln\left(\cos(t) + C_0\right).$$

A.1.2   *Initial condition and the initial value problem*

In a first course in calculus, the instructor usually takes great pain to emphasize the importance of constants of integration when we integrate. We see presently why such constants are critical in mathematical modeling. In the question we posed that led to the relation (2.1.1), we started with the knowledge that the Earth's current population being 7 billion in 2011. Whatever the model we choose to study the Earth's future population, its consequences must be consistent with that fact, i.e., $y(t = $ moment when population reaches 7 billions$) = 7$ billions. Since, the calendar label for time is merely a convention, we may work with a calendar that start at that moment (or year) so that we have instead

$$y(0) = 7 \text{ (billion)}.$$

(Note that such re-labeling of time may or may not be appropriate if the ODE system is non-autonomous without a corresponding change of t in $f(y,t)$.) With the model (2.1.1) and its consequence (A.1.2), we can choose the constant C to meet this requirement:

$$\int_0^7 \frac{dz}{f(z)} = C$$

so that

$$t = \int_7^y \frac{dz}{f(z)}.$$

More generally, if the initial population is Y_0 (instead of 7 billion) individuals, we would have

$$y(0) = Y_0 \qquad\qquad\qquad\text{(A.1.5)}$$

and

$$t = \int_{Y_0}^y \frac{dz}{f(z)}. \qquad\qquad\qquad\text{(A.1.6)}$$

Whether or not we can invert (A.1.6) to obtain y as a function t, $y = Y(t)$, explicitly, we can always plot the graph of t vs. y (by integrating the right-hand side of (A.1.6) numerically if necessary) and then rotate and flip the resulting graph to get y vs. t.

The ODE (2.1.1) and the initial condition (A.1.5) together constitute an *initial value problem* (or IVP for short). Had we forgotten to include a constant of integration in handling the calculus problem associated with (2.1.1), we would not have been able to meet the requirement posed by the *initial condition* (A.1.5). Since the study of the Earth's human population cannot proceed without either of these two items, the growth rate and the initial population, it is the IVP (and not just the ODE (2.1.1)) that constitutes the mathematical model for investigating the *evolution* of a population with time.

It should be noted that the choice of reference time may be changed provided that we make the corresponding change in the determination of the constant of integration. For example, if we take the time unit to be the calendar year and denote year 2011 by t_0 with $y(t_0) = Y_0$, then we have instead of (A.1.6)

$$t - t_0 = \int_{Y_0}^y \frac{dz}{f(z)}.$$

A.1.3 *Scale invariant first order ODE*

Before we extract more information from the exact solution (A.1.6), it should be noted that the separability of a first order ODE of the form (2.1.1) or (A.1.4) is important beyond yielding the exact solution of such a first order ODE. It is possible that other first order ODE, not separable as stated, may be transformed into a separable equation so that the same method of reduction to finding an anti-derivative applies. For example, the first order ODE

$$y' = \frac{t+y}{t-y} \tag{A.1.7}$$

is obviously not separable as it stands. However, if we set $y(t) = tv(t)$ with $y' = v + tv'$, the ODE (A.1.7) becomes

$$v' = \frac{1}{t}\left(\frac{1+v}{1-v} - v\right) = \frac{1}{t}\frac{1+v^2}{1-v}.$$

The ODE for $v(t)$ is separable and the method for such an equation now applies. (To become facile with the solution process, the reader should carry out the remaining steps to obtain the exact solution for the problem.)

The solution process for (A.1.7) may be extended to the more general class of *homogeneous* (also known as *scale invariant*) ODE defined as follows:

Definition 19 *The (scalar) first order ODE (A.1.3) is said to be homogeneous (of degree 1) or scale invariant if $f(y,t) = f(y/t)$.*

Proposition 45 *A homogeneous (first order) ODE can be transformed into a separable equation after a change of variable $y(t) = tv(t)$.*

Exercise 41 *Prove Proposition 45 by reducing the ODE $y' = f(y/t)$ to a separable ODE and hence a calculus problem.*

Exercise 42 *Obtain in the form of $y(t)$ the exact solution of the IVP*

$$y' = \frac{t-y}{t+y}, \qquad y(1) = 1.$$

A.2   First Order Linear ODE

A.2.1   *Integrating factor*

Definition 20 *A general (scalar) linear first order ODE is of the form*

$$y' + p(t)y = r(t) \tag{A.2.1}$$

where $p(t)$ and $r(t)$ are integrable functions in some interval $0 < |t| < L$.

Proposition 46 *The linear first order ODE (A.2.1) with the initial condition $y(t_0) = Y_0$ is solved by the integrating factor*

$$E(t) = e^{\int_{t_0}^{t} p(z)dz} \tag{A.2.2}$$

which helps to recast the linear ODE as

$$(Ey)' = E(t)r(t) \tag{A.2.3}$$

so that

$$y(t) = \frac{1}{E(t)}\left\{y^0 + \int_{t_0}^{t} E(z)r(z)dz\right\}. \tag{A.2.4}$$

Proof If $E(t)$ is an integrating factor for the general linear ODE (A.2.1), then we have

$$\{Ey\}' = Ey' + E'y = Ey' + p(t)Ey$$

which requires

$$\frac{E'(t)}{E(t)} = p(t)$$

The relation above can be integrated to give (A.2.2). With $E(t)$ in hand, we integrate (A.2.3) to get the exact solution (A.2.4). Note that our choice of constant of integration for $E(t)$ was to ensure $E(t_0) = 1$; other choices for the constant are also possible. □

Applying the results to (2.6.1), we obtain $E(t) = e^{a_0 \sin(t)}$ and therewith

$$M(t) = e^{-a_0 \sin(t)} \left\{ M_9 + \int_0^t e^{a_0 \sin(z)} r(z) dz \right\}.$$

Exercise 43 *Apply the method of Proposition 46 to the following IVP:*

$$
\begin{array}{lll}
a) & y' = ay - b(t), & y(0) = Y_0 \\
b) & y' = -\dfrac{1}{t}y + t, & y(1) = 1 \\
c) & y' = \cot(t)y + \sin t, & y(\pi/2) = \pi \\
d) & y' = \cot(t)y + \sin t, & y(\pi) = 0 \\
e) & y' = \cot(t)y + \sin t, & y(0) = 1.
\end{array}
$$

A.2.2 The Bernoulli equation

Just as a first order homogeneous (*aka* scale invariant) ODE can be reduced to a separable equation, many ODE can be reduced to a linear first order ODE. One example is the Bernoulli equation defined as follows:

Definition 21 *A (scalar) first order ODE is a Bernoulli equation if it is of the form*

$$y' + p(t)y = q(t)y^\alpha \tag{A.2.5}$$

for any constant α (which is real-valued in most applications).

For $\alpha = 0$, the Bernoulli equation (A.2.5) is a linear first order ODE (with $r(t) = q(t)$). For $\alpha = 1$, the first order ODE (A.2.5) is both linear (with $r(t) = 0$ and $p(t)$ replaced by $p(t) - q(t)$) and separable. Hence, we need to consider only the case $\alpha \neq 1$ and $\alpha \neq 0$ for which (A.2.5) is clearly not of the form (A.2.1) and hence not linear. For this case, we have the following result involving an "integrating factor" that depends only on y (and not on t).

Proposition 47 *A Bernoulli equation can be transformed into a linear first order ODE by an integrating factor $\tilde{E}(y) = y^{-\alpha}$.*

Exercise 44 *a) Prove Proposition 47. b) Solve the IVP $t^2 y' + 2ty = y^3$, $y(1) = 1$. c) Repeat b) but with the initial condition $y(0) = 0$. d) What happens if the initial condition is $y(0) = 1$?*

A.3   An Exact First Order ODE

A.3.1   *Test for an exact equation*

The methods of solution for four different kinds of scalar first order ODE have one feature in common. They all aim to reduce the given ODE to

$$\frac{d}{dx}\left[\varphi(x, y)\right] = 0. \tag{A.3.1}$$

In this section, we investigate what first order scalar ODE is of the form (A.3.1) in disguise so that they can be solved as a calculus problem immediately without additional manipulations. Note that we have switched the independent variable from t to x to convey the idea that it is not always time and may be something else including a space variable.

Definition 22 *A first order ODE of the form (A.3.1), possibly after some re-arrangements, is said to be an **exact equation**.*

Evidently, an exact equation can be integrated immediately to get an implicit form of the solution of the IVP

$$\varphi(x, y) = C_0 \tag{A.3.2}$$

which may be solved to get

$$y = Y(x, C_0).$$

On the other hand, the implicit form (A.3.2) is also an acceptable form of the actual solution.

Now, the general ODE (A.1.3) is not an exact equation; nor is the autonomous ODE (2.1.1). After relabeling the independent variable t as x, neither can be written in the form (A.3.1). While this seems intuitive, we will develop a proof of this claim so that we have a procedure to determine if another first order ODE is exact.

Suppose the non-autonomous ODE (A.1.3) can be written as (A.3.1). Then upon differentiating φ with respect to x using the chain rule to handle the fact that y is a function of x, we get from (A.3.1)

$$\frac{d}{dx}\left[\varphi(x, y)\right] = \frac{\partial \varphi}{\partial x} + \frac{\partial \varphi}{\partial y}\frac{dy}{dx} = 0.$$

If this is in fact the given ODE (A.1.3) written as

$$-f(x, y) + \frac{dy}{dx} = 0, \tag{A.3.3}$$

we must have

$$\frac{\partial \varphi}{\partial x} = -f(x, y) \qquad \text{and} \qquad \frac{\partial \varphi}{\partial y} = 1.$$

But with

$$\frac{\partial}{\partial y}\left(\frac{\partial \varphi}{\partial x}\right) = -\frac{\partial f}{\partial y}, \qquad \frac{\partial}{\partial x}\left(\frac{\partial \varphi}{\partial y}\right) = 0,$$

these requirements are inconsistent unless $f(x, y)$ does not depend on y so that $\partial f/\partial y = 0$ (in which case the given ODE is already a calculus problem).

The method for demonstrating whether an ODE is an exact equation applies to any first order ODE of the form

$$N(x, y)\frac{dy}{dx} + M(x, y) = 0 \qquad (A.3.4)$$

(which contains (A.3.3) as a special case with $N = 1$ and $M = -f(x, y)$). For this more general form of ODE to be the same as (A.3.1), we need

$$\frac{d}{dx}\left[\varphi(x, y)\right] = \frac{\partial \varphi}{\partial x} + \frac{\partial \varphi}{\partial y}\frac{dy}{dx} = N(x, y)\frac{dy}{dx} + M(x, y)$$

or

$$\frac{\partial \varphi}{\partial x} = M(x, y), \qquad \frac{\partial \varphi}{\partial y} = N(x, y). \qquad (A.3.5)$$

The two "matching" requirements above lead to the following important *necessary* condition for any such first order ODE to be *exact*:

Proposition 48 *For a first order scalar ODE in the general form (A.3.4) to be an exact equation, it is necessary that the consistency condition*

$$\frac{\partial M(x, y)}{\partial y} = \frac{\partial N(x, y)}{\partial x} \qquad (A.3.6)$$

be satisfied for all point (x, y) in some region of the (x, y)-plane containing the initial point (x_0, Y_0).

A.3.2 Reduction to a calculus problem

We now show how Proposition 48 can be used to solve an exact equation.

Example 36 $\qquad (x + y)\frac{dy}{dx} - (x - y) = 0.$

This ODE is actually scale invariant and can be solved by Proposition 45. With $M = x + y$ and $N = -(x - y)$, it is also straightforward to verify that the consistency condition (A.3.6) for an exact ODE is satisfied. We can therefore try to find a function $\varphi(x, y)$ with

$$\frac{\partial \varphi}{\partial x} = M(x, y) = -x + y, \qquad \frac{\partial \varphi}{\partial y} = N(x, y) = x + y. \qquad (A.3.7)$$

Now, we want to determine φ. This can be done by integrating either equation with respect to relevant argument of the unknown while holding the other argument fixed. Let us integrate the first equation in (A.3.7) with respect to x to get

$$\varphi(x,y) = -\frac{1}{2}x^2 + xy + C(y) \tag{A.3.8}$$

where we have allowed the "constant" of integration to depend on y since we held y fixed when we differentiated φ and integrated $M(x,y)$ to get (A.3.8). To determine $C(y)$, we make use of the fact that $\varphi(x,y)$ must satisfy the second condition in (A.3.7). That condition requires

$$\frac{\partial \varphi}{\partial y} = x + \frac{dC}{dy} = N(x,y) = x + y$$

or $dC/dy = y$ so that we have after integration

$$C(y) = \frac{1}{2}y^2 + c_0 \qquad \text{and} \qquad \varphi(x,y) = -\frac{1}{2}x^2 + xy + \frac{1}{2}y^2 + c_0.$$

The quantity c_0 in the expression for $C(y)$ is now a true constant of integration but can be omitted since the given exact ODE can be written as

$$(x+y)\frac{dy}{dx} - (x-y) = \frac{d}{dx}(-\frac{1}{2}x^2 + xy + \frac{1}{2}y^2 + c_0) = 0$$

which becomes

$$\varphi(x,y) = C_0. \tag{A.3.9}$$

Hence, the constant c_0 in $C(y)$ plays no role in the final solution.

Remark 17 *It is important to emphasize that $\varphi(x,y)$ is not the solution of the exact ODE. The solution is in the implicit form (A.3.9) where $\varphi(x,y)$ is determined by the two relations in (A.3.7) (and not y as an explicit function of x).*

We summarize the development above in the following proposition:

Proposition 49 *If the ODE (A.3.4) is exact so that (A.3.6) is satisfied, we can determine $\varphi(x,y)$ by integrating one of the two relations in (A.3.5) with respect to one independent variable while holding the other independent variable fixed and then use the other relation in (A.3.4) to find the "function of integration" associated with that integration.*

Exercise 45 *Show that the ODE $[1 - \sin(x) - x^2 e^y].y' = [y\cos(x) + 2xe^y]$ is an exact ODE and its solution is given by*

$$\frac{1}{2}y^2 - \frac{1}{2}x^2 + xy = C_0. \tag{A.3.10}$$

A.4 When an ODE Is Not Exact

We recall from Proposition 44 that a *Bernoulli* equation can be reduced to a linear first order ODE. Since the resulting linear ODE can be further reduced to (A.3.1) by the integrating factor $E(x)$ (known from (A.2.2) with $p(x)$ replaced by $\bar{p}(x) = (1-\alpha)p(x)$), a *Bernoulli* equation can be transformed into an exact equation after we multiply the given ODE through by the following integrating factor:

$$\bar{E}(x,y) = y^{-\alpha}E(x), \qquad E(x) = e^{\int^x \bar{p}(t)dt} = e^{(1-\alpha)\int^x p(t)dt}.$$

After multiplication by the $\bar{E}(x,y)$, the original Bernoulli equation becomes (A.3.4) with

$$N(x,y) = \bar{E}(x,y) = E(x)y^{-\alpha},$$
$$M(x,y) = \bar{E}(x,y)\left[p(x)y - q(x)y^{\alpha}\right] = E(x)\left[p(x)y^{1-\alpha} - q(x)\right].$$

which can be verified to be exact since

$$\frac{\partial M(x,y)}{\partial y} = E(x)(1-\alpha)p(x)y^{-\alpha},$$

$$\frac{\partial N(x,y)}{\partial x} = E'(x)y^{-\alpha} = (1-\alpha)E(x)p(x)y^{-\alpha}.$$

As such, it can be written as

$$\frac{d\varphi}{dx} = \left[E(x)y^{1-\alpha} - Q(x)\right]' = 0$$

with $Q'(x) = q(x)E(x)$ so that

$$\varphi(x,y) = y^{1-\alpha}e^{\int^x \bar{p}(t)dt} - Q(x) = y^{1-\alpha}E(x) - Q(x),$$

where α, $p(x)$, and $q(x)$ are prescribed quantities in the given *Bernoulli* equation.

Remark 18 *We emphasize that $\varphi(x,y)$ is not the solution of the given ODE. The solution is given by the implicit relation $\varphi(x,y) = c_0$ which defines y as a function x up to a constant of integration (with the constant determined by the initial condition). In the case of a Bernoulli equation, this relation gives formally*

$$y(x) = [E(x)]^{1/(\alpha-1)}\left[c_0 + Q(x)\right]^{1/(1-\alpha)},$$

where $E(x)$ and $Q(x)$ are as previously defined.

For a general ODE already in the form (A.3.4) which is not exact, i.e., the given factors M and N do not satisfy the consistency condition (A.3.6), there is no assurance that an appropriate integrating factor $\bar{E}(x,y)$ can be found unless one that is independent of x or y happens to work. While there are some special techniques that may lead to the desired results (see [1] for example); their success rate is not sufficiently high to deserve further discussion in a review such as this.

A.5   Summary of Methods for First Order ODE

Given a first order ODE in the form (2.1.1) or (A.3.4), we should check to see if it is one of the following five types of equations:

- a separable equation — by the form (A.1.4)
- a scale invariant (also known as homogeneous of degree 1) equation — by Definition 19
- a linear equation — by Definition 20
- a Bernoulli equation — by Definition 21
- an exact equation — by the consistency condition (A.3.6).

If it is, then the equation can be solved by the method of solution described in the relevant section in this chapter. If it is not one of the five types, there are still a few other specialized techniques (such as integrating factor that depends on both x and y) which we can also try. Because of the rather technical nature of these other techniques, we will not pursue a discussion of them in an introductory course. Instead, we will move on to discuss methods for finding accurate approximate solutions for problems not tractable by the five methods above.

It should be emphasized that many important problems in science and engineering have been solved by the five methods introduced in this chapter and many more will continue to be solved by them. It behooves all of us to become adroit in identifying these equations and in the method of their solutions. As we will see, an analytical solution in terms of elementary and special functions are more efficient for analyzing the natural or social phenomena of interest than the more broadly applicable numerical simulations which will be discussed in the next chapter.

Appendix B

Basic Numerical Methods

B.1 Simple Euler

As useful as the various known analytical methods discussed in the previous chapter may be, they do not begin to cover a significant number of the ODE that occur in applications. When an exact solution of the IVP in terms of elementary (and special) functions is not possible, we will have to settle for an approximate solution of the problem. While this can again be accomplished by other analytical approaches such as perturbation and asymptotic methods, we will focus on obtaining approximate solutions by one class of numerical methods (or *numerical solutions* for short) widely used in computational software, such as MatLab, Mathematica and Maple, for general applications in science and engineering. Most modern texts for a first course in ODE at the level of [1] discuss this class of numerical methods for IVP in ODE. It is our purpose here only to show the basic idea of this group of methods by describing the first order (or Simple) Euler method, the simplest approach for such problems. To avoid unnecessary complications in the description of this method (and other related ones), we will consider the IVP

$$y'(x) = f(x, y), \qquad y(x_0) = Y_0 \tag{B.1.1}$$

where the function $f(x, y)$ is continuously differentiable in both of its arguments in some bounded region of the (x, y)-plane containing the initial point (x_0, Y_0). Here $(\ )'$ denotes $d(\ )/dx$.

The Simple Euler method exemplifies a broad approach to numerical solutions based of Taylor's theorem taken in the form (see any calculus text at the level of [25]):

$$y(x + h) = y(x) + hy'(x) + \frac{h^2}{2}y''(\xi) \tag{B.1.2}$$

for some value ξ in the interval $(x, x+h)$. It is important to note that we generally do not know the exact value of ξ; but Taylor's theorem guarantees that it is somewhere inside the interval $(x, x + h)$ as long as $y(x)$ has a continuous second derivative.

Suppose we wish to compute the value of the unknown $y(x)$ of the IVP (B.1.1) for x (up to and) at the value x_T. If the terminal point x_T is very close to the initial point x_0, then we have from Taylor's theorem

$$y(x_T) = y(x_0) + hy'(x_0) + \frac{h^2}{2}y''(\xi) \qquad (\text{B.1.3})$$

where $h = x_T - x_0$ and $x_0 < \xi < x_T$. Since we know $y(x_0) = Y_0$ and, by the given ODE, $y'(x_0) = f(x_0, Y_0)$ as well, the first two terms in (B.1.3) are known from the initial condition. While we do not know the *local truncation error* (or simply *truncation error*)

$$E_t = y(x_0 + h) - [y(x_0) + hy'(x_0)] = h^2 y''(\xi)/2 \qquad (\text{B.1.4})$$

in using the two known terms as an approximate solution for $y(x_T)$, we do know that this error decreases to zero as h tends to zero. This is so because, given our restriction that $f(x, y)$ is continuously differentiable in both of its arguments in some bounded region in the (x, y)-plane, $y''(\xi)$ is bounded by a finite number $C(h)$ for very small mesh size h. Hence, the two known terms provide a good approximation for the unknown $y(x_T)$ if $h = x_T - x_0$ is sufficiently small.

When $x_T - x_0$ is not so small, a good approximate solution for $y(x_T)$ can still be obtained with a little more work. Consider an evenly spaced set of (*mesh*) points in the interval $[x_0, x_T]$ generated by

$$h = \frac{1}{N}(x_T - x_0), \qquad x_k = x_{k-1} + h = x_0 + kh \qquad (\text{B.1.5})$$

for some choice of positive integer N and $k = 1, 2, 3, \ldots, N$. Note that we have $x_N = x_0 + Nh = x_T$. Starting with the known initial value $y_0 = Y_0$, the Simple Euler method calculates $x_1, x_2, \ldots, x_N$ successively by the algorithm

$$y_k = y_{k-1} + hf(x_{k-1}, y_{k-1}), \qquad k = 1, 2, \ldots, N. \qquad (\text{B.1.6})$$

Once y_{k-1} is found (and we do know $y_0 = Y_0$), the right-hand side of (B.1.6) is just the first two (known) terms in the Taylor's theorem (B.1.2) for $y(x_k)$ and therefore enables us to calculate a good approximation for $y(x_k)$ when $\left|h^2 y''(\xi)/2\right|$ is small.

Starting with the known values x_0 and Y_0, the sequence of formulas (B.1.6) determines successively approximate solutions $\{y_1, y_2, \ldots, y_N\}$ for $\{y(x_1), \ldots, y(x_N)\}$. The omission of (at least) a local truncation error term proportional to h^2 at each step toward $x_N = x_T$ (each analogous to $h^2 y''(\xi)/2$ in (B.1.2)) notwithstanding, we generally expect $y_k \simeq y(x_k)$, for $k = 1, 2, \ldots, N$ when N is sufficiently large (so that h is sufficiently small). This is in fact true as we shall see in some examples in the next few sections. These examples will also bring out some of the deficiencies of the Simple Euler algorithm.

B.2 Slowly

The following example is a special case of the exponential growth problem of Chapter 1. It shows that Simple Euler works but is slow if you want an accurate solution.

Example 37

$$y' = y, \qquad y(0) = 1. \tag{B.2.1}$$

Suppose we want to compute an approximate solution for $y(1)$ by the Simple Euler algorithm. With $f(x, y) = y$, $x_0 = 0$, $x_T = 1$ and $y^o = 1$, we have $h = 1/N$ and the Simple Euler algorithm gives successively

$$y_1 = 1 + h, \qquad y_2 = (1 + h) + h(1 + h) = (1 + h)^2, \qquad \ldots,$$

and

$$y_N = (1 + h)^N = \left(1 + \frac{1}{N}\right)^N. \tag{B.2.2}$$

Table B.1 reports some numerical results produced by these formulas:

Table B.1   Simple Euler for $y' = y$, $y(0) = 1$.

N	h	y_N	$E_c = e^1 - y_N$
2^0	1	2	$0.7183..$
2^1	2^{-1}	2.25	$0.4683..$
2^2	2^{-2}	$2.4414..$	$0.2769..$
2^3	2^{-3}	$2.5658..$	$0.1525..$
2^4	2^{-4}	$2.6379..$	$0.0804..$
2^5	2^{-5}	$2.6770..$	$0.0413..$
∞	0	$e = 2.71828..$	0

We note the following important observations on these results:

- The analytical expression for y_N in (B.2.2) tends to the exact solution e^1 as $N \to \infty$.
- For a sufficiently large N, $E_N = e^1 - y_N$ given in the last column of Table B.1 decreases numerically roughly by a factor of 2 as we half the *mesh size* h (or equivalently double the number N of evenly spaced mesh points $\{x_k\}$ in $[0, 1]$).
- For a fixed N, we have

$$E_N = e^1 - y_N = e - e^{N \ln(1 + 1/N)}$$

$$= e - e^{N\left(\frac{1}{N} - \frac{1}{2N^2} + \frac{1}{3N^3} - \cdots\right)} = e - e^{\left(1 - \frac{1}{2N} + \frac{1}{3N^2} - \cdots\right)}$$

$$= e - e\left(1 - \frac{1}{2N} + \frac{c_2}{N^2} + \frac{c_3}{N^3} \cdots\right)$$

$$= \frac{e}{2N}\{1 + C(h)\} = \left\{\frac{e}{2}[1 + C(h)]\right\} h,$$

where $\{c_j\}$ are known constants and $C(h) \to 0$ as $N \to \infty$ (and $h = 1/N \to 0$).

The three observations are simply different aspects of the following general property of the Simple Euler algorithm applied to the particular problem:

For the simple IVP (B.2.1), the accumulation error E_N for y_N (as an approximation for the exact value $y(1)$) for the Simple Euler method after N steps is proportional to the (evenly spaced) mesh size $h = 1/N$ for sufficiently small h.

The conclusion above for a particularly simple problem turns out to be true more generally. We state this as the following proposition:

Proposition 50 *Suppose $f(x,y)$ is continuously differentiable in both arguments. The accumulation error E_N for y_N of the IVP (B.1.1) by the Simple Euler method is proportional to $h = (x_T - x_0)/N$.*

With the local truncation error $E_t(h)$ proportional to h^2 (see (B.1.4)), the accumulated error after N steps being proportional to h is not surprising. This is proved rigorously in most texts that discuss numerical solutions for ODE (see [32] for example). Informally, we note that there will be (at least) two (different) truncation errors after two steps so that we have (for some known constants C_1 and C_2) $E_2(h) = C_1(h)h^2 + C_2(h)h^2 \leq 2\bar{C}_2(h)h^2$ where $\bar{C}_2(h) = \max\{C_1, C_2\} \to$ a constant as $h \to 0$. Similarly, we have after N steps

$$E_N(h) = (N-1)\bar{C}_{N-1}h^2 + C_N h^2 \leq N\bar{C}_N h^2 = \bar{C}_N(h)(x_T - x_0)h$$

with $Nh = (x_T - x_0)$ and $\bar{C}_N(h) = \max\{\bar{C}_{N-1}, C_N\} \to$ a constant as $h \to 0$. We state this observation as a proposition (with a rigorous proof to be found in most text of numerical solutions for IVP of ODE such as [32]):

Proposition 51 *For the IVP (B.1.1) with $f(x,y)$ continuously differentiable in both arguments, the accumulation error for the simple Euler method after N steps is proportional to the mesh size $h = (x_T - x_0)/N$.*

It is important to emphasize that any approximate solution is meaningful only if we have a handle on the error incurred in the approximation and a mechanism for reducing it. (This requirement distinguishes mathematics from other sciences and imposes a more difficult task on mathematicians.) In applications, that the accumulated error decreases with mesh size h is the least we should expect from any approximate method of solution. Few users of approximate methods would be satisfied with this type of error behavior. Most would want to have the accumulation error decreases in proportional to a higher power of h. For example, if E_N is proportional to h^2, then halving the mesh size would reduce the accumulation error by a factor of 4. In practice, few users would be satisfied with a method slower than $E_N = C_N(h)h^4$ with $C_N(h) \to$ a constant as $h \to 0$. In this sense, the *Simple Euler* method is a slow method and used only if there is no faster method available because of some special features of the problem of interest (including the limited differentiability of $f(x,y)$). We will comment briefly about faster methods at the end of this chapter.

B.3   But Not So Surely

Simple Euler and other numerical methods for IVP can only be as good as the solution of the problem. The following example shows why a numerical method may not get us anything useful under some circumstances.

Example 38

$$y' = \tan(x)y + \frac{\sin(x)}{\cos^2(x)}, \qquad y(1) = 0. \qquad \text{(B.3.1)}$$

Suppose we want to compute $y(5/4)$, $y(3/2)$ and $y(2)$ by the Simple Euler algorithm using various choices of mesh size h and taking $N = (x_T - x_0)/h = (2-1)/h = 1/h$ given $x_T = 2$. With $f(x,y) = \tan(x)y + \sin(x)/\cos^2(x)$, the Simple Euler algorithm is used to obtain the results in Table B.2 with the last row being the values of the exact solution

$$y(x) = \frac{1}{\cos(x)} \ln\left(\frac{\cos(1)}{\cos(x)}\right) \qquad \text{(B.3.2)}$$

at the three values of x obtained by using an integrating factor $E(x) = \cos(x)$ for the given linear ODE. For a given h which determines N and $x_k = x_{k-1} + h = x_0 + kh$, the columns $y_{N/4}$, $y_{N/2}$ and y_N are the approximate solutions by the Simple Euler algorithm (B.1.6) for $y(5/4)$, $y(3/2)$ and $y(2)$, respectively. The two columns, $y(5/4) - y_{N/4}$ and $y(3/2) - y_{N/2}$ are the accumulation error incurred in the two approximate solutions for the different choices of the mesh size.

Table B.2   Simple Euler for $y' = \tan(x)y + \sin(x)/\cos^2(x)$, $y(1) = 0$.

h	N	$y_{N/4}$	$y(5/4) - y_{N/4}$	$y_{N/2}$	$y(3/2) - y_{N/2}$	y_N ($\approx y(2)$?)
2^{-2}	4	0.7206..	0.9873..	3.6489..	25.094..	-17.4767..
2^{-3}	8	1.0612..	0.6467..	7.5653..	21.177..	0.4474..
2^{-4}	16	1.3275..	0.3804..	12.8502..	15.892..	-20.9431..
2^{-5}	32	1.4997..	0.2082..	18.2286..	10.514..	-8.2575..
2^{-6}	64	1.5987..	0.1092..	22.4927..	6.250..	0.0225..
2^{-7}	128	1.6519..	0.0560..	25.2952..	3.447..	-39.7132..
0	∞	1.7079..	0	28.7424..	0	?

First, it is gratifying to see that $y_{N/4}$ becomes increasingly close to the exact value $y(5/4)$ as h reduces from 2^{-2} to 2^{-7} with the accumulation error only a little more than 3% of the exact value. The accumulation errors themselves decrease monotonically, pretty much in proportional to h for the last four choices of the mesh size.

The approximate solutions $y_{N/2}$ for $y(3/2)$ are not as satisfying. It is also true that $y_{N/2}$ becomes increasingly close to the exact value $y(3/2)$ as h reduces from 2^{-2} to 2^{-7}. However, the accumulation error for $h = 2^{-7}$ is 3.4472..., still more than 10% of the exact solution and the error decreases more or less in proportional to h only for the two smallest mesh sizes.

The approximate solutions y_N intended for $y(2)$ for different h not only do not tend to the exact value as the mesh size decreases, it seems to actually oscillates about $y = 0$. Having the exact solution (B.3.2), we now see that $y(x)$ is 1) unbounded at $x = \pi/2$ where $\cos(x)$ vanishes, and 2) undefined for larger values of x such as $x = 2$ since $\cos(x)$ is negative for x in the interval $(\pi/2, \pi)$.

Definition 23 *A function $g(x)$ is said to have a singularity (or to be singular) at $x = x_s$ if it is either unbounded at the x_s or does not have the same limiting value as x approaches x_s from the left and the right, i.e., $x \uparrow x_s$ and $x \downarrow x_s$.*

Definition 24 *A linear ODE (of any order) such as (A.2.1) is said to have a singularity at $x = x_s$ if any one of its coefficients (including the forcing term $r(x)$) has a singularity at x_s.*

With the exact solution (B.3.2) at our disposal, it is easy to see that $y(x)$ has a singularity at $x = \pi/2$. But what if we do not have the exact solution for a more complicated ODE? How then would we know when the solution has a singularity or ceases to be meaningful? In fact, how do we know if there is a meaningful solution for a finite interval $x_0 < x < x_T$ before the appearance of a singularity? For example, it is not possible to find a solution for the simple IVP

$$y' = \cot(x), \ y(0) = 0. \tag{B.3.3}$$

The exact solution of the ODE is

$$y(x) = \ln(\sin(x)) + c_0. \tag{B.3.4}$$

Given $\sin(0) = 0$, no finite constant c_0 can be found to satisfy the initial condition; hence the IVP has no solution for any finite interval $(0, x_T)$ of the independent variable.

Fortunately, for the general linear ODE (A.2.1) of Appendix A (which includes (B.3.1) as a special case), we always have the explicit exact solution (A.2.4) of Appendix A in terms of the coefficients $p(x)$ and $r(x)$ of the ODE. Since we know the location of all the singularities of these coefficients, once prescribed, we have the following re-assuring theorem:

Theorem 25 *Any singularity in the exact solution $y(x)$ for a (scalar) linear first order ODE must also be a singularity of the ODE.*

It is important to note what the theorem does not say. A singularity of either or both coefficients need not be a singularity of the solution. The IVP

$$y' + \frac{1}{x}y = x, \qquad y(0) = 0$$

has as its exact solution

$$y(x) = \frac{x^2}{3}.$$

While the ODE has a singularity at $x = 0$, the solution does not. Hence, a singularity of a linear ODE does NOT automatically become a singularity of the solution of the corresponding IVP.

B.4 No Clue

Linear ODE is an important but only a small fraction of the ODE in applications. While we know precisely where are the singularities in the solution of an IVP involving linear ODE, the same cannot be said about those for nonlinear ODE. Below is an example showing that a nonlinear ODE giving no apparent clue to the existence or location of any singularity in its solution.

Example 39

$$y' = 1 + y^2, \qquad y(1) = 0. \tag{B.4.1}$$

Suppose we want to compute $y(3/2)$, $y(2)$ and $y(3)$ by the Simple Euler algorithm for different mesh size h. With $f(x,y) = 1 + y^2$, and $N = 2/h$, we obtained the results in Table B.3. The exact values $y(1.5) = 0.5463\ldots$ and $y(2) = 1.5574\ldots$ given in the last row (with $h = 0$) are obtained from the exact analytic solution $y(x) = \tan(x - 1)$ since the ODE is separable.

Table B.3   Simple Euler for $y' = 1 + y^2$, $y(1) = 0$.

h	N	$y_{N/4}$	$y(3/2) - y_{N/4}$	$y_{N/2}$	$y(2) - y_{N/2}$	$y_N \ (\simeq y(3)?)$
2^{-1}	2	0.5	0.0463..	1.125	0.4325..	5.3067..
2^{-2}	4	0.5156..	0.0307..	1.2552..	0.3023..	13.7940..
2^{-3}	8	0.5284..	0.0179..	1.3669..	0.1905..	141.2236..
2^{-4}	16	0.5366..	0.0097..	1.4472..	0.1102..	5.8482×10^9
2^{-5}	32	0.5412..	0.0051..	1.4974..	0.0600..	6.4954×10^{625}
0	∞	0.5463..	0	1.5574..	0	?

The approximate solutions $y_{N/4}$ for different mesh sizes tend to $y(1.5)$ monotonically as $h \to 0$ with the accumulation error proportional to the mesh size for sufficiently small h. The actual accumulation error is less than 1% of the exact solution for $h = 1/2^5$. The same is true for the approximate solutions $y_{N/2}$ in relation to $y(2)$. The principal difference is the accumulation error for $h = 1/2^5$ is larger at about 4%.

The behavior of the sequence of y_N is dramatically different. If we had expected that members of the sequence to be better and better approximation of the $y(3)$, the numerical results from the Simple Euler algorithm show an opposite trend, the sequence of y_N becomes increasingly larger with decreasing mesh size h with no indication of converging to a finite value. Yet, unlike the linear ODE of the previous section, the nonlinear ODE (B.4.1) gives no sign of any complication. The ODE in question has no singularity; there is no term in $f(x,y) = 1 + y^2$ that would become unbounded or discontinuous in the finite (x, y)-plane.

Fortunately, the ODE in (B.4.1) is separable and the IVP can be solved exactly to get

$$y(x) = \tan(x - 1).$$

We now see from this exact solution (plotted in Figure B.1 of this chapter) the reason for the nonconvergence of the sequence $\{y_N\}$ as N increases. The function $\tan(x-1)$ has a singularity at $x_s = 1 + \frac{1}{2}\pi$ and its graph is discontinuous at that point. Starting at the initial point $(x_0, Y_0) = (1, 0)$, the ODE drives the point forward with y increasing monotonically with x tending to ∞ as $x \to \frac{1}{2}\pi$. The vertical line $x = \frac{1}{2}\pi$ acts as a barrier to the graph of $\tan(x-1)$ preventing it from crossing that line. If we evaluate (the formal solution) $\tan(x-1)$ for $x > \frac{1}{2}\pi$, we would find the graph below the x-axis. Hence $x_s = \frac{1}{2}\pi$ is a singularity of the solution $y(x)$ as it meets both criteria for a singular point in Definition 23: unboundedness and discontinuity. Because of the unboundedness and discontinuity, the (graph of the) actual solution $y(x)$ of the IVP can never get to $x_s = \frac{1}{2}\pi$ not to mention the interval of x beyond the singularity. As such, the actual solution of the IVP exists only for $x < \frac{1}{2}\pi$.

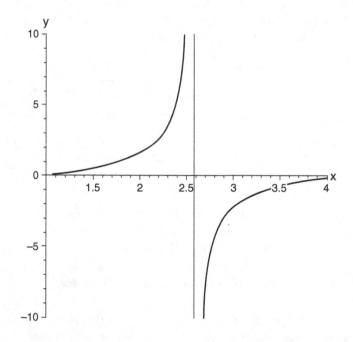

Fig. B.1   Formal solution of the IVP (B.4.1).

On the other hand, when we compute approximate solutions for $y(3)$, we do not encounter the effect of the singularity as long as none of the mesh points is x_s. For $h = 2, 1, \ldots$, we have $y(3) \approx y_1 = 2$, $y(3) \approx y_2 = 3, \ldots$, showing that in discretizing the problem (without x_s as a mesh point), we bypassed the singularity and neglected the information associated with that location. But as h decreased, mesh points adjacent to x_s became closer and closer to the singular point so that the approximate solutions experience more and more the effects of the larger and larger value of y_k (and hence steeper and steeper slope) near x_s. The steeper and steeper slopes at the mesh points just before the singularity for successively smaller mesh sizes in turn generated the ever larger approximate values for $y(3)$ in the last column of Table B.3. In the limit as $h \to 0$, it gives every indication that the approximate value for y at x_k^s, the value of x_k just before x_s, denoted by y_k^s, would become unbounded as it should.

Except for finer details, the situation described above would not be different even if we use a more efficient numerical method. Intrinsically, the ODE in (B.4.1) requires the solution $y(x)$ to grow faster than the exponential rate of the simpler ODE $y' = y$ in Example 37 in Section 2 of this chapter given that for any fixed y, we have $y < 1 + y^2$ so that the slope of the solution for (B.4.1) is always steeper than that of Example 37. With e^x becoming unbounded as x becomes unbounded, it is not surprising that $y(x)$ for the IVP (B.4.1) that grows faster would become unbounded after a finite interval of x and no efficient numerical method can change that intrinsic property of the solution.

B.5   Clueless?

Unlike linear differential equations, the nonlinear ODE in (B.4.1) gives no apparent clue in advance of the solution process that there would be a singularity in $y(x)$ and that there is only a solution for the IVP for x in the interval $[x_0, x_s)$. It was fortuitous that the IVP has an exact solution in terms of an elementary function whose mathematical properties we know well. For other nonlinear ODE, we would not be so fortunate; there would not be an exact solution. (If there were, we would not need to use the Simple Euler or any other approximate method to begin with!) Still, we are not always completely in the dark, especially if we are willing to engage in some mathematical gymnastics. The following example illustrates what we can learn about the solution singularities when there is no obvious clue from (or an exact solution for) the ODE in the IVP.

Example 40

$$y' = x^2 + y^2, \qquad y(1) = 0. \tag{B.5.1}$$

Before we compute approximate solutions by Simple Euler (or another more efficient method), we would like to know whether there is a singularity in the solution beyond which we cannot compute the solution. The existence of a singularity would enable us to avoid unnecessary computations and/or spurious approximate solutions that may appear reasonable. But similar to (B.4.1), the form of the ODE in (B.5.1) suggests that its solution would also grow faster than an exponential rate. In fact, a more careful comparison between the two ODE gives for a fixed y

$$1 + y^2 \le x^2 + y^2$$

since $x \ge 1$ for these two IVP. Moreover, the inequality is strict for $x > 1$. So starting with the same initial condition $y(1) = y^o$, the ODE in (B.5.1) requires the

solution to grow faster than that of (B.4.1) after the first step. Since the solution for
(B.4.1) has a singularity at $\pi/2$, the solution for (B.5.1) should have a singularity
x_s before that, i.e., $x_s < \pi/2$.

Table B.4   Simple Euler for $y' = x^2 + y^2$, $y(1) = 0$.

h	N	$y_{N/2}$	$y(3/2) - y_N$	y_N	$y(2) - y_N$	$y_{3N/2}$ $(\simeq y(5/2)?)$
2^{-1}	2	0.5	0.3900..	1.75	4.9539..	5.2813..
2^{-2}	4	0.6563..	0.2337..	2.5319..	4.1720..	12.9909..
2^{-3}	8	0.7571..	0.1329..	3.3920..	3.3120..	78.2845..
2^{-4}	16	0.8180..	0.0720..	4.2823..	2.4217..	$6.6278.. \times 10^5$
2^{-5}	32	0.8523..	0.0377..	5.0952..	1.6087..	3.5800×10^{150}
0	∞	0.8900..	0	6.7039..	0	?

The numerical solutions by Simple Euler shown in Table B.4 suggest that the
exact solution $y(x)$ should have a singularity at some x_s located somewhere between
$x = 2$ and $x = 2.5$. We can obviously narrow down the location of the x_s by
computing $y(x_T)$ for some terminal value x_T closer and closer to $x = 2$ (instead
of $x_T = 5/2$). This requires a lot of computing and in any case it is not possible
to computing that location precisely since y is unbounded at x_s (and we cannot
compute infinity). But we need not despair because we can compute $1/y(x_s) = 0$.

Instead of solving for $y(x)$, we can solve for $z(x) = 1/y(x)$ by setting up an IVP
for $z(x)$. Upon observing $dy/dx = d(1/z)/dx = -z^{-2}dz/dx$, we can write the ODE
(B.5.1) as

$$\frac{dz}{dx} = -(1 + x^2 z^2), \qquad z(\bar{x}) = \bar{z} \qquad \text{(B.5.2)}$$

for some starting condition $z(\bar{x}) = \bar{z}$. Note that we cannot use the original initial
condition for $y(x_0)$ written in terms of $z(x_0)$ because $z(1) = 1/y(1) = 1/0 = \infty$. But
we have already computed $y(3/2)$ quite accurately (with an accumulation error of
the order of 10^{-6} which is achieved with a mesh size less than 2^{-8}) to be $0.8900\ldots$
and we can compute the solution for $z(x)$ from that point on. With $z(\bar{x} = 3/2) =
1/0.8900\ldots \equiv \bar{z}$, the resulting $z(x)$ is plotted in Figure B.2, with the graph showing
a zero $x_s \approx 2.15$.

With a little more creativity, we can actually pin down x_s once we have $z(\bar{x}) = \bar{z}$.
Instead of finding $z(x)$, we can calculate $x(z)$ from the ODE

$$\frac{dx}{dz} = \left(\frac{dz}{dx}\right)^{-1} = (-1 - x^2 z^2)^{-1}, \qquad x(\bar{z}) = \bar{x} \qquad \text{(B.5.3)}$$

(with $\bar{x}$ taken to be $x(1./0.8900\ldots) = 3/2$). The desired location of the solution
singularity is $x_s = x(z = 0)$. An accurate numerical solution for the IVP for $x(z)$
gives $x_s = 2.1448\ldots$. A graph of $x(z)$ of the IVP for dx/dz (or more accurately
the *terminal value problem* since we are interest in x at $z = 0$ with x prescribed at
$\bar{z} > 0$) is given in Figure B.3.

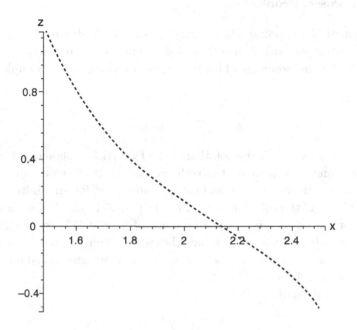

Fig. B.2   Graph for the solution of IVP for $z(x)$.

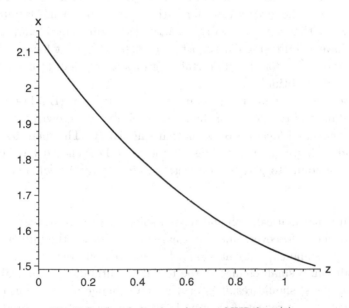

Fig. B.3   Graph for the solution of IVP for $x(z)$.

B.6   Well-Posed Problems

The mathematical reasoning which served us so well in identifying the singularity in the last section would fail to pick up a different kind of problem in the solution of other initial value problems. This is seen from the following example:

Example 41

$$y' = \sqrt{y}, \qquad y(0) = 0. \tag{B.6.1}$$

Since $\sqrt{y} < y$ for $y > 1$, the solution of the IVP (B.6.1) should not grow as fast as the corresponding exponential growth problem (B.2.1), which does not have a finite singularity. Hence, $y(x)$ should not be unbounded for any finite x_s. It is true that for $0 \le y < 1$, the solution $y(x)$ of the IVP (B.6.1) should increase no slower than that for $y' = y$ as $\sqrt{y} \ge y$ in this range of y. But if it should increase past $y = 1$, the growth rate would be slower thereafter. Thus, as before, a comparison of the relevant solution slopes enabled us to eliminate any possibility of a finite singularity in the solution of (B.6.1).

Now, the solution of the IVP

$$y' = y, \quad y(0) = 0 \tag{B.6.2}$$

is $y(x) \equiv 0$ for all $x > 0$ and no others (exercise). This appears to pose no inconsistency with the previous conclusion on the relative growth rates of the two solutions since $y(x) \equiv 0$ is also a solution of the new IVP (B.6.1). With both solutions identically zero for all x, the growth rates for both are the same. Unfortunately, it is not difficult to verify that $y(x) = x^2/4$ is also a solution of new problem (B.6.1). While it does not have a finite singularity just as $y(x) \equiv 0$ does not for the IVP (B.6.2), the new solution $x^2/4$ for (B.6.1) certainly grows faster than that for $y' = y$ with the same initial condition!

The existence of the additional solution $y(x) = x^2/4$ for (B.6.1) poses a serious problem well beyond the failure of the method of "relative growth rate" in estimating the existence and location of a solution singularity. The mere fact that there are two (and perhaps more) solutions for the same IVP should be alarming. The situation tantamount to getting completely different results when you repeat an experiment. This is unacceptable if the IVP is expected to describe a natural or social phenomenon. It is expected that an IVP for ODE that model an observed phenomenon mathematically should have a solution just as an experiment should have an outcome. Moreover, under normal circumstances, the outcome should be the same if the same experiment should be repeated without any change whatsoever. In addition, a small change of the initial condition(s) (or the values of some system parameters) should result in only a small change in the solution. After all, no experiment can be repeated exactly; but the outcome of a repeated experiment should not be too different if the difference in the experimental setup is insignificant. In the parlance of the mathematics of ODE, the IVP should have a solution (*existence*), only one solution for the same problem (*uniqueness*) and only a slight change in the solution for a small change in the parameter values in the IVP (*continuity with respect to system parameters*). An IVP problem that meets these three criteria is called a *well-posed problem*.

When an IVP is not well-posed so that the mathematical results are not scientifically meaningful, we should re-examine how we model the phenomenon mathematically, i.e., how we use mathematics to describe the scientific principles and/or processes underlying the phenomenon being investigated. To some, this task is the responsibility of the relevant group of scientists (physicists, chemists, biologists, economists, or a combination thereof) interested in the quantitative analysis of the phenomenon. This would require the scientist to have a good command of the mathematics relevant to the quantitative investigation. To others, a good applied mathematician should know enough about the relevant science to undertake the task. To mathematicians, there is also the problem of tracking down the mathematical pitfall of a mathematical problem that is not well-posed. One way of addressing the mathematical issue of well-posedness is to tell us when we are assured of a well-posed IVP. The following theorem is one of the first results in this direction:

Theorem 26 *Let $f(x,y)$ be continuously differentiable in both arguments for the region $D = \{|x - x_0| < a. \ |y - y^o| < b\}$ of the (x,y)-plane and M be the finite upper bound for $f(x,y)$ on D so that $|f(x,y)| \leq M$. Then the IVP (B.1.1) has only one solution $y(x)$ for all x in the interval $|x - x_0| < \delta$ where*

$$\delta = \min\{a, b/M\}. \tag{B.6.3}$$

The proof of this theorem can be found in most books on ODE including [8]. Here, we will limit our discussion to some important observations about what this (seemingly less than concrete) theorem tells us and how it can be useful in applications.

(1) As a starter, the theorem is conservative in several ways. It gives the condition under which you are guaranteed a solution. A solution may still exist when a stipulated condition (e.g., differentiability of $f(x,y)$) is not satisfied. For example, $f(x,y) = |y|$ is not differentiable in y at $y = 0$ and hence not in any region D of the (x,y)-plane containing any part of the x-axis. However, it is not difficult to show (exercise) that

$$y(x) = \begin{cases} y^o e^{-x} & (y^o < 0) \\ 0 & (y^o = 0) \\ y^o e^x & (y^o > 0) \end{cases}.$$

On the other hand, the stipulated condition could not possibly be satisfied if a solution of the IVP does not exist.

(2) The theorem is also conservative in another way. A solution exists for at least the interval $|x - x_0| < \delta$ with δ given by (B.6.3). The solution may exist for

a larger interval but the theorem is silent about that. For example, an ODE with $f(x, y) = -y/(x + x_0)$ gives a minimum interval of $|x - x_0| < 2x_0$ but $y(x) = y^o/(x + x_0)$ is a well-defined solution of the IVP for all $x > -x_0$.

(3) The theorem does not tell you what the solution is; but you can find as good an approximation of the exact solution as you would like by Simple Euler with a sufficiently small h once you know it exists.

(4) The theorem assures us that there is only one solution if the hypothesis is met which allows us to meet the second requirement of a well-posed problem. In that case, once we have found a solution of the IVP (even by guessing), we can stop searching for others since there is only one solution possible.

(5) Again, the hypothesis is only a sufficient condition; the problem may still have a unique solution even if it does not satisfy the sufficient condition. On the other hand, the IVP (B.6.1) has at least two solutions; hence, it must not have met the hypothesis (and it does not).

(6) The bound δ on the length of interval $|x - x_0| < \delta$ in which we are guaranteed a unique solution for our IVP may already be conservative (as shown in item (2) of this list). Naturally, we would like δ to be as big as possible. For this purpose, we need the maximum of b/M, keeping in mind that M generally depends on both a and b. This can be found once we have $f(x, y)$. We illustrate with the IVP (B.4.1) where $f(x, y) = 1 + y^2$, $x_0 = 1$ and $y^o = 0$. The given $f(x, y)$ is continuously differentiable for all $|x| < \infty$ and all $|y| < \infty$. With $|x - 1| \le a < \infty$ and $|y| < b$, we get $|1 + y^2| \le 1 + b^2 = M$ which is independent of a. Now the maximum of $b/M = b/(1+b^2)$ is $1/2$ which occurs at $b = 1$. Since a is unbounded, we have $\delta = 1/2$ so that a unique solution is guaranteed in the interval $|x - 1| < 1/2$. This is considerably smaller than the actual interval $|x - 1| \le \pi/2$ where we have found a well-defined exact solution $\tan(x - 1)$.

B.7 Higher Order Numerical Schemes for IVP

The Simple Euler method is slow in that the accumulation error after N steps, $E_N(h)$, is only proportional to h or $O(h)$ for short. If we half the mesh size h (and thereby double the number of mesh points), the accumulation error of the approximate solution at the same x value would be reduced only by half. To get a highly accurate approximate solution, we would have to use a very fine mesh and thereby a great deal of computing, measured by the number of multiplications and evaluations of functions needed to get the solution (since these operations are substantially more time consuming than others such as additions). Given the speed of the current generation of computers, no one would bat an eye to a requirement of ten thousand operations to solve a single first order ODE over a moderately long time interval. However, for a system of a thousand ODE (which is far from massive in current applications), we would be talking about 10 million multiplications which is more serious but still a long way from heavy duty computing. To cut down on the computing time and the computing cost associated with repeated runs of the same massive computing, considerable effort has been made to develop more efficient numerical algorithms for IVP in ODE. (Time and cost are not the only factors that have stimulated this effort; those interested should take an advanced course in numerical analysis and scientific computing.) One kind of more efficient algorithms is to have an accumulation error after N steps proportional to higher power of the mesh size, i.e., $E_N(h) = O(h^m)$, with $m > 1$.

To illustrate this kind of improvement, consider the following two Taylor's formulas for $y(x)$:

$$y(x + h) = y(x) + hy'(x) + \frac{h^2}{2}y''(x) + \frac{h^3}{6}y'''(\xi)$$

and

$$y(x) = y(x + h) + (-h)y'(x + h) + \frac{(-h)^2}{2}y''(x + h) + \frac{(-h)^3}{6}y'''(\eta)$$

from which we get

$$y(x + h) - y(x) = [y(x) - y(x + h)] + h[y'(x) + y'(x + h)]$$
$$+ \frac{h^2}{2}[y''(x) - y''(x + h)] + \frac{h^3}{6}[y'''(\xi) - y'''(\eta)]$$

or

$$y(x + h) = y(x) + \frac{h}{2}[y'(x) + y'(x + h)] + \frac{h^3}{12}[y'''(\xi) - y'''(\eta) - 3y'''(\varsigma)].$$

For the ODE (B.1.1) and with the notations already adopted in the previous sections, an algorithm such as

$$y_{k+1} = y_k + \frac{h}{2}[y'_k + y'_{k+1}] = y_k + \frac{h}{2}[f(x_k, y_k) + f(x_{k+1}, y_{k+1})] \qquad (B.7.1)$$

would have a local truncation error proportional to h^3 and an accumulation error after N steps proportional to h^2. As it stands, (B.7.1) is not a formula that gives y_{k+1} explicitly in terms of x_k and y_k as in the case of Simple Euler. The right-hand side involved y_{k+1}, the unknown we are trying to calculate at this step, so that (B.7.1) is generally a nonlinear equation to be solved for y_{k+1}. But by Taylor's theorem, we have

$$f(x_{k+1}, y_{k+1}) = f(x_{k+1}, y_k + hy'_k + \beta h^2)$$
$$= f(x_{k+1}, y_k + hy'_k) + \beta h^2 f_y(x_{k+1}, \sigma)$$
$$= f(x_{k+1}, y_k + hf(x_k, y_k)) + \beta h^2 f_y(x_{k+1}, \sigma)$$

so that (B.7.1) may be replaced by the *Improved Euler method*

$$y_{k+1} = y_k + \frac{h}{2}[f(x_k, y_k) + f(x_{k+1}, y_k + hf(x_k, y_k))]$$

without changing the order of accuracy. The accumulation error after N steps E_N remains $O(h^2)$. When we halve the mesh size h, we reduce the accumulation error after N steps by a factor of 4.

Numerical schemes with higher-order accuracies than the Improved Euler method have also been developed. One of the more popular higher-order schemes today is the fourth-order Runge-Kutta method with

$$y_{n+1} = y_n + \frac{h}{6}\left(k_{n1} + 2k_{n2} + 2k_{n3} + k_{n4}\right) \tag{B.7.2}$$

where

$$k_{n1} = f(x_n, y_n), \tag{B.7.3}$$

$$k_{n2} = f\left(x_n + \frac{h}{2},\, y_n + \frac{h}{2}k_{n1}\right), \tag{B.7.4}$$

$$k_{n3} = f\left(x_n + \frac{h}{2},\, y_n + \frac{h}{2}k_{n2}\right), \tag{B.7.5}$$

$$k_{n4} = f\left(x_n + h,\, y_n + hk_{n3}\right). \tag{B.7.6}$$

The accumulation error after N steps for this algorithm is $O(h^4)$.

Appendix C

Assignments

C.1 Assignment I

READING: Chapters 1 and 2.
SUGGESTED EXERCISES: Do any seven (7) problems.

1. Do Exercise 1 in Chapter 2.

2. Do Exercise 2 in Chapter 2.

3. Plot the four data points and their straight line fit obtained in Problem 1.

4. Do Exercise 42 in Appendix A.

5. Do Exercise 43 (b, c) in Appendix A.

6. Do Exercise 4 of Chapter 2.

7. A population of bacteria is growing but its growth rate diminishes with time (measured in hours). Suppose the bacteria is adequately characterized by a linear growth rate model and the "growth rate constant a" for the bacteria is estimated to be $1/(1 + 2t)$. Suppose the initial bacteria population is $y(0) = 1$ (micromole of biomass).

   a) What will the bacteria population be after 40 hours?
   b) What would be the bacteria population after 40 hours if the "growth rate constant a" is 1 instead?

8. An infectious disease is introduced to a small fraction of a large population at time $t = 0$. Let $y(t)$ be the fraction of the population infected at time t (measured in months).

   a) Suppose the fraction of the infected population increases at a rate equal to $1/36$ of the fraction of uninfected. Formulate the differential equation for the rate of increase of the infected fraction $y(t)$.

   b) Suppose $y(0) = 10^{-6}$. What will $y(t)$ be for $t = 12$ months? 60 months?

C.2   Assignment II

READING: Chapter 3.

SUGGESTED EXERCISES: Do any seven (7) problems.

1. The mosquito eradication model from Assignment I may have larva population becoming negative eventually. Formulate an improved (and more realistic) model that does not have this deficiency. (Hint: One possible and realistic modification is to take reduction by bio-agent proportional to the larva population.)

2. Starting with $y(0) = 2$, find the population as a function of time determined by the logistic growth model with emigration $y' = y(2 - y) - 1$. Does the population evolve into a steady state? Justify your answer.

3. One of the perceived shortcoming of the logistic model for population growth is its average growth rate (or more correctly, growth rate per unit population), y'/y, being large when the population is small. Some populations of fish and other species do become extinct if their population falls below a certain critical size. In addition to the depensation model, the following model has been proposed to reflect this reality:

$$y' = \alpha(Y_c - y)(y - Y_e) \equiv f(y),$$

   where α, Y_e and $Y_c(> Y_e > 0)$ are three parameters for the "modified" logistic growth rate function $f(y)$.

   a) Identify the coefficients $f(0)$, $f'(0)$, $f''(0)$, etc., of the Taylor series expansion for $f(y) = \alpha(Y_c - y)(y - Y_e)$ about the origin.

   b) Does the model allow for immigration (or emigration)? Justify your answer.

   c) Introduce new variables to eliminate at least two of the three parameters α, Y_e and Y_c.

   d) Solve the (original or normalized) IVP with $y(0) = Y_0$ and show that $y(t) \to 0$ if $Y_0 < Y_e$.

4. Set $\alpha = 1$ in the modified logistic growth model of the previous problem.

   a) Determine all the critical points.

   b) Determine by the graphical method the stability of each critical point.

c) Determine by the linear stability analysis the stability of each critical point.

d) Starting with $y(0) = Y_0$, determine the behavior of $y(t)$ as $t \to \infty$.

5. The modified logistic growth models of the previous two problems are not the only realistic models to allow for extinction when there is no immigration.

a) Determine a more appropriate growth model using a third degree Taylor polynomial of $f(y)$ that allows for extinction (below a threshold population Y_e) and limited growth but no immigration.

b) Identify all the critical points of your model.

c) Use the graphical method to determine the stability of each critical point.

6. The Gompertz growth model $y' = -ay \ln(y/Y_c)$ also captures the feature of limited growth.

a) With $z = y/Y_c$, sketch the graph of $f(z) = -az \ln(z)$ for $a = 1$ and $a = 10$.

b) Determine the critical points and their stability.

c) If the population characterized by the Gompertz model is harvested at a constant rate $h \geq 0$, use the graphical method to investigate the location and stability of the critical points as h increases from 0.

d) Draw a bifurcation diagram and identify the type of bifurcation.

7. Consider the normalized growth model $z' = z(1-z)(z-z_e) - \mu$ with a constant harvest rate μ and an extinction size parameter z_e, $0 < z_e < 1$.

a) Show by the graphical method the location and stability of all the critical points of the growth model.

b) With μ as the bifurcation parameter, plot a bifurcation diagram for the problem.

c) What is the bifurcation type (including negative critical points)?

8. Consider the ODE $y' = \ln(y) + m(y-1)$ not tied to any population growth so that we may have $-\infty < y < \infty$.

a) For $m = 1$, locate all critical points of the ODE and determine their stability by the graphical method.

b) Confirm the stability of all critical points by a linear stability analysis.

c) With m as the bifurcation parameter, determine the bifurcation point(s) and type(s) by plotting the corresponding bifurcation diagram for the critical points of the problem as m varies over $(-\infty, \infty)$.

354

C.3   Assignment III

READING: Chapter 4.

SUGGESTED EXERCISES: Do any seven (7) problems.

1. Deduce by the method for separable first order ODE the exact solution

$$y(t) = \frac{Y_0 Y_c}{Y_0 + (Y_c - Y_0)e^{-a_0 t}}$$

of the IVP

$$\frac{dy}{dt} = a_0 y \left(1 - \frac{y}{Y_c}\right), \qquad y(0) = Y_0.$$

2. $$z'(t) = z(z - \mu)(2 - z) \equiv f(z; \mu)$$

   a) Plot $f(z; \mu)$ to determine critical points and their stability for i) $\mu < 0$, ii) $\mu = 0$, iii) $0 < \mu < 2$, iv) $\mu = 2$ and v) $\mu > 2$.

   b) Plot the bifurcation diagram and identify the bifurcation points and the bifurcation type for each.

3. Consider the ODE $y' = \mu y + y^3 - y^5$ with μ as the bifurcation parameter:

   a) Restricting y to be real (but allowing negative y), obtain all possible critical points of the ODE as functions of μ for different range of μ values. From these expressions, determine the number and location of the bifurcation points for the problem.

   b) Determine the stability of the critical point at the origin by a linear stability analysis.

   c) Give an argument why the two non-zero critical points with the largest equal magnitude are asymptotically stable.

   d) What is the stability of the two critical points of equal intermediate magnitude (which exist only for μ in the range $(-1/4, 0)$)?

4. A huge dense forest comprising of red oaks and white pines with one or the other tree type found at any particular tree location inside the forest and both having roughly the same life span. Available data shows that death red oaks are lost and replaced by white pines 50% of the time while dead white pines are replaced by red oaks 75% of the time.

   a) Formulate a linear growth rate model for the evolution of the two species with time.

   b) Determine the eventual distribution of the two tree types given the initial distribution (R_0, W_0).

5. $\mathbf{y}' = A\mathbf{y} - \mathbf{b}$, $\mathbf{y}(0) = \mathbf{0}$, with

$$A = \begin{bmatrix} 0 & 1 & 1 \\ 1 & 0 & 1 \\ 1 & 1 & 0 \end{bmatrix}, \qquad \mathbf{b} = \begin{pmatrix} 3 \\ 2 \\ 1 \end{pmatrix}.$$

a) Find all eigen-pairs (eigenvalues and their associated eigenvectors) for the matrix A (Hint: the three eigenvalues are $\lambda_1 = \lambda_2 = -1$ and $\lambda_3 = 2$ with three independent eigenvectors).

b) De-couple the linear system by diagonalizing A and obtaining the exact solution of the IVP.

6. Clinical data from an isolated village show that a fraction p_i of the healthy villagers is infected by an infectious disease in unit time (and the rest remains healthy). Among the infected, a fraction p_r recovers and the rest die.

a) Formulate a linear (growth rate) model for the evolution of the three groups of individuals (the fraction of healthy $H(t)$, the fraction of infected $I(t)$, and the fraction of dead $D(t)$) in the village (with $H + I + D = 1$).

b) Show that all villagers die eventually for any initial distribution (H_0, I_0, D_0) with $H_0 + I_0 + D_0 = 1$.

c) Modify your model if in addition to a fraction p_r of infected recovers, another fraction p_d dies while the rest remains sick (infected). Determine the eventual fate of the villagers.

7. The conclusion of all villagers die eventually is not realistic. In looking for the cause, we note that those recover from the infectious disease should be immune from being infected again for a long time. For the time horizon of the model application, we should add a fourth group of those recovered, $R(t)$, distinct from the susceptible and not to be infected again.

a) Formulate an improved linear model for the four groups of individuals (assuming that all infected either recover or die).

b) With $p_i = 1/4$ and $p_r = 9/16$, determine the eventual distribution of the villagers among the four groups given the initial distribution (H_0, I_0, D_0, R_0) with $H_0 + I_0 + D_0 + R_0 = 1$.

8. A drug is injected intravenously (and hence continuously) into a patient's blood stream at the constant rate D units/day. The drug is partially absorbed by the patient's body tissue through some blood-tissue exchange process, from the blood stream concentration, $B(t)$, to the body tissue concentration, $T(t)$, at the rate $k_{bt}B$ and gaining from the concentration in the tissue at the rate $k_{tb}T$. The drug concentration in the blood stream at time t is also excreted through urination at the rate $\mu B(t)$. In addition to blood-tissue exchange, the drug concentration in the body tissue at time t is partially absorbed by tissue cells and partially excreted through sweat at a combined rate of $\gamma T(t)$.

Without additional information on other biochemical processes, proceed with
the accounting of the gain and loss rates of both $B(t)$ and $T(t)$ to formulate a
linear (growth rate) model of the two interacting concentrations. [Comments:
The exchange rate constants as defined are normally non-negative. Since we
are interested in the case of drug absorption by tissues (possibly of a particular
organ), we would have $k_{bt} > k_{tb}$.]

C.4 Assignment IV

READING: Chapters 4 and 5.
SUGGESTED EXERCISES: Do any seven (7) problems.

$$A_{RW} = \begin{bmatrix} -p & q \\ p & -q \end{bmatrix}, \qquad A_{JC} = \begin{bmatrix} -3p & p & p & p \\ p & -3p & p & p \\ p & p & -3p & p \\ p & p & p & -3p \end{bmatrix}$$

1. A general form of the Red Oak-White Pine problem has as its ODE $\mathbf{y'} = A_{RW}\mathbf{y}$
   with a 2×2 coefficient matrix A_{RW} as given above where $0 < p, q < 1$.
   a) Show that the tree site total, $R(t) + W(t)$, is conserved.
   b) Obtain the unique non-trivial (not zero) solution for the homogeneous ODE
       by determining the complementary solutions of the ODE.
   c) Obtain the non-trivial solution again, this time by making use of the linear
       dependence of the matrix rows of A_{RW}. (Without obtaining the eigen-
       values, the solution by this approach may or may not be asymptotically
       stable.)

2. For the (transition) matrix A_{JC} of the Jukes-Cantor (equal opportunity) model
   $\mathbf{y'} = A_{JC}\mathbf{y}$ of DNA base substitution given above, it is impractical to determine
   all the eigenvalues of the coefficient matrix by hand calculation.
   a) From linear dependence of the rows of A_{JC}, obtain the eigenvector for the
       one obvious eigenvalue.
   b) Note that A_{JC} is a symmetric matrix. Show that $-4p$ is a (multiple)
       eigenvalue and determine all the associated linear independent eigenvectors.
   c) Determine the steady state solution in terms of the initial distribution of
       the four elements.

3. Obtain the exact solution of the IVP below:
$$\mathbf{y'} = \begin{bmatrix} -2 & 1 \\ 1 & -2 \end{bmatrix}\mathbf{y} + \begin{pmatrix} e^{-t} \\ 0 \end{pmatrix}, \qquad \mathbf{y}(0) = \begin{pmatrix} 0 \\ 0 \end{pmatrix}.$$

4. Obtain (by reduction of order) the exact solution of the following IVP:

   a) $y'y'' - t = 0,$ $y(1) = 2,$ $y'(1) = 1.$

   b) $y'' + y(y')^3 = 0,$ $y(0) = 0,$ $y'(0) = 1.$

   c) $y'' + y(y')^3 = 0,$ $y(0) = 0,$ $y(1) = 2.$

5. $x' = x - y,$ $y' = 1 - e^x,$ $(\ )' = \frac{d(\ )}{dt}$

   a) Find all the critical points of the systems above.

   b) Sketch the eigenvectors of the Jacobian matrix, the null-clines (the locus of $x' = 0$ or $y' = 0$) and the phase portrait for the problem.

6. Find all the critical points and perform a linear stability analysis for each critical point of the following problems:

   a) $x' = \sin(y),$ $y' = x - x^3,$

   b) $x' = 1 + y - e^{-x},$ $y' = x^3 - y.$

7. A possible linear model of the drug uptake problem of a previous assignment is $B' = D - (u + k_{21})B + k_{12}T,$ $T' = -(s + k_{12})T + k_{21}B,$ $B(0) = 0,$ $T(0) = 0$

   a) Interpret each term on the right-hand side of these two equations in the context of the information given in Problem 8 of the last assignment.

   b) In practice, it is desirable not to have much drug returned to the blood stream from tissue. For the extreme case of $k_{12} = 0$, solve the IVP.

   c) What is the solution behavior for large t?

8.

$$A_3 = \begin{bmatrix} -2 & 1 & 0 \\ 1 & -2 & 1 \\ 0 & 1 & -2 \end{bmatrix}$$

   a) Find the eigen-pairs for the matrix A_3 above.

   b) Construct the matrix exponential $e^{A_3 t}$.

   c) Obtain the exact solution of the IVP: $\mathbf{y}' = A_3\mathbf{y},$ $\mathbf{y}(0) = (1, 0, -1)^T.$

9. Repeat Problem 8 for the matrix A_4 below and the initial condition $\mathbf{y}(0) = (1, 0, 0, 1)^T.$

$$A_4 = \begin{bmatrix} -2 & 1 & 0 & 0 \\ 1 & -2 & 1 & 0 \\ 0 & 1 & -2 & 1 \\ 0 & 0 & 1 & -2 \end{bmatrix}, \qquad B = \begin{bmatrix} 2 & -2 & 0 \\ -1 & 2 & -1 \\ 0 & -2 & 2 \end{bmatrix}$$

10. Repeat Problem 8 for the matrix B above. (What is the difference between B and A_3?)

C.5   Assignment V

READING: Chapters 5 and 6.

SUGGESTED EXERCISES: Do any seven (7) problems.

1. A more general version of the rabbit-fox problem:

$$R' = R - \mu R^2 - FR, \qquad F' = -F + RF + \alpha F^2.$$

   a) For the special case of $\alpha = 0$, locate the three critical points in the finite phase plane.

   b) Determine the linear stability of each of the three critical points as a function of μ.

   c) Sketch the null-clines of the ODE system (with $\alpha = 0$) and explain why trajectories that start in the first quadrant do not leave that quadrant.

   d) Locate the four critical points for this problem in terms of the two general parameters μ and α (≥ 0).

2. Find the potential $V(x, y)$ for the following gradient systems:

   a) $x' = y^2 + y\cos x, \qquad y' = 2xy + \sin x,$

   b) $x' = 3x^2 - 1 - e^{2y}, \qquad y' = -2xe^{2y}.$

3.
$$x' = y + \frac{x(x^2 + y^2 - \mu)}{\sqrt{x^2 + y^2}}, \qquad y' = -x + \frac{y(x^2 + y^2 - \mu)}{\sqrt{x^2 + y^2}}.$$

   a) Find all periodic solutions (if any) of this system.

   b) For any periodic solution found, determine its stability.

4. Decide where each of the following systems can or cannot have a limit cycle:

   a) $x' = x - y^2, \qquad\qquad y' = x - 2y + x^2.$

   b) $x' = x^2 + xy^2 - 1, \qquad y! = x^2 y - 2y.$

   c) $x' = (1 + x)\sin(y), \qquad y' = y - x + \cos(y).$

5. Show that the linear center at $(1, 1)$ of the special case with $\alpha = \mu = 0$ of Problem 1 is a true center. (Hint: About what direction is the ODE system reversible?)

6.
$$x' = \sigma(y - x), \quad y' = x(\gamma - z) - y, \quad z' = xy - \beta z \ (\beta > 0, \ \gamma > 0, \ \sigma > 0).$$

   a) Locate all critical points of the third order ODE system above (known as the Lorenz system for the discovery of the mathematical chaos phenomenon).

   b) Determine the stability of the critical point at the origin by a linear stability analysis.

7.
$$x' = \mu - x + xy, \qquad y' = y - xy.$$

   a) Find all critical points of the system above.

b) Determine the linear stability of all critical points.

c) By way of a bifurcation diagram with μ as the bifurcation parameter, determine the location of the bifurcation point and the type of bifurcation.

8. We determine in this problem how the critical points found for Problem 1 change with the parameter μ:

a) For fixed $\alpha = 0$, plot the three $R^{(c)}$ as functions of μ and do the same for the three $F^{(c)}$.

b) Deduce from the changes of the critical points with μ the location of a bifurcation point μ_c and the type of bifurcation.

c) Discuss the long time behavior of the two populations for the case $\mu > 1$ (small carrying capacity for the rabbit population) and give a possible reason for the seemingly unlikely outcome of the two limiting populations.

C.6 A Typical Midterm Examination

INSTRUCTION
Do any three (3) of the four (4) problems.
(If not crossed out, the first three will be counted toward your in-class score.)
The score of Bonus problem will be an add-on but you must have a completely correct solution to get credit.

1. (34 points) Determine the two constants a and b in $y(t) = a + bt$ for a least square fit of the following four pieces of data: $y(1) = 4$, $y(2) = 2.5$, $y(3) = 1$, $y(4) = 0$.

2. (33 points) The synchronization of the blinking/flashing of male fireflies to attract a mate has been modeled by the first order autonomous ODE: $y' = \mu - \sin(y)$.

a) Find all the critical points of the first order ODE above.

b) Determine the stability of all critical points for $\mu > 0$.

c) Plot a bifurcation diagram of all critical points as the parameter μ ranges over $-\infty < \mu < \infty$.

d) Deduce from the bifurcation diagram the type of bifurcation for the problem.

3. (33 points) Mathematical models of social mobility have been formulated based of available census data on generational family wealth. Coarsely separated into Rich (R), Middle Class (M) and Poor (P), it is seen from available data that the same fraction of each class changes its class status after one generation. For illustrative purposes, suppose 30% of the rich and 30% of the poor become middle class, while 15% of the middle class becomes poor and another 15% becomes rich.

a) For the long time evolution of upward social mobility, formulate a linear model of rate of changes of the three classes of families.
b) Show that there is a non-zero steady state for this system.
c) Show that this nontrivial steady state is asymptotically stable. (Hint: What are the eigenvalues of the relevant coefficient matrix?)
d) What is the long term steady state class distribution of the families?

4. (33 points) $x' = \mu - x - xy, \qquad y' = y(-1).$

a) Locate the two critical points of this autonomous ODE system.
b) Determine their stability by a linear stability analysis.
c) Determine all bifurcation points, if any, and the type of bifurcation for each.

Bonus (10 points) Let $\mu = 1.5$ in Problem 4.

a) Show that one critical point is a saddle point and the other is an asymptotically stable node.
b) If $x(t)$ is the normal CD4+T cell population and $y(t)$ is the HIV infected CD4+T cell population, interpret the clinical implications of your mathematical results for this case.

C.7   Assignment VI

READING: Chapter 5 and (optional) sections* of Chapter 7 on bi-stability and cell lineage.
SUGGESTED EXERCISES: Do any seven (7) problems

(In principle, you can do all problems involving phase portraits by looking for null-clines and critical point types. But free feel to use mathematical software to plot the direction fields.)

1. Two predator populations $x(t)$ and $y(t)$ share the same habitat and compete for the same finite food source. Without the other, the evolution of each is governed by normalized logistic growth ODE in the form: $z' = z(1-z)$. The growth rate of each cohort is reduced by the presence of the other by a factor proportional to $x(t)y(t)$. For simplicity, we take their growth rates to be modeled by

$$x' = x - x^2 - \alpha xy, \qquad y' = y - y^2 - \beta xy.$$

a) For $\alpha \neq 1$ and $\beta \neq 1$, obtain the four critical points of this system: $P_1 = (0,0)$, $P_2 = (1,0)$, $P_3 = (0,1)$ and $P_4 = (1-\alpha, 1-\beta)/(1-\alpha\beta)$.
b) Show that P_1 is always unstable and P_2 is asymptotically stable if $\beta > 1$ and unstable if $\beta < 1$. (By symmetry, P_3 is asymptotically stable if $\alpha > 1$ and unstable if $\alpha < 1$.)
c) Show that P_4 is unstable when P_2 and P_3 are asymptotically stable.

2. For the same system when both P_2 and P_3 are asymptotically stable, to which (stable) critical point does the competing populations approach as t tends to infinity? (Hint: The answer depends on the initial conditions for the two populations. One way to infer the answer is to draw null-clines and direction fields in the phase (x, y)-plane (by StreamPlot if you use Mathematica).

3. This problem investigated cases with $\alpha = 1$ but $\beta \neq 1$ of the same system. (We get also from symmetry the solution for $\beta = 1$ and $\alpha \neq 1$.)
   a) Show that two of the critical points coalesce, leaving only P_1, P_2 and P_3.
   b) Show that the stability of P_1 and P_2 remain the same as in Problem 1.
   c) Show that the critical point P_3 is nonhyperbolic (suggesting the possibility of bifurcation at $\alpha = 1$.
   d) Infer from the information above and what you can about the null-clines the limiting behavior of trajectories in the first quadrant for $\beta > 1$ and $\beta < 1$. (Feel free to use MatLab or any computer software.)

4. For the special case $\beta = 1$ and $\alpha = 1$ of the same system,
   a) determine all critical points;
   b) the stability of $(0,0)$, $(1,0)$ and $(0,1)$ by a linear stability analysis, and
   c) the limiting behavior of the two populations for large t. (Hint: There are some non-isolated critical points. You do not need to but may get some idea for part c by doing a StreamPlot using Mathematica.)

5. Human bodies break down glucose to provide energy for activities. This process of glycolysis may oscillate in some cases and has been analyzed by the following simple model:
$$x' = -x + \alpha y + x^2 y, \qquad y' = b - \alpha y - x^2 y \qquad (a, b > 0)$$
   a) Find a bounded region in the first quadrant (including the positive x and y axes) with trajectories along the boundary not leaving the region enclosed by the boundary.
   b) Can you conclude that there is a limit cycle inside the region found?
   c) Show that the ODE system has only one critical point at $(b, b/(a+b^2))$ and it is asymptotically stable.
   d) Does Poincaré-Bendixson theorem apply? Why or why not?

6. Show that the following planar system has no periodic orbit in the (x, y)-plane:
$$x' = -x - y - x(x^2 + 2y^2), \qquad y' = x - y - y(x^2 + 2y^2).$$

7. Show that the following planar autonomous system has at least one periodic solution:
$$x' = x - y - x(x^2 + 2y^2), \qquad y\prime = x + y - y(x^2 + 2y^2).$$

8. Discuss possible bifurcation for the following planar autonomous system:

$$x' = \mu x - y - x(x^2 + 2y^2), \qquad y' = x + \mu y - y(x^2 + 2y^2).$$

C.8   Assignment VII

READING: Chapters 8–10, 11*.
SUGGESTED EXERCISES: Do any seven (7) problems.

1. $y' = -y \ln(y) - qEy.$

a) Obtain the analytical expressions for the two critical points of the ODE in terms of the fishing effort E and use graphical method to determine their stability.

b) Determine the maximum sustainable yield h_{MSY}, and the corresponding effort E_{MSY} and steady state fish population y_{MSY} for this harvested population.

c) If the fishery is a price taker and the cost of harvesting effort E is cE, determine the effort E_0 for zero profit (break-even point) and the effort E_M for maximum profit.

d) What is the significance of $c/pq < 1$?

2. $J[y] = \int_0^1 \left[(y')^2 - 2\alpha y y' - \beta y' \right] dx, \qquad y(0) = 0, \qquad y(1) = 1.$

a) Determine the unique admissible extremal for the basic problem above. (Hint: It may not be an advantage to use the first integral for this problem.)

b) If instead of the basic problem, only $y(0) = 0$ is prescribed and the condition at the terminal end is not specified. Determine the unique extremal for this *modified* problem.

c) Obtain the extremal for the modified problem with only the terminal condition $y(1) = 1$ prescribed instead.

3. The flight path of an aircraft is required to be between two points of distance D apart on a level desert. For simplicity, assume the cost of flying the aircraft a unit (arclength) distance of the path at a height y to be $e^{-y/H}$ for some positive constant H. Find a smooth flight path for minimum total cost over the entire flight. (You do not have to prove the extremal is a minimum.)

4. (A Sufficiency Theorem) For the special Lagrangian $F(x, y, y') = F(y')$, show that any admissible extremal of the basic problem minimizes the performance index if the strengthened Legendre condition

$$\frac{d^2 F(v)}{dv^2} > 0$$

holds for all real values of the argument. (Hint: Examine $J''(\varepsilon)$.)

5. $J[y] = \int_0^1 [xy'' + y' + y]dx$, $y(1) = A$, $y'(1) = U$, $y(2) = B$, $y'(2) = V$.

a)   Obtain the Euler DE for the basic problem above. (Hint: You have to integrate by parts some term(s) more than once.)

b)   If only $y(1) = A$ and $y'(1) = U$ are prescribed, obtain the two Euler BC at the terminal end $x = 2$.

6. Consider the performance index $J[u(x), v(x)]$ with a multivariate Lagrangian $F(x, u(x), v(x), u'(x), v'(x))$ with

$$J[u, v] = \int_a^b F(x, u(x), v(x), u'(x), v'(x))dx$$

subject to the end conditions: $u(a) = A_u$, $v(a) = A_v$, $u(b) = B_u$ and $v(b) = B_v$. Deduce two Euler DE for the two unknowns $u(x)$ and $v(x)$. (Hint: Let $u(x) = U(x) + \varepsilon z(x)$ and $v(x) = V(x) + \varepsilon w(x)$ so that $J[u, v] = J(\varepsilon)$ and make use of the technique employed for the basic problem with only one unknown $y(x)$.)

7. Suppose the curve $y(x)$ in the (x, y)-plane is given parametrically by $x = \varphi(t)$ and $y = \psi(t)$, $\alpha \le t \le \beta$. The performance index for the basic problem may be written by the chain rule in calculus as

$$J[x, y] = \int_\alpha^\beta F\left(x, y, \frac{y^\cdot}{x^\cdot}\right) x^\cdot dt = \int_\alpha^\beta f(x, y, x^\cdot, y^\cdot)dt, \qquad (\ )^\cdot = \frac{d(\ )}{dt}$$

where $f(x, y, x^\cdot, y^\cdot) = F(x, y, y^\cdot/x^\cdot)x^\cdot$ does not depend on the independent variable t. Deduce the two Euler DEs for the extremal pair $(X(t), Y(t))$.

8. a)   Rewrite the two Euler DEs of Problem 7 in terms of $F(x, y, y')$ with no trace of the parameter t in them.

b)   Use the result of part a) to explain (the Erdmann second corner condition) why the quantity $\hat{F} - Y'\hat{F}_{,y'}$ is continuous along an extremal (even when the Lagrangian F depends on x).

b)   Use result in part a) to show $\hat{F} - Y'\hat{F}_{,y'} = C_0$ when the Lagrangian F does not depend on x.

C.9   Assignment VIII

READING: Chapters 9, 10, 11*, 12.

SUGGESTED EXERCISES: Do any five (5) of the eight (8) problems.

1. a)   If $F(x, y, y') = F(y')$, i.e., if the Lagrangian depends only on y' and not on x or y, show that the extremal must be a linear function of x, i.e., $Y(x) = c_0 + c_1 x$ for two arbitrary constants c_0 and c_1.

b) If $F(x, y, y') = F(y, y')$, multiply the Euler DE by Y' and re-arrange the resulting expression so that it can be integrated once to give the first integral $\hat{F} - Y'\hat{F}_{,y'} = C_0$.

2. $$\min_y J[y] = \int_a^b y^2(1 - y')^2 dx, \qquad y(a) = A, \qquad y(b) = B.$$

a) Obtain the exact solution for $a = 2$, $A = 1$, $b = 3$ and $B = \sqrt{3}$.
b) Show that there is no C_1 solution for $a = 0$, $A = 0$, $b = 2$ and $B = 1$.
c) Obtain a PWS solution for Part b.

3. Do Exercise 30 in Chapter 10.

4. In this and the next problem, we show that the Euler DE in calculus of variations can be rewritten as a Hamiltonian system. We begins the process by introducing the momentum variable $p = F_{,y'}(x, y, y')$ and solve this relation for y' to get $y' = v(x, y, p)$. Determine $v(x, y, p)$ for the following two Lagrangians:
a) $F(x, y, y') = \frac{1}{2}[m(y')^2 - ky^2]$,
b) $F(x, y, y') = f(x, y)\sqrt{1 + (y')^2}(f(x, y) > 0)$.

5. Keeping in mind that $y' = v(x, y, p)$ from Problem 4, define the Hamiltonian function to be $H(x, y, p) = pv - F(x, y, v)$ (known as the Legendre transformation) for the given Lagrangian.
a) Verify that an extremal of a general Lagrangian satisfies

$$\frac{\partial H}{\partial p} = y', \qquad -\frac{\partial H}{\partial y} = p'.$$

b) Determine the Hamiltonian for the two Lagrangians in Problem 1 and verify that an extremal of each satisfies the Hamiltonian system with the specific Hamiltonian found for the corresponding Lagrangian.

6. A firm is contracted to produce B units of flu vaccine units to be delivered before the flu season next year ($t = T$). More resources will be needed for a higher rate of vaccine production. Let $y(t)$ be the total vaccines produced by time t and $y' = u$ with $y(0) = 0$ where $u(t)$ is the resource investment rate for production. No vaccine is produced until the firm begins its investment. Vaccines produced will have to be stored at a cost of α per unit vaccine per unit time. There is a trade-off between producing faster near the end to lower total storage cost and the higher cost to produce more in a shorter period near delivery time.

a) Formulate a performance index for the total cost over the planning period $[0, T]$ to fulfill the contract. The total cost per unit time consists of the sum of a unit production cost per unit time equal to u^2 and unit storage cost of the vaccine produced equal to αy.

b) Use $y' = u$ to eliminate u from the performance index and solve the resulting calculus of variations problem.

7. a) If in the basic problem of calculus of variations with $y(a) = A$ and $y(b) = B$ where a, A and B are prescribed but b is not, use the result of Problem 7 of Assignment VII to deduce a condition for determining b for an optimal (maximum or minimum) $J[y]$ where

$$J[y] = \int_a^b F(x, y, y')dx + G(b, B).$$

b) Use the result above to obtain the solution of Problem 6 when T is not prescribed but there is a salvage value of $G(T, B) = -T/4$.

8. A firm buys and sells animal stock (cattle for its beef). Let $y(t)$ and $z(t)$ be the firm's cash and cattle holding, respectively, at time t. Suppose $u(t)$ is the firm's rate of cattle purchase at time t, then the rates of change of the two holdings are given by

$$y' = -pu - \alpha z, \qquad z' = u, \qquad y(0) = Y_0, \qquad z(0) = Z_0,$$

where $p(t)$ is the unit cattle stock cost at time t and α the constant unit storage cost. The firm wants to find a cattle purchase/sale policy $U(t)$ to maximize its total asset at the end of a prescribed planning period $[0, T]$,

$$J[y, z] = pz(T) + y(T),$$

subject to the financial and trading constraints $-u_m \leq u(t) \leq u_M$. (Negative u corresponds to (rate of) selling the cattle holding.) Historical record shows the cattle price as a function of time to be given by

$$p(t) = \beta t + p_0 \cos(2\pi t),$$

where p_0 and β are constant with $\beta T/2 \leq p_0 \leq \beta T$. Obtain the optimal purchase/sale policy $U(t)$ for the firm.

C.10   A Typical Final Examination

INSTRUCTION: The points awarded to each problem on this take-home final are as indicated. Do no more than seven problems to get 100 points. Any work on an eighth problem will be discarded. Exam is due within one week of the starting date.

1. (15 points) Recall that a planar system $x' = f(t, x, y)$ and $y' = g(t, x, y)$, where $(\ )' = d(\ )/dt$, is a Hamiltonian system if

$$x' = H_{,y} \quad \text{and} \quad y' = H_{,x}$$

with $H_{,z} = \partial H/\partial z$ for some (sufficiently differentiable) function $H(t, x, y)$.

a) Show that if a planar autonomous system, $x` = f(t, x, y)$ and $y` = g(t, x, y)$, is Hamiltonian, we must have

$$f_{,x}(x, y) + g_{,y}(x, y) = 0.$$

b) Show that the following autonomous systems satisfy the necessary condition for a Hamiltonian system above and find the appropriate Hamiltonian for each system:

   i) $x` = y, \quad y` = -x + x^2$

   ii) $x` = a_{11}x + a_{12}y + Ax^2 - 2Bxy + Cy^2$, $y` = a_{21}x - a_{11}y + Dx^2 - 2Axy + By^2$

c) Show that the ODE $x`` = g(x)$ can be converted to a Hamiltonian system. (Hint: Let $x` = y$.)

2. (15 points)

   a) Show that if a planar autonomous system is Hamiltonian, then a critical point of the system with a nonvanishing Jacobian can only be either a saddle point or a linear center.

   b) Show that along a solution trajectory $X(t)$ and $Y(t)$ of the autonomous system, the Hamiltonian is a constant independent of t, i.e., $H(X(t), Y(t)) = H_0$. (Hint: Differentiate $H(X(t), Y(t))$ and show that it must vanish.)

3. (15 points) In the well-known model of budworms depleting a spruce forest (D. Ludwig, 1978), the early stage dynamics of the budworms as preys for birds shortly after budworm's arrival may be simplified in two ways: i) the forest's carrying capacity for the budworms is essentially constant and may be normalized to 1, and ii) the budworm loss rate to predators changes with budworm population size $y(t)$ (ignored by the birds when it is small and reaching a limit by the size of the bird population when $y(t)$ is large). For our purpose, the budworm growth dynamics at this early stage may be modeled by

$$y' = y(1 - y) - \frac{\mu y}{1 + y}$$

   a) Determine all the critical point as a function of μ.

   b) Determine the stability of each critical point, by graphing $y(1 - y)$ and $\mu y/(1 + y)$ and examining their intersections.

   c) Determine the stability of the critical points by a linear stability analysis.

   d) Identify the bifurcation point and the type of bifurcation.

4. (15 points)

$$x` = x(\mu - 4x^2 - y^2) - \frac{1}{2}y, \qquad y` = y(\mu - 4x^2 - y^2) + 2x$$

   a) Show that $(0, 0)$ is the only critical point of this system and determine its stability.

b) Obtain two uncoupled ODEs for the polar angle θ and an appropriate distance variable ρ to show the existence of a limit cycle for $\mu > 0$. (Hint: Form $cxx' + yy'$ for some $c \neq 1$.)

c) Deduce from the uncoupled system the bifurcation point μ_c and the type of bifurcation there.

5. (15 points)

a) Formulate the calculus of variations problem that minimizes the surface area generated by revolving a curve $y(x)$ around the x-axis between $x = 0$ and $x = b$ with $y(0) = y(b) = A$.

b) Obtain the extremal for the problem and show that there may be two, one or no solution depending on the number of intersections between the straight line $2Az/b$ and the curve $\cosh(z)$.

c) What are the possible Euler boundary conditions at $x = b$ if $y(b)$ is not prescribed? (You do not have to solve for the extremal, just give the possible conditions and justify your answers.)

6. (20 points) The simple model for medical use of stem cells was seen to be appropriate only for a very special allocation for medical use. In the steps below, formulate a more realistic model and obtain the corresponding optimal allocation for maximum medical benefit:

a) If $u(t)$ is the fraction of the stem cell population $y(t)$ assigned for further reproduction so that the corresponding stem cell growth rate is αuy as before, the actual growth rate of the stem cell pool should reduced by the fraction allocated for medical application (per unit time). Use this assumption to formulate a new optimization problem (with control constraints) over the planning period of $[0, T]$ that maximizes the performance index:

$$J[u] = \int_0^T (1 - u)ydt + \beta y(T)$$

for the special case $T = 1$, $\beta = 1/2$ and $\alpha = 1/2$ (not 3/2).

b) Find the optimal allocation policy by the Non-improvement (i.e., Do Your Best) Principle.

c) Solve the problem by the Maximum Principle, this time for $T = 1$, $\beta = 1/2$ and $\alpha = 3/2$.

7. (25 points) $\min_{u \geq 0} \left\{ J[u] = \int_0^T (u^2 + 4y)dt \mid y' = u, \ y(0) = 0, \ y(T) = B \right\}$.

a) This problem was solved in Assignment VIII by the method of calculus of variations. In order to see that the solution is inappropriate for the case $B < T^2$, obtain the following formal solution by eliminating the investment rate $u(t)$ from the performance index and find the following extremal for the resulting problem in calculus of variations:

$$Y(t) = \frac{B - T^2}{T} t + t^2.$$

For $B < T^2$, show that the "optimal policy" is unacceptable requiring negative investment for some interval $0 \leq t < t_0$. Determine t_0 in terms of B and T.

b) Suppose we take $u(t) = 0$ ($0 \leq t \leq t_0$) and $u(t) = Y'(t)$ for ($t_0 \leq t \leq T$) with $Y'(t_0) = 0$ to avoid negative investment. Explain why the resulting solution is not acceptable.

c) With the constraint $u \geq 0$, obtain the solution of the minimum production cost problem by the Non-improving (Do Your Best) Principle with a non-negative optimal control $U(t)$.

8. (20 points) Solve the same Problem 7 using the Maximum Principle with the same non-negative constraint on the optimal control $U(t)$.

a) Write down an appropriate Hamiltonian $H(t, y, \lambda, u)$ and generate the state and adjoint DE in the standard form of a Hamiltonian ODE system.

b) Obtain the same interior (stationary) solution for $y_i(t)$, $\lambda_i(t)$ and $u_i(t)$ and show $u_i(t) < 0$ for $0 \leq t < t_s$ with t_s determined by $u_i(t_s) = 0$. (We are just reproducing the results obtained by the Non-improving Principle but emphasizing a different aspect of the solution.)

c) Knowing that the stationary solution fails to meet the constraint $y \geq 0$ and therefore not acceptable, deduce from the minimization condition

$$H(t, Y(t), \Lambda(t), U(t)) = \min_{0 \leq u \leq u_{\max}} [H(t, Y(t), \Lambda(t), u)]$$

that we must have $\Lambda(t_s) = 0$ (and change sign at this switch point).

d) Show that the interior control applies for $t > t_s$ and obtain $U(t)$ and $Y(t)$ for that interval.

Bibliography

[1] W.E. Boyce and R.C. DiPrima, *Elementary Differential Equations*, 7th ed., Wiley, 2001.

[2] A. Bryson and Y.C. Ho, *Applied Optimal Control*, Ginn and Company,Waltham MA, 1969.

[3] C.W. Clark, *Mathematical Bioeconomics: The Optimal Management of Renewable Resources*, Wiley, 1976.

[4] L. Edelstein-Keshet, *Mathematical Models in Biology*, SIAM ed., 2005.

[5] C.P. Fall, E.S. Marland, J.M. Wagner and J.J. Tyson, *Computational Cell Biology*, Springer, 2002.

[6] F.B. Hildebrand, *Advance Calculus for Appications*, 2nd ed., Prentice Hall, 1976.

[7] H.B. Keller, *Numerical Methods for Two-Point Boundary-Value Problems*, Ginn Blaisdell, 1968.

[8] W. Kelley and A. Peterson, *The Theory of Differential Equations, Classical and Qualitative*, Pearson Prentice Hall, 2004.

[9] N.L. Komarova and D. Wodarz, The optimal rate of chromosome loss for the inactivation of tumor suppressor genes in cancer. Proc. Natl. Acad. Sci. USA 101(18), (2004) 7017-7021.

[10] N.L. Komarova, A. Sadovsky and F.Y.M. Wan, Selective pressures for and against genetic instability in cancer: a time-dependent problem, J. Royal Society, Interface, 5, (2008) 105-121. (Online: June 19, 2007, doi: 10.1098/rsif.2007.1054).

[11] A.D. Lander, K.K. Gokoffski, F.Y.M. Wan, Q. Nie and A.L. Calof, Cell lineages and the logic of proliferative control, PLoS, Vol. 7(1), (2009) 84-100 (DOI:10.1371/journal.pbio.1000015).

[12] P.A. Larkin, Canada and the International Management of the Oceans, Institute of International Relations 1977.

[13] P.A. Larkin, Pacific Salmon, Scenarios for the Future, 7, Washington Seagrant Publ. No. 80-3, 1980.

[14] J. Lee, G. Enciso, D. Boassa, C. Chandler, T. Lou, S. Pairawan, M. Guo, F. Wan, M. Ellisman and M. Tan, Replication-dependent size reduction precedes differentiation dProgressive decrease in Chlamydia, (2016) submitted for publication.

[15] W.-C. Lo, C.-S. Chou, K.K. Gokoffski, F.Y.M. Wan, A.D. Lander, A.L. Calof and Q. Nie, Feedback regulations in multistage cell lineages, Math. BioSci & Engr., 6 (1), (2009) 59-82.

[16] M. Marden, *The Geomety of the Zeros of a Polynomial in a Complex Variable*, Mathematical Surveys, No. 3, Amer. Math. Soc., Providence RI, 1966.

[17] L. Perko, *Differential Equations and Dynamical Systems*, 3rd ed., Springer, 2001.

[18] A.N. Phillips, Reduction of HIV concentration during acute infection: independenc from a specific immune response. Science, 271, (1966) 497-499.

[19] L.S. Pontryagin *et al.*, *The Mathematical Theory of Optimal Control Processes*. Interscience Publishers, New York, 1962.

[20] Q.I. Rahman and G. Schmeisser, Analytic Theory of Polynomials, London Mathematical Society Monographs. New Series 26. Oxford: Oxford University Press, 2002.

[21] C. Sanchez-Tapia and F.Y.M. Wan, Fastest time to cancer by loss of tumor suppressor genes, Bull. Math. Biol. 76, (2014) 2737-2784.

[22] L. Segel, *Modeling Dynamic Phenomena in Molecular and Cellular Biology*, Cambridge University Press, 1984.

[23] J.M. Smith, *The Theory of Evolution*, London, Penguin Books, 1058.

[24] J.M. Smith, *Mathematical Ideas in Biology*, Cambridge University Press, 1968.

[25] R.T. Smith and R.B. Hinton, *Calculus*, 3rd ed., McGraw-Hill, 2008.

[26] G. Strang, *Linear Algebra and Its Applications*, 3rd ed., Harcourt, 1988.

[27] S.H. Strogatz, *Nonlinear Dynamics and Chaos*, Perseus Publishing, 2000.

[28] H.C. Tuckwell, *Introduction to theoretical neurobiology*, Vols. 1 & 2, Cambridge University Press, 1988.

[29] H.C. Tuckwell and F.Y.M. Wan, Nature of equilibria and effects of drug treatments in some viral population dynamical models, IMA J. Math. Appl. Med. & Biol. 17, (2000) 311-327.

[30] H.C. Tuckwell and F.Y.M. Wan, On the behavior of solutions in viral dynamical models, BioSystems 73, (2004) 157-161.

[31] U.S. Census Bureau, *U.S. and World Population Clocks*, http://www.census.gov/popclock/.

[32] C.F. Van Loan, *Introduction to Scientific Computing*, 2nd ed., Prentice Hall, 2000.

[33] F.Y.M. Wan, *Mathematical Models and Their Analysis*, Harper & Row, 1989.

[34] F.Y.M. Wan, *Introduction to the Calculus of Variations and Its Applications*, Chapman and Hall, 1995.

[35] F.Y.M. Wan, A. Sadovsky and N.L. Komarova, Genetic instability in cancer: An optimal control prolbem, Studies in Appl. Math. 125(1), (2010) 1-10.

[36] F.Y.M. Wan and G. Enciso, Optimal proliferation and differentiation of Chlamydia Trachomatis, Studies in Appl. Math. (2017) doi:10.1111/sapm12175.

[37] D.P. Wilson, P. Timms, D.L. McElwain and P.M. Bavoil, Type III secretion, contact-dependent model for the intracellular development of Chlamydia. Bull. Math. Biol. 68, (2006) 161-178.

Index

A

accumulation error, 337

Acquired Immune Deficiency Syndrome (see AIDS)

activation of an oncogene, 294

adenine, 68

adjoint, 234, 254
    boundary condition, 302
    functions, 234, 241, 254, 264
    ODE, 249, 263, 302
    variable, 239, 244

admissible comparison functions, 172, 173, 231, 235

admissible control, 294

aging adult case, 57

AIDS, 117, 119
    AIDS cocktail treatment, 121
    ART, cART, HAART, 121
    multi-drug regiment, 121

almost linear, 102

anchiovy, 167

anemia, 54

angular momentum, 82

animal stock, 364

antelope family, 144

anti-derivative, 325

apoptosis, 151, 293

arclength, 361

assets in the common, 166

asymptomatic phase, 118

asymptotes, 92

attractor, 33

autonomous system, 13, 30, 82, 87, 93, 96

auxiliary condition, 206

axon, 319

AZT, 130

B

B cells, 117

bacteria-antibody, 106, 110, 352

Baltimore, David, 117

bang-bang control, 246, 265, 272, 289, 308, 314

base sequence, 69

base substitution, 69
    deletion, 69
    insertion, 69
    inversion, 69
    transition, 69
    transversion, 69

basic problem, 172, 173, 177–179, 191, 204

Bendixson, 125

Bernoulli, 219
    Bernoulli ODE, 219, 329, 333, 334

bifurcation, 27, 39, 126, 141, 161
    basic bifurcation types, 43, 110, 353
    bifurcation diagram, 41, 43, 45, 112, 353
    bifurcation parameters, 41, 353
    bifurcation point, 40, 48, 110, 114, 124, 162, 164, 354
    Hopf bifurcation, 113, 114
    pitchfork bifurcation, 43, 46, 113, 141
    saddle-node bifurcation, 41, 43, 45, 113, 141, 143
    transcritical bifurcation, 43, 45, 112, 128, 132, 151, 164

bio-agent, 352

bio-control, 25
bistability, 137, 141, 143, 145, 148, 149
blood clotting, 57
blood-tissue exchange, 355
bone marrow, 54, 55, 84
bounds, 266, 311
    lower, 312
    upper, 312
boundary conditions (BC), 184, 209, 214
    Dirichlet, 176, 224, 227
    Euler BC, 194, 195
    mixed condition, 216
    Neumann, 224, 227, 228
boundary value problem, 116, 209
Brahe, Tycho, 318
break even, 165, 361
budworm, 365
BVP, 116, 176, 177, 252, 264, 282–293

C

C^1, 177, 178, 191
C^2, 191, 205
calculus of variations, 171
cancer, 293, 301
carcinogenesis, 249, 293
carrying capacity, 28, 37, 38, 81, 86, 123,
    162, 266, 316, 358
    catch-dependent cost, 184
CD4+T cells, 117, 118, 359
cell division, 69, 266, 293
cell lineage, 151
    progenitor, 151, 153, 156
    stem cells, 151, 154, 156
    terminally differentiated, 151
    transit-amplifying, 151
census data, 16
center, 92, 107, 137
    linear, 137
    stable, 93
    true, 137
central dogma of biology, 68, 117
central planning board, 168, 171
chaos
    chaotic phenomenon, 358
Chlamydia, 249, 261
    Chlamydia Trachomatis, 249, 261
chromosomal, 294
    instability (CIN), 295
    loss, 294
closed contour, 284

coefficient matrix, 58
commutative, 232
comparison functions, 173
competitive, 53
complementary solutions, 56, 59
complex conjugates, 62
compound interest, 168
concave, 306
    strictly, 310
consistency condition, 331, 333, 334
constant coefficients, 25, 58
constrained optimization, 186, 249, 289
constraints, 198, 236, 251
    inequality, 251
    non-negative, 238, 239
continuity, 238, 257, 263, 269, 274, 311,
    347
control, 162, 241, 246, 251, 275, 286, 293
convergence, 227, 342
    quadratic, 227
    super-linear, 227
conversion, 261, 265, 289
convexity, 297
    non-convex, 295
    strictly convex, 295, 314
cooperative, 53
corner control, 241, 257, 303
corner solution,
    lower, 243, 251
    upper, 243, 251
costs, 202
    production, 202
    storage, 202
Crick, Francis, 68
critical point, 30, 31, 85, 96, 131, 214
    critical point analysis, 127
    hyperbolic critical point, 100, 102
    isolated critical point, 34
    node, 124
    non-hyperbolic critical point, 108, 111
    non-isolated critical point, 34, 102,
        107, 109
crowding, 289
cure, 133
cylindrical cable, 318
cytoplasmic calcium, 144
cytosine, 68

D

Darwinian microevolution, 294

death rate, 307, 310, 314
de-coupling, 63
degradation rate, 54
deletion, 69
depensation growth model, 352
determinant, 95
differentiable, 37
    continuously differentiable, 37
differential, 231
differential equations (ODE), 13
    autonomous ODE, 13, 30, 82, 218
    canonical ODE, 44
    fundamental matrix solution, 59, 60,
        64
    homogeneous, ODE, 56, 66
    inhomogeneous equation, 59, 61, 74
    linear ODE, 328
    non-autonomous ODE, 13, 31, 220
    nonlinear ODE, 341
    particular solution, 25, 59, 64
diffusion, 211
    diffusive phenomena, 210
dikdiks, 144
dimer, 140
diminishing return, 28, 196, 267
direction field, 103, 110, 360
Dirichlet, 176
discontinuities, 204, 342
discounting, 167, 168
    discounted net revenue, 171, 259
discriminant, 95
divergence theorem, 107
DNA, 68, 117
    Influenza A virus DNA, 71
    mitochondrial DNA, 71
Do Your Best Principle, 231, 241, 250, 260
double helix, 68
dragonflies, 26
drug uptake, 357
dsolve, 223
Dulac, 125
dynamical systems, 12

E
Earth, 11, 326
    Earth's population, 12
economics, 161
effective index, 134
effort cost, 165
eigen-pairs, 60, 357

eigenvalue multiplicty, 66, 67
eigenvalue problem, 211
eigenvalues, 59, 116, 132, 356
eigenvector normalization, 67
eigenvectors, 59, 356
    multiple eigenvalues, 109
elementary bodies (EB), 261
emigration, 23, 352
end conditions, 172, 179, 192
endocytosis, 271
enzymes, 117, 140
equilibrium, 5, 11
Erdmann conditions, 204, 205, 362
estimation of parameters, 17
Euler, 177, 178, 195
    Euler BC, 181, 194, 195, 205, 240, 243,
        249, 253, 263, 274, 302
    Euler DE, 176, 179–203, 205, 207, 233,
        363
    inhomogeneous Euler BC, 199
    Simple Euler method, 335, 336
Evolution, 11, 293, 327, 359
exact equation, 330, 332, 334
existence, 211, 221, 226, 228, 347
extinction, 162
extremal, 176, 180, 192, 204–206, 233, 235,
    241, 246, 252, 361

F
feedback, 137, 151, 153, 156
    control, 153
finite difference, 227
finite jump discontinuities, 178
first integral, 82, 183, 192, 204, 238
first order ODE, 325
fish harvesting, 37
fishery management, 38
fishing effort, 161
fishing habitat, 171
fishing moratorium, 167
fixed point (see critical point), 141, 196
flight path, 361
focus, 93
Fourier's law, 213
foxes, 81
free end problem, 203, 205
frictional resistance, 33
Franklin, Rosalind, 68
free boundary, 237, 298
fundamental lemma, 176, 197, 241

fundamental set, 222, 223
future return, 171

G
gag gene, 130
gazelles, 144
GDF11, 153
genetic, 69
  code, 69
  instability, 249, 293
  mutation, 293
genomic, 293
geometrical method, 30
glucose, 22, 360
  deficiency, 24
  glycolysis, 360
  infusion, 22
Gompertz growth, 322, 353
gp41, 117
gp120, 117, 130
gradient system, 106, 357
graphical method, 34, 35, 353
gravitational constant, 82
Green's theorem, 188, 189, 284
growth, 11
  constant growth rate model, 21
  cyclical growth, 26
  depensation growth, 48, 163, 352
  economics of growth, 161
  exponential growth, 15
  Gompertz growth, 353
  growth dynamics, 172, 249, 255
  growth postulate, 12
  growth rate, 11
  growth rate constant, 16, 161
  linear growth rate, 15, 184
  logistic growth, 27, 38, 81, 144, 161, 164
  Lotka-Volterra, 81
  marginal growth rate, 323
  modified logistic growth, 352
  seasonal growth rate, 26
  size-dependent growth, 11, 144
guanine, 68

H
half life, 55
halibut, 167
Hamilton, 174, 302, 363
  Hamiltonian, 249, 256, 263

Hamiltonian system, 264, 303, 363
  Hamilton's principle, 176
Hartman-Grobman, 102, 112
harvest cost, 185, 255
harvest rate, 171, 240
harvesting strategy, 38, 188
  fixed effort harvesting, 45
heat, 211
  heat conduction, 211
  heat equation, 213
  heat flux, 212, 215, 225
  specific heat, 212
helper cells, 118
hemolysis, 54
higher derivatives, 199
higher order, 318
  rate of change, 318
  schemes, 348
Herz, *et al.*, 122, 126
Hill's function, 154–156
HIV, 117, 119, 359
  cocktail treatment, 120
  early stage, 123, 126
HIV dynamics, 122
  homeostasis, 293
  homogeneous ODE, 56, 66, 88, 328, 334
  inhibitors, 120
  RT blocker, 134
  treatment, 120, 130
Hopf bifurcation, 113
  subcritical, 115
  supercritical, 113
horizontal tangency, 41, 42, 44, 47
host cell, 270
Human Immunodeficiency Virus (see HIV)
hydrogen bond, 68
hyenas, 144
hyperbola, 91, 192
hyperbolic critical point, 48, 49, 100
  non-hyperbolic critical point, 49
hypoglycemia, 22

I
immigration, 11, 23, 353
immune system, 117, 123
impalas, 144
Improved Euler Method, 349
inclusion membrane, 289
index theory, 137

inequalities, 293
inequality constraints, 198, 234, 230, 239, 251, 255, 262, 301
infectious diesease, 352
inhomgeneous ODE, 25, 224
initial-boundary value problem (IBVP), 217
initial condition, 14, 55, 301, 326
initial distribution, 356
initial value problem, 14
INP (immediate neuronal precursor), 153, 157
insertion, 69
instability, 293
instantaneous rate, 12, 317
insulated, 212, 214
integrase, 117, 120
integrating factor, 23, 328, 333, 339
interior, 236, 251
    control, 236, 241, 251, 268, 303, 367
    solution, 240
intravenous infusion, 21
invariant, 204
inverse, 67
inverse square law, 315, 319
inversion, 69
investment, 235
IVP, 14, 15, 326

J
Jacobian, 97, 109, 110, 113, 128, 132, 139, 141, 145, 155, 356, 365
Jordan block, 74, 77
Jordan form, 73, 75
Jukes-Cantor model, 71, 355
jump discontinuities, 204, 237, 302

K
Kepler, J., 318, 319
killer cells, 118
Kimura models, 72
kinetic energy, 174, 175

L
Lagrange, 174
    multipliers, 200
Lagrangian, 173, 174, 179, 182, 205, 239, 361
least square method, 17
    least square fit, 17, 20, 358

least squares solution, 20
Legondre, 303
    condition, 361
    transformation, 303
leopard, 144
life cycle, 118, 261
limit cycle, 103, 105, 106, 114, 115, 125, 139, 140, 358, 360
limit of integration, 325
linear dependence, 356
linear independence, 56, 61, 78
    linearly independent solution, 56, 66
linear center, 115, 358, 365
linear differential operator, 210
linear growth rate, 184, 261
linear interaction, 53
linear model, 70
    equal opportunity model, 71
    equal opportunity substitution, 70, 72
    Jukes-Cantor model, 71
    Kimura models, 72
    linear growth rate model, 17
linear oscillator, 174
linear stability analysis, 36, 85, 97, 98, 116, 137, 155, 156, 353
linear systems, 58
    almost linear systems, 102
    linear autonomous systems, 87
    homogeneous linear systems, 59, 66, 89
local optimum, 231
logistic growth, 27, 38, 81, 144, 164, 257, 295, 352
    modified logistic growth, 352
loop integral, 284
Lotka-Volterra model, 81
lower corner control, 271, 278
Ludwig, D., 365
Lyapunov stability 137
lymphocytes, 117

M
macrophage, 117
macroscopic, 316
Maple, 31, 58, 219, 223, 226
Mars, 318
mass-spring-dashpot system, 36, 174–177
Mathematica, 31, 58, 103, 219, 223, 226
mathematical modeling, 1, 315
MatLab, 31, 58, 219, 223, 226

matrices
  defective, 63, 73
  diagonalization, 63, 66, 68
  eigenvalues, 59
  eigenvectors, 59
  fundamental matrix solution, 59, 60, 64
  Inverse, 67
  Jacobian matrix, 97, 109, 124, 128, 139, 145, 155, 356
  Jordan block, 74
  Jordan form, 73, 110
  matrix eigenvalue problem, 58
  matrix exponential, 58
  matrix inversion, 58
  matrix norm, 65
  modal matrix, 67, 87
  nondefective, 63, 65, 68, 73, 87
  pseudo modal matrix, 76, 79
  symmetric matrix, 66, 68, 71, 356
  transpose, 67
  upper triangular, 73
mature adult case, 56, 84
Maximum, 161
  Maximum Principle, 231, 239, 249, 253, 256, 263, 273, 302, 366
maximum sustainable yield, 161–164, 186, 361
Mercury, 316
mesh points, 336
mesh size, 337
method of elimination, 55
method of IVP, 225
method of least squares, 18, 320
method of undetermined coefficients, 25
Michaelis-Menten, 140
minimum
  global, 239, 299
  local, 239, 299
  minimum cost, 234
  strong, 239
  weak, 239
mitosis, 293
model, 1, 293
  exponential growth model, 16
  linear (growth rate) model, 16
  mathematical models, 1, 293
  model analysis, 4
  model applications, 4
  model formulation, 4

model improvement, 4, 24, 30
model validation, 4, 24, 30
modeling
  mathematical modeling, 2
  modeling cycle, 1, 324
morphogenesis, 144
mosquito, 25
  eradication, 25, 352
  fish, 26
  lavrae, 26
  predators, 26
Most Rapid Approach, 187, 244, 259, 283, 286
multicellular, 293
multipotent, 151
multi-scale, 6
mutation, 68, 295
  genetic, 68

N

NDSolve, 223
nerve axon, 318
  cable model, 319
neurons, 143
  cable neuron, 319
Newton's law, 82, 174, 175, 315, 319
Newton's method, 219, 220
Newton-Raphson, 220
nihilistic, 53
node, 89, 91, 113, 124, 139
non-convergence, 342
non-existence, 211
non-improvement principle, 366 (see Do Your Best Principle)
nonlinear systems, 96, 341
non-negative, 238
non-nucleoside inhibitors, 120
non-uniqueness, 211
normalization, 29
nucleobases, 68
nucleoside inhibitors, 120
nucleotide, 69
null-clines, 356
numerical methods, 335

O

obligopotent, 151
ODE, 13, 325
ode23, ode45, 223 (see differential equations)

offshore fishing limit, 166
offshore territorial rights, 167
olfactory epithelium, 153
oncology, 293
open access, 167, 171
optimal control, 186, 236, 237, 249, 251, 258, 262
optimality, 255, 269, 303
optimization, 5, 159, 171, 186, 196, 293
orbit, 89, 318
order of accuracy, 349
organism, 293
ORN (olfactory receptor neuron), 153, 157
orthogonality, 67
oscillatory motion, 36, 174
overfishing, 37, 162, 164, 166

P

pancreatic beta cells, 143
parameter estimation, 320, 321
parametric form, 203
parametric plot, 84
parametric representation, 203
parasitic, 53
partial derivatives, 179
partial differential equations (PDE), 209
partial fractions, 29
particular solution, 25, 59, 64
pendulum, 82, 96, 98, 103, 107, 108
performance index, 173, 191, 203, 232, 235, 239
performance objectives, 153
perihelion, 316
periodic orbit/solution, 106, 114, 139
Peruvian anchiovy, 162
phase portrait, 84, 103
phenotypes, 294
piece-wise smooth (PWS), 191
    PWS comparison functions, 193
    PWS functions, 172, 178, 193
planar systems, 93
planetary motion, 315
    inverse square law, 315
    Mercury, 316
    Mars, 318
planning period, 167, 172, 255
platelets, 57
pluripotent, 151
Poincaré index, 137
Poincaré-Bendixson, 106, 360

polar coordinates, 101, 114, 115
polycythemias, 54
Pontryagin, 249, 260
population, 11
    competing 144
    human population, 11
    interacting population, 51, 81
    population clock, 11
potential energy, 174, 175
potential function, 106, 357
precession, 316
predator-prey, 53, 81, 107, 144, 359
present value, 171
price taker, 164, 171, 361
production cost, 202, 363
profit, 165, 171, 255, 361
progenitor, 151, 153, 156
progressive catch tax, 167
proliferation, 261
proof of concept, 262
protease, 117, 120, 130, 131, 134, 135
protein, 117
purines, 69
PWS, 178, 235, 237, 302
pyrimidines, 69

R

rabbits, 81
    rabbit-fox problem, 84, 86, 96, 357
red blood cells, 54, 84
    anemia, 54
    hemolysis, 54
    polycythemias, 54
    Sickle-cell anemia, 54
red oaks, 354
reduction of order, 82, 83, 85, 218, 356
regeneration, 153, 156
regression analysis, 18
    linear regression, 21
relativity, 316
replication data, 70
reproductive ratio, 125–127
reticulate bodies (RB), 261
retrovirus, 117
REU, 324
revenue, 164, 165, 168
revenue-effort curve, 165
reverse transcriptase, 117, 120, 130, 131
    RT blocker, 134
reverse transcription, 117

reversible systems, 107–109
RNA, 68, 117
roots, 31
root finding, 31
Routh-Hurwitz, 129, 132
row echelon reduction, 58
Runge-Kutta, 350

S

saddle-node, 41, 43, 45, 113, 163
saddle point, 91, 92, 98, 107, 110, 124, 125, 137, 139, 140
salvage value, 248, 250
sardine, 167
scales, 6
    scale-invariant, 327, 334
    anatomical, 6
    cellular, 6
    developmental, 6
    ecosystem, 6
    evolutionary, 6, 69
    molecular, 6
    organismic, 6
    population, 6
    societal, 6
    species, 6
    temporal scales, 6
scientific issues, 4
    bifurcation, 5
    chaos, 5
    control, 5
    diffusion, 5
    equilibrium, 5
    evolution, 5
    feedback, 5
    optimization, 5
    periodicity, 5
    propagation, 5
    randomness, 5
    robustness, 5
    stability, 5
    stochasticity, 5
secant method, 227
self-adjoint, 210
separable, 29, 325, 334, 354
separatrix, 92, 103, 110
Serengeti National Park, 144
shadow worth (price), 286
shooting method, 225, 228
shortest distance, 195, 205, 207

shortest time, 294
sign function, 240, 249
similarity transformation, 63, 66, 79
Simple Euler, 335
simply-connected, 106
single pass, 300
singular solution, 187, 244, 247, 251, 253, 256, 263, 268, 282, 303
singularity, 340
sink, 90
size, 6
size limiting growth, 144
smooth functions, 177
    smooth comparison functions, 172
social mobility, 359
source, 91
space curve, 89
specific heat, 212
speed of light, 317
spiral point, 93, 104, 110, 113–115
spruce, 365
stability, 27, 30, 32, 132
    asymptotically stable, 33
    semi-stable, 37, 42, 44
    stable critical point, 33
    structural stability, 48, 102
    unstable critical point, 33
state variable, 239, 254, 302
stationarity, 175, 195, 233, 236, 239, 241, 257, 275, 285
stationary condition, 197, 207, 234, 240, 252, 300
stationary point, 48, 175
steady state, 56
    non-isolated, 71
    temperature distribution, 214
    time-dependent, 115
stem cell, 151, 154, 156, 248, 365
stochastic models, 291
stock, 172, 188
storage cost, 202, 235, 363
straight line fit, 351
StreamPlot, 103, 360
subspace, 88, 109
substrate, 140
sufficiency, 102
super-diagonal, 74
superposition principle, 56, 59, 222, 224
sustainable yield, 161, 163
switch, 237, 238, 259

switch condition, 270, 303
switch point, 270, 286, 309
switch time, 237, 265
two-switch control, 275, 283, 286
synchronization, 358
system parameters, 347

T

T cells, 117
Taylor
Taylor expansion, 42, 102
Taylor polynomials, 27, 30, 353
Taylor's series, 27, 65, 323, 353
Taylor's theorem, 12, 30, 37, 48, 97, 335
Temin, Howard, 117
temperature, 211
distribution, 214
terminal condition, 301
terminal payoff, 196–198
terminal time, 267
terminal value problem, 273, 344
thermal, 212
conductivity, 213
diffusivity, 214
energy, 212
thymine, 68
time-invariant, 59, 215, 299
time optimal problem (TOP), 302
time-reversal, 108, 109
torque, 82
trace, 95
trajectory, 89, 284
transcritical bifurcation, 43, 45, 112, 128, 132, 151
transient, 217
transition, 69
transversality, 206, 314
transversion, 69
true center, 358
truncation error, 336
local, 336

U

ultra-differentiated form, 180
unipotent, 151
unique solution, 215, 224
uniqueness, 108, 211, 217, 221, 225, 347
uniqueness theorem, 217, 226
upgrade, 316
upper bound, 266–347
upper corner control, 245, 251, 254, 256, 270, 283, 304
U.S. Census Bureau, 18

V

vaccine, 196, 234, 363
Validation, 3
variational notation, 231
variations, 231, 232
first variation, 232, 233, 237
powers, 232
products, 232
quotients, 232
smooth variations, 233
sums, 232
vector control, 26
VectorPlot, 103
virions, 117, 119
virion envelop, 130
virus enzyme protease, 130
virus saturation, 123

W

Watson, James, 68
well-posed problems, 346
white blood cell, 57
white pines, 354
Wilkins, Maurice, 68

Y

yield, 171
Yield-effort, 161

Z

zero, 31, 263

Printed in the United States
By Bookmasters